Analysis of Genes and Genomes

Richard J. Reece

University of Manchester, UK

John Wiley & Sons, Ltd

Other Wiley Editorial Offices

John Wiley & Sons Inc., 111 River Street, Hoboken, NJ 07030, USA

Jossey-Bass, 989 Market Street, San Francisco, CA 94103-1741, USA

Wiley-VCH Verlag GmbH, Boschstr. 12, D-69469 Weinheim, Germany

John Wiley & Sons Australia Ltd, 33 Park Road, Milton, Queensland 4064, Australia

John Wiley & Sons (Asia) Pte Ltd, 2 Clementi Loop #02-01, Jin Xing Distripark, Singapore 129809

John Wiley & Sons Canada Ltd, 22 Worcester Road, Etobicoke, Ontario, Canada M9W 1L1

Wiley also publishes its books in a variety of electronic formats. Some content that appears
in print may not be available in electronic books.

Library of Congress Cataloging-in-Publication Data

Reece, Richard J.
    Analysis of genes & genomes / Richard J. Reece.
        p. ; cm.
Includes bibliographical references and index.
    ISBN 0-470-84379-9 (cloth : alk. paper) – ISBN 0-470-84380-2 (paper : alk. paper)
  1. Molecular genetics – Research – Methodology. 2. Genetic engineering – Research – Methodology.
    [DNLM: 1. Genetic Techniques. 2. DNA–analysis. 3. Genome. QZ 52 R322a 2003]
I. Title: Analysis of genes and genomes. II. Title.
    QH442.R445 2003
    572.8′6 – dc21
                                                                              2003012937

British Library Cataloguing in Publication Data

A catalogue record for this book is available from the British Library

ISBN 0-470-84379-9 (HB)
        0-470-84380-2 (PB)

Typeset in 11/14pt Sabon by Laserwords Private Limited, Chennai, India
Printed and bound in Italy by Conti Tipocolor SpA, Florence
This book is printed on acid-free paper responsibly manufactured from sustainable forestry
in which at least two trees are planted for each one used for paper production.

For Judith

Contents

Preface

There are few phrases that can elicit such an emotive response as 'genetic engineering' and 'cloning'. Newspapers and television invariably use these phrases to describe something that is not quite right – even perhaps against nature. Genetic engineering and the modification of genes invariably conjures up images of Frankenstein foods and abnormal animals. During the course of reading this book, however, I hope that readers will appreciate that genetic engineering, and the techniques of molecular biology that underpin it, are essential components to understanding how organisms work. Man has been playing, often unwittingly, with genes for thousands of years through selective breeding to promote certain traits that were seen as desirable. We are currently at a watershed in the way in which we look at genes. Behind us is 50 years of knowledge of the structure of the genetic material, and ahead is the ability to see how every gene that we contain responds to other genes and environmental conditions. Determining the biochemical basis of why certain people respond differently to drug treatments, for example, may not be possible yet, but the techniques to address the appropriate questions are in place. The excitement of entering the post-genome age will go hand-in-hand with concerns over what we have the ability to do – whether we actually do it or not.

The analysis of genes and genomes could easily fall into a list of techniques that can be applied to a particular problem. I have tried to avoid this and, wherever possible, I have used specific examples to illustrate the problem and potential solutions. I have relied heavily on published works and have endeavoured to reference all primary material so that interested readers can explore the topic further. This has also allowed me to place many of the ideas and experiments into a historical context. It seems a common misconception that Watson and Crick were solely responsible for our understanding of how genes work. Their contribution should never be underestimated, but the work of many others should not be discounted. The full sequence of the human genome and, equally or even more importantly, the genomes of experimentally amenable organisms provide exceptional opportunities for advances in biological sciences over the coming years. More and more experiments can now be performed on a genome-wide scale and we are just beginning to understand the consequences of this.

One of the main problems that I have encountered during the writing of this text is attaining a balance between depth and coverage. I have purposefully concentrated on more amenable experimental systems – *E. coli* for prokaryotes and yeast for eukaryotes. In addition, I have treated higher eukaryotes as being almost exclusively mammals, and especially humans. This is intended to give readers a flavour of the ideas and experiments that are currently being undertaken, but also to give a historical framework onto which today's experiments may be hung. We ignore the past at our peril. This approach has, however, led to the exclusion of some other systems, e.g. *Drosophila* and prokaryotes other than *E. coli*, but is by no means meant as a slight to these neglected fields. Rather than either covering all fields in scant detail or explaining the intricate details and nuances of only a few, I have attempted to provide a broad overview that is punctuated with specific examples. Whether I have succeeded in getting the balance right I will leave to individual readers. I can say for certain, however, that there has never been a more exciting time to study biology, and I hope that this is reflected in this text.

Richard J. Reece
The University of Manchester
October 2003

Acknowledgements

I have had a great deal of help in writing this book. Of course, omissions and inaccuracies are entirely my responsibility, but I thank those who have (hopefully) kept these to a minimum – David Timson, Noel Curtis, Cristina Merlotti, Chris Sellick, Carolyn Byrne, Ray Boot-Handford and Ged Brady. I am also very grateful to Robert Slater (University of Hertfordshire) and to Mick Tuite (University of Kent) for their immensely helpful comments and suggestions. I thank the many friends and colleagues, mentioned in the text, who have so generously provided both figures for the book and for permission to cite their work. I am also deeply indented to Jordi Bella for showing me that molecular graphics programmes are usable by idiots. Nicky McGirr at John Wiley persuaded me that this project was a good idea. Her boundless enthusiasm and encouragement saw me through the times when I was not so sure and, of course, she was right. The 'guinea pigs' for many of the ideas presented here have been successive years of Genetic Engineering students at The University of Manchester. I thank the many of them who read parts of the manuscript, and all of them for challenging me, and many of my preconceived ideas. Judith, Daniel and Kathryn have been incredibly patient throughout the inception and writing of this book. Readers who find it useful should be thanking them, not me. Finally, I want to thank my teachers – Tony Maxwell and Mark Ptashne – who, each in his own way, have true passion for science and an insistence that the right experiments are done.

Abbreviations and acronyms

AAT	α_1-antitrypsin
AAV	adeno-associated virus
AD	activation domain
BAC	bacterial artificial chromosome
CaMV	cauliflower mosaic virus
CAP	catabolite activator protein
CBD	chitin binding domain
CDK	cyclin-dependent kinase
cDNA	complementary DNA
CFI	cleavage factor I
CFII	cleavage factor II
CHEF	contour-clamped homogeneous electric field
ChIP	chromatin immunoprecipitation
CMV	cytomegalovirus
CPSF	cleavage and polyadenylation specificity factor
CStF	cleavage stimulation factor
CTD	carboxy-terminal repeat domain
DBD	DNA binding domain
DEAE	diethylaminoethanol
DHFR	dihydrofolate reductase
DNA	deoxyribonucleic acid
DTT	dithiothreitol
ECM	extra-cellular matrix
EMS	ethyl methane sulphonate
ER	endoplasmic reticulum
ES	embryonic stem
EST	expressed sequence tag
FIGE	field inversion gel electrophoresis
FISH	fluorescent *in situ* hybridization
FRET	fluorescence resonance energy transfer
GST	glutathione S-transferase
HAC	human artificial chromosome
HAT	histone acetyltransferase
H-DAC	histone deacetylase

HSV	herpes simplex virus
IMAC	immobilized metal ion affinity chromatography
IMPACT	intein mediated purification with an affinity chitin binding tag
ITR	inverted terminal repeat
LTR	long terminal repeat
MBP	maltose binding protein
mRNA	messenger RNA
MCS	multiple cloning site
MLP	major late promoter
MSV	maize streak virus
NLS	nuclear localization signal
OD	optical density
ORF	open reading frame
PABII	polyA binding protein II
PAC	P1 artificial chromosome
PAP	polyA polymerase
PCR	polymerase chain reaction
PFGE	pulsed-field gel electrophoresis
RdRp	RNA-dependent RNA polymerase
RF	release factor
	replicative form
RFLP	restriction fragment length polymorphism
RIP	ribosome inactivating protein
RISC	RNA induced silencing complex
RNAi	RNA interference
rRNA	ribosomal RNA
RT	reverse transcription
	reverse transcriptase
RT-PCR	reverse transcription-polymerase chain reaction
SAM	S-adenosylmethionine
SDS	sodium dodecyl sulphate
siRNAs	small inhibiting RNAs
SNP	single-nucleotide polymorphism
snRNP	small nuclear ribonucleoprotein
SRB	suppressor of RNA polymerase B
STS	sequence tagged site
SV40	simian virus 40
TAF	TATA-box binding associated factor

TBP	TATA-box binding protein
TdT	terminal deoxynucleotidal transferase
TGMV	tomato golden mosaic virus
TK	thymidine kinase
tRNA	transfer RNA
VA RNAs	viral associated RNAs
VNTR	variable number tandem repeat
YAC	yeast artificial chromosome

1 DNA: Structure and function

Key concepts

♦ The genetic information is contained within nucleic acids

♦ DNA is a double-stranded antiparallel helix

♦ Base pairing (A to T and G to C) holds the two strands of the helix together

♦ DNA replication occurs through the unwinding of the DNA strands and copying each strand

♦ The central dogma of molecular biology:

  ○ DNA makes RNA makes protein

♦ Transcription is the production of an RNA copy of one of the DNA strands

♦ Translation is decoding of an RNA molecule to produce protein

Every organism possesses the information required to construct and maintain a living copy of itself. The basic concepts of heredity and, as a consequence, genes can be traced back to 1865 and the studies of Gregor Mendel – discussed by Orel (1995). From the results of his breeding experiments with peas, Mendel concluded that each pea plant possessed two **alleles** for each gene, but only displayed a single **phenotype**. Perhaps the most remarkable achievement of Mendel was his ability to correctly identify a complex phenomenon with no knowledge of the molecular processes involved in the formation of that phenomenon. Hereditary transmission through sperm and egg became known about the same time and Ernst Haeckel, noting that sperm consists largely of nuclear material, postulated that the nucleus was responsible for heredity.

Analysis of Genes and Genomes   Richard J. Reece
© 2004 John Wiley & Sons, Ltd   ISBNs: 0-470-84379-9 (HB); 0-470-84380-2 (PB)

1.1   Nucleic Acid is the Material of Heredity

The idea that genetic material is physically transmitted from parent to offspring has been accepted for as long as the concept of inheritance has existed. Both proteins and nucleic acid were considered as likely candidates for the role of the genetic material. Until the 1940s, however, many scientists favoured proteins. There were two main reasons for this. Firstly, proteins are abundant in cells; although the amount of an individual protein varies considerably from one cell type to another, the overall protein content of most cells accounts for over 50% of the dry weight. Secondly, nucleic acids appeared to be too simple to convey the complex information presumed to be required to convey the characteristics of heredity. **DNA** (deoxyribonucleic acid) was first isolated in 1869 by the Swiss chemist Johann Frederick Miescher. He separated nuclei from the cytoplasm of cells, and then isolated an acidic substance from these nuclei that he called nuclein. Miescher showed that nuclein contained large amounts of phosphorus and no sulphur, characteristics that differentiated it from proteins. In what proved to be a remarkable insight, he suggested that 'if one wants to assume that a single substance... is the specific cause of fertilization then one should undoubtedly first of all think of nuclein'.

In 1926, based on the idea that DNA contained approximately equal amounts of four different groups, called nucleotides, and by determining the type of linkage that joined the nucleotides together, Levene and Simms proposed a tetranucleotide structure (Figure 1.1) to explain the chemical arrangement of nucleotides within nucleic acids (Levene and Simms, 1926). They proposed a very simple four-nucleotide unit that was repeated many times to form long nucleic acid molecules. Because the tetranucleotide structure was relatively simple, it was widely believed that nucleic acids could not provide the chemical variation expected of the genetic material. Proteins, on the

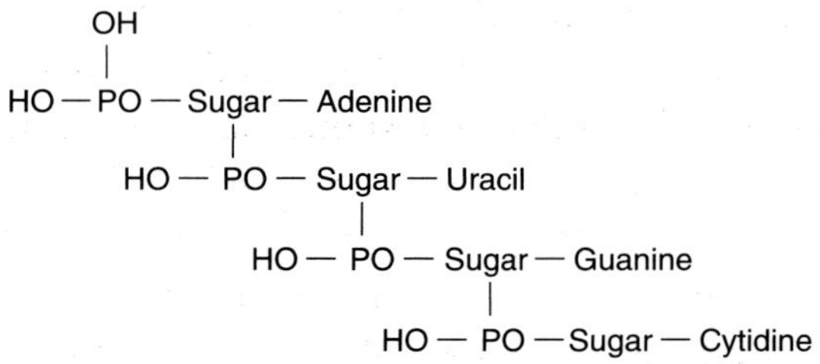

```
            OH
            |
      HO — PO — Sugar — Adenine
            |
        HO — PO — Sugar — Uracil
              |
          HO — PO — Sugar — Guanine
                |
            HO — PO — Sugar — Cytidine
```

Figure 1.1. The tetranucleotide model for nucleic acid structure proposed by Levene and Simms in 1926. At the time that this model was proposed, it was thought that plant and animal nucleic acid might be different, and the differences between DNA and RNA were not fully understood

other hand, containing 20 different amino acids, could provide the basis for substantial variation.

In 1928, Frederick Griffith performed experiments using several different strains of the bacterium *Streptococcus pneumoniae* (Griffith, 1928). Some of the strains used were termed virulent, meaning that they caused pneumonia in both humans and mice. Other strains were avirulent, and did not cause illness. Virulent and avirulent strains are morphologically distinct in that the virulent strains have a polysaccharide capsule surrounding the bacterium and form smooth, shiny-surfaced colonies when grown on agar plates. Avirulent bacteria lack the capsule and produce rough colonies on the same plates. The smooth bacteria are virulent because the polysaccharide capsule means that they are not easily engulfed by the immune system of an infected animal, and thus are able to multiply and cause pneumonia. The rough bacteria that lack the polysaccharide capsule do not have this protection and are consequently readily engulfed and destroyed by the host immune system.

Griffith knew that only living virulent bacteria would produce pneumonia when injected into mice. If heat-killed virulent bacteria were injected into mice, no pneumonia would result, just as living avirulent bacteria failed to produce the disease when similarly injected. Griffith's critical experiment (Figure 1.2) involved the injection into mice of living rough bacteria (avirulent) combined with heat-treated smooth bacteria. Neither cell type caused death in mice when they were injected alone, but all mice receiving the combined injections died. The analysis of blood of the dead mice revealed a large number of living smooth bacteria when grown on agar plates. Griffith concluded that the heat-killed smooth bacteria were somehow responsible for converting the live avirulent rough bacteria into virulent smooth ones. He called the phenomenon **transformation**, and suggested that the **transforming principle** might be some part of the polysaccharide capsule or some compound required for capsule synthesis, although he noted that the capsule alone did not cause pneumonia.

In 1944, Oswald Avery, Colin MacLeod and Maclyn McCarty published their work to show that the molecule responsible for the transforming principle was DNA (Avery, MacLeod and McCarty, 1944). They began by culturing large quantities of smooth *Streptococcus pneumoniae* cells. The cells were harvested from cultures and then heat-killed. Following homogenization and several extractions with detergent, they obtained an extract that, when tested by co-injection with live rough bacteria, still contained the transforming principle. Protein was removed from the extract by several chloroform extractions and polysaccharides were enzymatically digested and removed. Finally, precipitation of the resultant fraction with ethanol yielded a fibrous mass that still retained the ability to induce transformation of the rough avirulent cells. From the

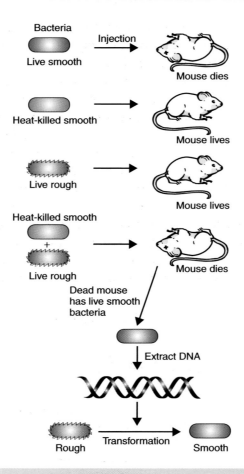

Figure 1.2. The key features of Griffith's experiments, combined with data of Avery, MacLeod and McCarty to identify DNA as the transforming principle from *Streptococcus pneumoniae*. Griffith noted that injecting live smooth bacteria into mice led to the formation of pneumonia and eventually to the death of the mouse. Heat-treating the bacteria before injection did not result in the formation of the disease. Non-virulent bacterial strains, which did not cause the disease on their own, could be transformed by co-injection with heat-treated virulent bacteria. Avery, MacLeod and McCarty identified DNA as the transforming principle

original 75 L culture of bacterial cells, the procedure yielded 10–25 mg of the 'active factor'. Further testing established beyond a reasonable doubt that the transforming principle was DNA. The fibrous mass was analysed for its nitrogen/phosphorus ratio, which was shown to coincide with the ratio expected for DNA. In order to eliminate all probable contaminants from their final extract, they treated it with the proteolytic enzymes trypsin and chyomtrypsin, and then digested it with an RNA digesting enzyme called ribonuclease. Such treatments destroyed any remaining activity of proteins and RNA, but still retained the transforming activity. The final confirmation that DNA was transforming principle came by digesting the extract with deoxyribonuclease, which destroyed the transforming activity.

The second major piece of evidence supporting DNA as the genetic material was provided by the study of the infection of the bacterium *Escherichia coli* by one of its viruses, bacteriophage T2. Often simply referred to as **phage**, the virus consists of a protein coat surrounding a core of DNA (see Figure 1.3). In the early 1950s, little was known about the early steps of phage infection. The phage was known to be adsorbed to the surface of the bacteria, after which there was a latent period of approximately ten minutes before infectious virus particles started to be made, ultimately leading to host cell lysis and phage release. Alfred Hershey and Martha Chase reasoned that if they knew the fate of the phage protein and the nucleic acid at the beginning of the infection process, they would understand more about the nature of those early steps.

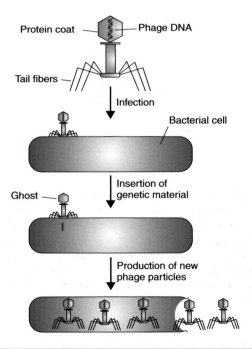

Figure 1.3. The life cycle of T2 phage. During the course of infection, the bacteriophage adheres to the surface of the *Escherichia coli* cell. The genetic information, but not the whole phage particle, is inserted into the bacterium, where it is replicated. The phage 'ghost', which lacks the genetic material, remains at the bacterial surface. Once the newly synthesized phage particles are produced, bacterial cell lysis occurs and the phage particles are released into the surrounding medium

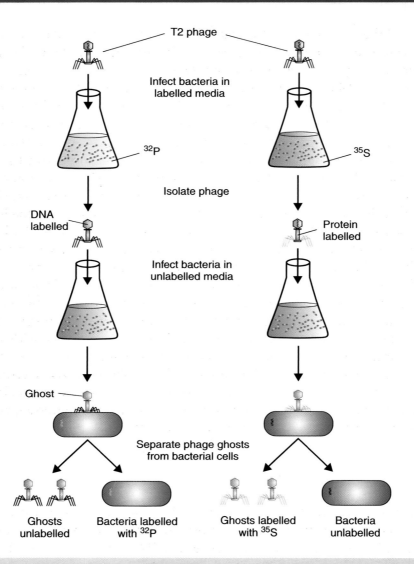

Figure 1.4. The Hershey–Chase blender experiment to show that nucleic acid was the genetic material. Hershey and Chase grew T2 bacteriophages on bacteria whose media contained either ^{32}P (to label the phosphorus of nucleic acid) or ^{35}S (to label the sulphur of proteins – the side chains of the amino acids methionine and cysteine both contain sulphur). They used their radio-labelled bacteriophages to infect a new culture of unlabelled bacteria. After a brief incubation, the bacteria were harvested by centrifugation and put into a blender to shear the bacteria away from the phage particles attached to their surface. They found that, when the DNA was labelled, the label was transferred to the bacterial cell, while the labelled protein remained with the phage ghosts. They concluded, therefore, that the material of heredity – i.e. the material passed on to make new offspring – was nucleic acid

They used the radioisotopes ^{32}P and ^{35}S to follow the molecular components of the phages during infection (Figure 1.4). Because DNA contains phosphorus but not sulphur, ^{32}P effectively labels DNA, and because proteins contain sulphur but not phosphorus, ^{35}S labels protein. If *E. coli* cells are grown in the presence of ^{32}P or ^{35}S and then infected with T2 virus, the newly synthesized phages will have either a radioactively labelled DNA core or a radioactively labelled protein coat, respectively. These labelled phages can be isolated from the medium of infected cultures and used to infect other unlabelled bacteria. Hershey and Chase labelled the T2 phages with either ^{35}S or ^{32}P, and allowed them to adsorb to unlabelled bacteria. The cells were then separated from the unadsorbed material by centrifugation. The cells were resuspended and the suspension was blended to separate the phage 'ghosts' from the infected bacteria. The ghosts should not contain the genetic material, which needs to be replicated inside the host bacteria. Hershey and Chase found that ∼80% of the ^{35}S label was removed from the bacteria whereas only ∼20% of the ^{32}P label was removed. They concluded that 'most of the phage DNA enters the cell, and a residue containing at least 80% of the sulphur-containing protein of the phage remains at the cell surface'. This work, together with that of Avery, MacLeod and McCarty, provided overwhelming evidence that DNA was the molecule responsible for heredity. Curiously, however, Hershey and Chase seemed somewhat skeptical about their findings, concluding their paper by saying that the '...protein has no function in phage multiplication, and that DNA has some function' (Hershey and Chase, 1952).

1.2 Structure of Nucleic Acids

As we have seen in the above section, the material of heredity is DNA, or more correctly nucleic acids. In most organisms, the hereditary material is DNA. However, a number of viruses use RNA (ribonucleic acid) as the building block for their genome. DNA and RNA are polymeric molecules made up of linear chains of subunits called nucleotides. Each nucleotide has three parts: a nitrogenous base, a five-carbon-atom sugar and a phosphate group (Figure 1.5). The combination of base and sugar is termed a **nucleoside**, while the base–sugar–phosphate is called a **nucleotide**. Since they contain the sugar 2′-deoxyribose, the nucleotides of DNA are termed deoxyribonucleotides, while those of RNA, which contain the sugar ribose, are known as ribonucleotides. The nucleotide bases can be either a double-ringed **purine** or a single-ringed **pyrimidine**. DNA and RNA are both built up from two purine containing nucleotides and two pyrimidine containing nucleotides. The purines of both DNA and RNA are the same – **adenine** (A) and **guanine** (G). The pyrimidine

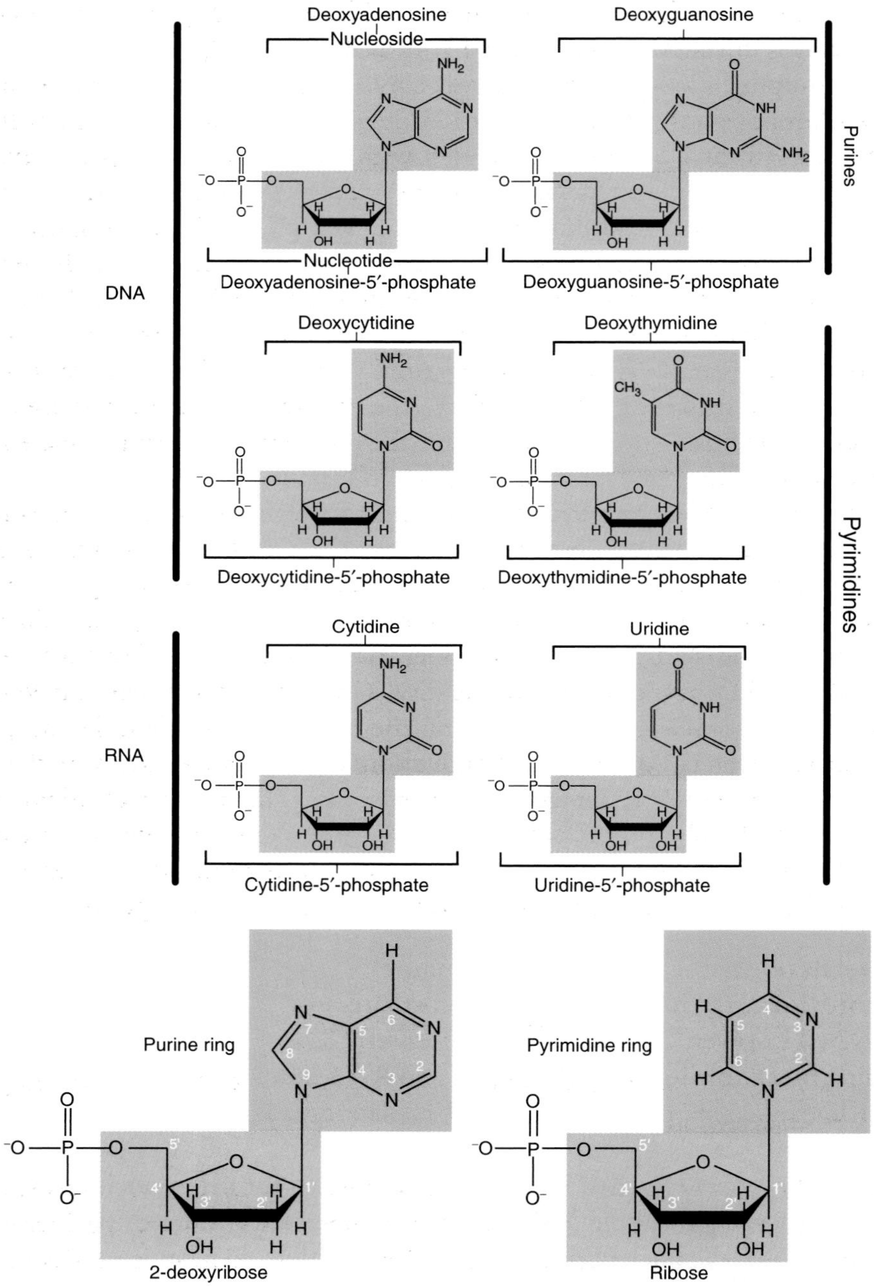

Figure 1.5. The structures of the purines and pyrimidines found in nucleic acids. The nitrogenous bases are highlighted in orange and the sugar groups are highlighted in blue. Beneath is the numbering system used throughout this text. The atoms of the purine ring are numbered from 1 to 9, and those of the pyrimidine ring are numbered from 1 to 6. The atoms of the sugar are numbered from 1' to 5'

cytosine (C) is also found in both nucleic acids, while the pyrimidine **thymine** (T) is limited to DNA, being replaced by **uracil** (U) in RNA.

The numbering system for nucleotides that is used extensively through this text is shown in Figure 1.5. Each of the carbon and nitrogen atoms in both the pyrimidine and purine rings is numbered from 1 to 6, or 1 to 9, respectively. The carbon atoms of the sugar ring – either ribose or deoxyribose – are numbered from 1′ to 5′ (spoken as 1-prime to 5-prime). Thus, 2′-deoxyribose lacks a hydroxyl group attached to the 2′ carbon of the sugar ring. Individual nucleotides are connected to each other in both DNA and RNA through sugar–phosphate bonds that connect the hydroxyl group on the 3′ carbon of one nucleotide with the phosphate group on the 5′ carbon of another nucleotide. See Figure 1.6. Two nucleotides connected to each other are called a **dinucleotide**, three are called a **trinucleotide** and numerous nucleotides connected in a long chain is termed a **polynucleotide**.

In the early 1950s, the chemist Erwin Chargaff was performing experiments to address the chemical composition of nucleic acids, and he realized that nucleic acids did not contain equal proportions of each nucleotide. Chargaff isolated DNA from a number of organisms, both prokaryotic and eukaryotic (Chargaff, Lipshitz and Green, 1952; Chargaff *et al.*, 1951; Zamenhof, Brawerman and Chargaff, 1952). He hydrolysed the DNA into its constituent nucleotides by treatment with strong acid, and then separated the nucleotides by paper chromatography. His experiments showed that the relative ratios of the four bases were not equal, but were also not random. The number of adenine (A) residues in all DNA samples was equal to the number of thymine (T) residues, while the number of guanine (G) residues equalled the number of cytosine (C) residues (Table 1.1). Chargaff's rules state that for any given species

- A = T and G = C

- sum of the purines = sum of the pyrimidines

- the percentage of (C + G) does not necessarily equal the percentage of (A + T).

These findings opened the possibility that it was the precise arrangements of nucleotides within a DNA molecule that conferred its genetic specificity, but the fundamental significance of the A = T and G = C relationships was not full realized until the three-dimensional structure of DNA was solved. As we will see later, in DNA A always pairs with T and G always pairs with C.

Between 1940 and 1953, many scientists were interested in solving the structure of DNA. X-ray diffraction as a method of determining protein structure was becoming an established technique. X-ray diffraction involves

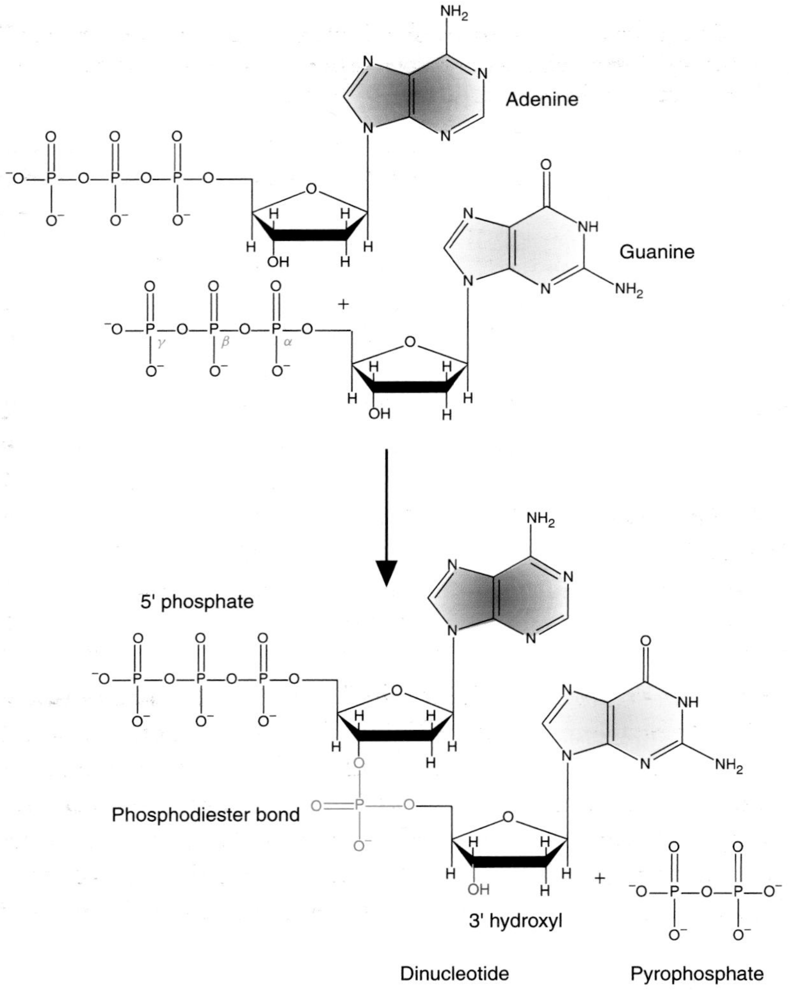

Figure 1.6. The joining of nucleotides. The joining of an adenine and a guanine nucleotide. The phosphates on the sugar ring of guanine are designated as α, β or γ. In the formation of the dinucleotide, pyrophosphate (representing the β and γ phosphates) is lost and the phosphodiester bond links the 3' hydroxyl to the phosphate on the 5' carbon atom of the sugar. DNA molecules invariably have a free 5' phosphate and 3' hydroxyl

firing a beam of X-rays at a regular array of molecules – either a crystal or a fibre. When the X-rays hit an atom in the array they will be diffracted, and the diffracted beams are detected as spots on X-ray film. Analysis of the diffraction patterns yields information about the structure and shape of the molecules in the array. As early as 1938 William Astbury applied the technique to fibres of DNA. By 1947, he had detected a periodicity (or repeating unit) within DNA of

Table 1.1. Chargaff's rules. The ratios of individual nucleotides isolated from DNA of various sources. While the ratios of purine:purine and pyrimidine:pyrimidine vary widely, the ratio of purine:pyrimidine was found to be a constant unity

Organism	A to G	T to C	A to T	G to C	Purines: pyrimidines
Ox	1.29	1.43	1.04	1	1.1
Human	1.59	1.75	1	1	1
Hen	1.45	1.29	1.06	0.91	0.99
Salmon	1.43	1.43	1.02	1.02	1.02
Sea urchin	1.83	1.80	1.02	1.00	1.01
Wheat	1.22	1.18	1	0.97	0.99
Yeast	1.67	1.92	1.03	1.2	1
Hemophilus influenzae	1.75	1.54	1.06	0.93	1.01
Escherichia coli	1.05	0.95	1.09	0.99	1
Serratia marcescens	0.76	0.63	1.03	0.85	0.92
Bacillus schatz	0.68	0.58	1.07	0.9	0.96

0.34 nm. Between 1950 and 1953, Rosalind Franklin obtained improved X-ray data from highly purified samples of DNA. Her work confirmed the 0.34 nm periodicity, and suggested that the structure of DNA was some sort of helix. Franklin, however, did not propose a model for the structure of DNA. Rather, Linus Pauling and Robert Corey used Franklin's data, together with that of others, to propose that DNA was a triple helix with the phosphates near the centre of the axis and the bases on the outside (Pauling and Corey, 1953).

1.3 The Double Helix

Franklin noted that DNA fibres could give two distinct types of diffraction pattern depending upon how the samples were prepared and stored. The first (termed Structure A) was composed of fibres that were relatively dehydrated, while the second (Structure B) was prevalent over a wide variety of conditions. She noted that the change from Structure A to Structure B was reversible, depending on the levels of sample hydration (Franklin and Gosling, 1953). It is thought that the B-form of DNA is the biologically significant conformation. Other forms of DNA (the right-handed A form and the left-handed Z form) certainly do exist under certain conditions, and may play significant roles in certain cellular processes. For example, a family of proteins that bind specifically to Z-DNA has recently been described (Schwartz *et al.*, 2001). Here, however, we will concentrate on the properties and interactions of B-form DNA.

In 1953, James Watson and Francis Crick attempted to build molecular models of DNA and realized that the Pauling–Corey structure was incorrect, with some atoms having to be closer together than was possible. By combining Franklin's X-ray diffraction patterns with Chargaff's rules, Watson and Crick proposed the, now famous, double-helix model in 1953 (Watson and Crick, 1953a). This model, shown in Figure 1.7, has the following major features, some of which have been updated slightly from the original model in the light of high-resolution crystal X-ray diffraction data.

(a) Two long polynucleotide chains coiled around a central axis, forming a right-handed double helix – this means that the turns are clockwise when looking down the helical axis.

(b) The two chains are antiparallel; that is, each chain has a specific orientation, and these run in opposite directions.

(c) The bases of both chains are flat structures, lying perpendicular to the axis. They are 'stacked' on one another, 0.34 nm apart, and are located on the inside of the helix.

(d) The nitrogenous bases of opposite strands are paired to one another by hydrogen bonds.

(e) Each complete turn of the helix is 3.4 nm long. This means that just over ten bases from each strand (10.4 bp) form one complete turn of the helix.

(f) Along the molecule, alternating larger major grooves and smaller minor grooves are apparent.

(g) The double helix measures approximately 2 nm in diameter.

The pairing of the nitrogenous bases in the centre of the helix is the most significant feature of the model by Watson and Crick. However, several other features are also important to understand the double helix.

1.3.1 The Antiparallel Helix

The antiparallel nature of the two polynucleotide chains is a key part of the double helix. Given the constraints of the bond angles of the bases and sugar phosphates, the double helix could not be constructed easily if both chains ran parallel to each another. One chain of the helix runs in the 5′ to 3′ orientation, and the other chain runs in the 3′ to 5′ orientation. This is illustrated in Figure 1.8. The 5′ and 3′ nomenclature is derived from the numbering system of the sugar ring that we saw in Figure 1.5. By convention, DNA sequences are

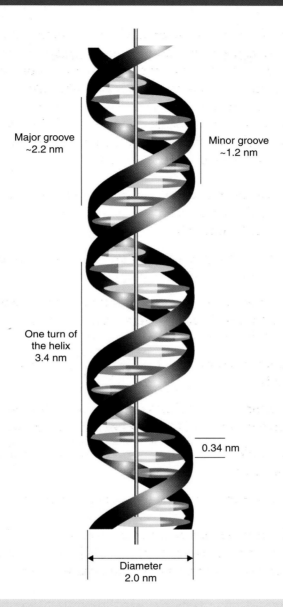

Major groove
~2.2 nm

Minor groove
~1.2 nm

One turn of
the helix
3.4 nm

0.34 nm

Diameter
2.0 nm

Figure 1.7. The Watson and Crick model of DNA

written in the 5′ to 3′ direction. This means that a single DNA chain begins
with a free phosphate group on the 5′ carbon of a deoxyribose ring. Additional
nucleotides are joined to the chain through phosphodiester bonds, which link
the hydroxyl group on the 3′ carbon atom of one sugar with the phosphate
on the 5′ carbon atom of an adjoining sugar. The chain terminates in a free
hydroxyl group on the 3′ carbon atom of the last sugar.

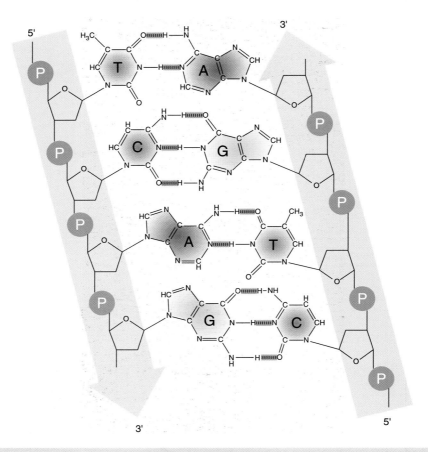

Figure 1.8. DNA base pairing and complementation. The two chains of the helix, arrowed in the 5′ to 3′ direction, are antiparallel. The bases on one strand of the helix are complementary to those on the opposite strand, A always base pairs with T and G always base pairs with C

1.3.2   Base Pairs and Stacking

The bases of both DNA chains are flat structures that lie approximately perpendicular to the helical axis. The bases themselves are stacked upon each other. The arrangement is best illustrated by inspection of a computer-generated model of high-resolution crystal X-ray diffraction data (Figure 1.9). It can be noted that the base pairs are not all perpendicular to the helical axis, and that some show **propeller twist**, where the purine and pyrimidine pair do not lie flat but are twisted with respect to each other, like the blades of a propeller (Dickerson, 1983). The pairing of a purine (A or G) with a pyrimidine (T or C) within the helix is important for the integrity of the helix.

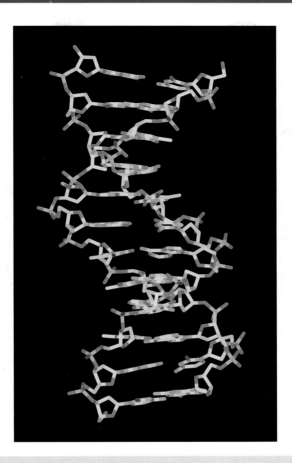

Figure 1.9. Computer generated model of DNA. The structure of double-stranded B-form DNA as derived from high-resolution X-ray diffraction of DNA crystals. Oxygen atoms are coloured red, phosphorus is orange, carbon is white and nitrogen is blue

The constant length of the purine–pyrimidine pairing would be disrupted if purine–purine (too large) or pyrimidine–pyrimidine (too small) pairings occurred. The purine–pyrimidine pairs are said to complement each other, and the two strands of a single DNA molecule are thus **complementary** to one another. Thus, if the sequence 5′-ATGATCAGTACG-3′ occurs on one strand of the DNA, the other strand must have the sequence 5′-CGTACTGATCAT-3′. These two sequences are complementary to each other:

```
Strand one: 5'-ATGATCAGTACG-3'
               ||||||||||||
Strand two: 3'-TACTAGTCATGC-5'
```

As we will we see in many of the subsequent chapters, the ideas of complementation between two strands of DNA form the basis of many genetic engineering experiments. The pairing of two DNA strands is very specific. Precise matches between two DNA strands, like those shown above, are highly stable and readily form helices. As we will see later, two DNA strands that are not precisely complementary to one another, but where there is still a high degree of complementation, retain the ability to interact with each other.

1.3.3 Gaining Access to Information with the Double Helix without Breaking it Apart

How can the information held within the sequence of DNA be read without having to unravel the double helix? The invariant nature of the sugar–phosphate backbone would seem to provide an almost impenetrable barrier to 'reading' the DNA base sequence. The grooves along the helical axis do, however, provide a mechanism whereby the bases can be distinguished from one another. As we can see in Figure 1.7, DNA is composed of alternating major and minor grooves along its axis. This is a result of the glycosidic bonds that attach a base pair to its sugar rings not lying directly opposite each other across the helical axis. As a result, the two sugar–phosphate backbones of the double helix are not equally spaced along the helical axis, and the grooves that form between the backbones are not of equal size. The major groove is wide ($\sim$0.22 nm) and shallow, while the minor groove is narrow ($\sim$0.12 nm) and deep. The floor of the major group is composed mainly of nitrogen and oxygen atoms that belong to the unique portions of each base pair. In contrast, the floor of the minor groove is filled with nitrogen and oxygen atoms that are generally common to either the purines or to the pyrimidines. Thus, the potential of the major groove for interactions shows a much greater dependence on base sequence than that of the minor groove. This finding led to the speculation that DNA sequence-specific binding proteins recognize DNA by forming hydrogen bonds predominantly to specific groups positioned within and along the major groove. One of the most common ways in which proteins can recognize specific DNA sequences is by the insertion of a protein α-helix into the major groove of DNA. The α-helix, originally postulated by Pauling and Corey in 1951, is a protein secondary structure motif in which a right-handed helix is formed by amino acids on a polypeptide chain (Pauling *et al.*, 1951). Each amino acid in the helix occupies a vertical distance of 0.15 nm, and there are 3.6 amino acid residues per turn of the helix (see Appendix 1). The diameter of the polypeptide backbone in an α-helix is approximately 0.5 nm; however, the amino acid side chains project away from the helical axis. This results in a protein α-helix being able to fit almost exactly into the major groove of double-stranded DNA. The amino acid side chains that project away from the α-helix are able to form hydrogen bonds with the DNA bases in the major groove. The type of protein–DNA interaction is shown in Figure 1.10.

1.3.4   Hydrogen Bonding

The Watson and Crick model of DNA structure predicts that the two polynucleotide chains are held together by non-covalent hydrogen bonds rather than by covalent interactions. This raises several important questions – what is a hydrogen bond, and is it sufficiently strong to maintain the integrity of the double helix?

To address the first question, a hydrogen bond is a weak electrostatic interaction between a covalently bonded hydrogen atom and an atom with

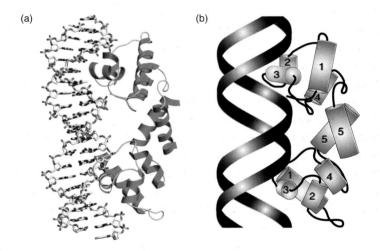

(a)   (b)

Figure 1.10. The interaction between the λ-repressor of bacteriophage λ and DNA. (a) Computer generated model of the interaction between DNA, shown with the same colouring scheme as in Figure 1.9, and λ-repressor, whose α-helices are shown in green and are connected by amino acids that do not adopt secondary structure. Two molecules of the λ-repressor (a dimer) interact with a 17 bp segment of double-stranded DNA. (b) The five helices of λ-repressor. Each monomer of the λ-repressor DNA binding domain has five helices, numbered 1–5 from the amino terminal end of the protein. Helix 3 lies in the major groove and the side chains (not shown) extend to the edges of the major groove and make contacts with the DNA bases. Helices 2 and 3 form a 'helix–turn–helix' DNA binding motif that is found in many DNA binding proteins. Helix 3 – the recognition helix – forms DNA sequence-specific contacts in the major groove, while helix 2 – the stabilization helix – interacts non-specifically with the DNA backbone to provide stability to the DNA–protein interaction

an unshared electron pair – a characteristic of covalently bonded oxygen and nitrogen atoms. The hydrogen atom assumes a partial positive charge, while the unshared electron pair assumes a partial negative charge. These opposite charges are responsible for the weak chemical attraction.

In general terms, a chemical bond is an attractive force that holds atoms together. The spontaneous formation of a bond between two free atoms involves the release of internal energy of the unbonded atoms and its conversion to another energy form, e.g. heat. The strength of a particular bond is measured by the amount of energy released upon the formation of the bond. The stronger the bond, the more energy is released. The change in energy (ΔG) that accompanies bond formation is used to describe the strength of a bond. The value of ΔG is calculated according to the following equation:

$$\Delta G = -RT \ln K_{eq}$$

where R is the universal gas constant (8.314 J K^{-1} mol^{-1}), T is the absolute temperature and $\ln K_{eq}$ is the natural log of the equilibrium constant between the bonded and the unbonded forms of the molecule. Covalent bonds are short (0.095 nm) and are very strong, with ΔG values in the range of -100 to -500 kJ mol^{-1}. Hydrogen bonds on the other hand, are longer (approximately 0.3 nm) and are much weaker, with ΔG values in the range of -10 to -30 kJ mol^{-1}.

As found in the double helix, adenine forms two hydrogen bonds with thymine, and cytosine forms three hydrogen bonds with guanine (see Figure 1.8). While a single hydrogen bond is itself very weak, the 2500 or so hydrogen bonds that hold together every kilobase of DNA provide an extraordinary level of stability to the helix. This also means, as we will see below and in later chapters, that an AT base pair is less stable in thermodynamic terms than a GC base pair.

1.4   Reversible Denaturing of DNA

Double-stranded DNA is an immensely stable molecule. In a dehydrated sample, it can survive virtually intact for thousands of years. Fragments of DNA have been isolated from Egyptian mummies that date back some 3000 years. In such samples, the double helix is still intact, but the DNA is fragmented. That is, rather than existing as a chain of millions of nucleotides joined as a contiguous unit, the DNA is composed of short, 500–1000 base pair (bp) fragments, with some of the covalent linkages forming the phosphodiester backbone having been broken.

A major consequence of the non-covalent forces that hold the double helix together is that the two constituent strands of DNA may be separated (denatured) simply by heating. The thermal energy provided by heating a DNA sample will break the relatively weak hydrogen bonds connecting the two strand of the helix, but will not affect the covalent linkages that hold each strand together (Lin and Chargaff, 1966). The separation of the two DNA strands in a helix is accompanied by changes in the physical properties of DNA. One of these changes is the way in which DNA absorbs UV light. DNA absorbs light at a wavelength of 260 nm due to the presence of alternating single and double bonds in the DNA bases (Figure 1.5). Each of the four nucleotide bases has a slightly different absorption spectrum, and the overall absorption spectrum of DNA is the average of them. A pure DNA solution appears transparent to the eye, and absorption does not become measurable until approximately 320 nm. Moving further into the UV region, there is an absorption peak at about 260 nm, followed by a dip between 220 and 230, and then the solution becomes essentially opaque in the far UV. A solution of double-stranded, native DNA, with a concentration of 0.05 mg/mL, has an absorbance (or optical density, OD) of about 1.0 at the 260 nm peak.

When a DNA helix is denatured to become single strands, e.g. by heating, the absorbance increases. The same 0.05 mg mL double-stranded DNA solution that we used above will have an absorbance of 1.5 when in the single-stranded form. This type of absorbance data is often expressed in the form shown below:

$$1.0 \; A_{260 \text{ nm}} \text{ double-stranded DNA} = 50 \; \mu\text{g/mL}$$

$$1.0 \; A_{260 \text{ nm}} \text{ single-stranded DNA} = 33 \; \mu\text{g/mL}$$

The increase in absorbance as double-stranded DNA becomes single stranded, called the **hyperchromic effect** (Thomas, 1993), reveals the interaction between the electronic dipoles in the stacked bases of the double helix and is shown diagrammatically in Figure 1.11. The stacking of the base pairs in double-stranded DNA has the effect of dampening the absorption of individual nucleotides due to the competing interactions of adjacent base pairs in the stack. In the single-stranded form, no such competition occurs and consequently single-stranded DNA absorbs light to greater extent than its double-stranded counterpart.

When a DNA sample is heated, the helix begins to separate, in a process sometimes referred to as **melting**. Since AT base pairs have only two hydrogen bonds holding them together, regions of DNA with a high level of A and T residues will separate first. As the temperature continues to increase, more and more of the DNA will assume a single-stranded nature, until eventually complete strand separation occurs. The temperature at which half of the DNA is single stranded is called the **melting temperature** (T_m). The melting temperature

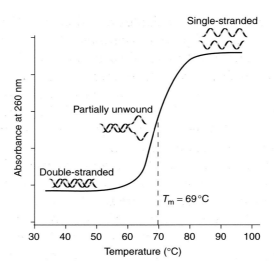

Figure 1.11. The effect of heating double-stranded DNA. As a double-stranded DNA molecule is heated above the ambient temperature, the non-covalent hydrogen bonds that hold the two strands together begin to break. This is accompanied by an increase in absorbance at 260 nm. At higher temperatures – above 90 °C – the two DNA strands will completely separate from each other. This process is reversible. If the temperature is reduced relatively slowly, the two complementary strands of DNA will reform to produce a correctly paired double-stranded DNA molecule. The melting temperature (T_m) is the temperature at which half the strands are separated

of all DNA molecules is not the same, and depends upon the length of DNA, and the proportion of GC and AT base pairs that it contains. Generally, short DNA molecules have low T_m values, as do those that contain a high proportion of AT base pairs. The melting temperature of a long DNA molecule can be calculated according to the following equation:

$$T_m = 16.5(\log[Na^+]) + 0.41(\%GC) + 81.5\,°C$$

where $[Na^+]$ is the concentration of sodium ions in the DNA solution and ($\%GC$) is the percentage of GC residues in the duplex. Thus, for a DNA molecule that contains 40% GC residues (typical for a mammalian genome) in a solution of 50 mM NaCl, the melting temperature is 76.4 °C. As we will see in later chapters, short DNA molecules (15–30 bases long) are often used in genetic engineering experiments. For short regions of complementation, the equation above breaks down, and the following estimation (sometimes called the Wallace rule) is used to determine the T_m (Wallace *et al.*, 1979):

$$T_m = 2(AT) + 4(GC)$$

That is, every AT base pair in the duplex contributes 2 °C of stabilization the double helix, while every GC pairing contributes 4 °C of stabilization. Therefore, an oligonucleotide of 20 bases of single-stranded DNA containing five A residues, six T residues, three C resides and six G residues will have an annealing temperature to its complementary DNA sequence of approximately $2(11) + 4(9) = 58$ °C. The importance of the annealing temperature will become apparent when we look at polymerase chain reaction (PCR) and nucleic acid hybridization in later chapters.

The heating of double-stranded DNA to produce the single-stranded version is an entirely reversible process. Single-stranded DNA made by heating a duplex will reform when the temperature is reduced. As the temperature falls, the thermal energy that was used to break the hydrogen bonds between the strands is reduced, and random collisions between the complementary strands will result in their re-association. Providing that the temperature is reduced relatively slowly (1–2 °C per second), complete duplex formation will result. The two complementary single-stranded DNA molecules will come together and reform the exact duplex that was present before the sample was heated. If the temperature is dropped rapidly, for example by plunging a DNA sample at 94 °C directly into ice, correct base pairing will not occur and the DNA will remain in a relatively single-stranded form.

1.5 Structure of DNA in the Cell

Different types of nucleic acid are used to form the genome of organisms depending on the organism itself. For example, viruses have a genome composed of either double-stranded DNA, single-stranded DNA or RNA, depending on the type of virus. The chromosomes of most eukaryotes are composed of single linear double-stranded DNA molecules. The genomes of prokaryotic organisms are generally composed of a circular DNA molecule. That is, rather than having free 5′- and 3′- ends, the ends are joined to each other to form a continuous ring of double-stranded DNA. A number of extra-chromosomal DNA molecules, called **plasmids**, are found in prokaryotic cells. As we will see in Chapter 3, understanding how cells deal with these plasmids has played a pivotal role in advances in molecular biology and genetic engineering. Plasmid DNA molecules are usually closed circles of either single-stranded or double-stranded DNA.

When looking at the structure of DNA as presented in Figure 1.7, it is easy to get the impression that the double helix is a rather inflexible solid rod. This, however, is simply not the case. Electron microscopy images of DNA molecules (Figure 1.12) indicate that DNA is more string-like than rod-like, and will wrap around itself to form a variety of irregular structures. Inside a eukaryotic

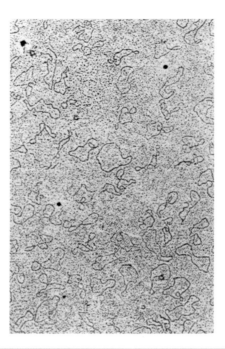

Figure 1.12. Images of double-stranded DNA generated by electron microscopy. Circular DNA isolated from polyoma virus shows many crossings of the DNA double strands, indicating that the DNA is supercoiled. Reproduced from Vinograd *et al.* (1965)

cell, DNA is associated with a vast array of proteins – for example, proteins required for its replication, its transcription into RNA and its packaging within the cell. Many of these proteins wrap DNA around themselves, or in other ways constrain or bend the DNA molecule. For short linear DNA molecules, this type of constraint is not a major problem. Stresses placed in one part of a DNA molecule by, for example, twisting the double helix can be relieved by untwisting another part of the DNA. The free ends of the linear DNA allow relatively free rotation of the DNA strands. For circular DNA molecules, however, these stresses can prove extremely problematic. Since the DNA has no free ends, it cannot simply untwist to counteract a twist elsewhere in the molecule. Twisting DNA molecules results in the formation of DNA **supercoils** (Figure 1.13).

In 1965 Jerome Vinograd and his colleagues suggested that closed-circular DNA molecules could adopt a 'twisted circular form' (Vinograd *et al.*, 1965). Such a form would result if, before joining the ends of a linear duplex DNA into a closed circle, one end was twisted relative to the other to introduce some strain into the molecule. Such coiling of the DNA helix upon itself is called **supercoiling**. Supercoiling can only be introduced into or released from

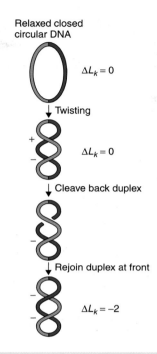

Relaxed closed
circular DNA

$\Delta L_k = 0$

↓ Twisting

$+$

$-$

$\Delta L_k = 0$

↓ Cleave back duplex

$-$

↓ Rejoin duplex at front

$-$

$-$

$\Delta L_k = -2$

Figure 1.13. The formation of DNA supercoiling. A relaxed closed-circular DNA molecule containing no supercoils is shown. The two halves of the molecule have been arbitrarily coloured red and blue to aid the visualization of DNA strands crossing over each other. A twist in the molecule will result in the formation of a positive (+) and a negative (−) supercoil. This process does not result in a change in the linking number. If, however, the DNA backbone is broken and subsequently resealed after a segment of DNA has passed through the break, then the linking number is reduced by 2. The molecule is now negatively supercoiled

a closed-circular DNA molecule by breaking at least one of the phosphodiester backbones.

The level of supercoiling within a particular DNA molecule can be described by its linking number (L_k). This number corresponds to the number of double-helical turns in the original linear molecule. The process of introducing supercoils into a DNA molecule reduces or increases the number of helical turns trapped when the circle is formed, and results in changes in the linking number to the final closed-circular species. The linking difference (ΔL_k) gives an indication of the overall levels of supercoiling within a molecule (Figure 1.13). Interested readers should refer to dedicated texts for a fuller description of DNA topology and its associated problems (Bates and Maxwell, 1993).

Closed-circular DNA isolated from prokaryotic cells is negatively super-coiled. That is, the DNA is relatively underwound in comparison to its linear state. Although eukaryotic cells contain linear DNA molecules, the length of DNA within a chromosome means that constraints placed in one part of the DNA molecule cannot be easily relieved by transferring the strain to one of the ends. An appropriate analogy here is to think about how you would unkink a twisted garden hosepipe. You would transfer the kink to the end of the pipe by untwisting it. At the end, the rotation of the pipe itself will result in the release of the kink. This type of action to release a kink or twist in a linear DNA molecule would be relatively straightforward if the DNA were short. Long chromosomal DNA molecules, however, suffer from a number of difficulties that make this type of constraint release impractical. The sheer length of DNA involved would make it difficult to move twists from the middle of the DNA to the ends, and the association of the DNA with proteins will provide a barrier to this type of movement. Coming back to the hosepipe analogy, another way to remove the kinks (although less practical in the garden) would be to cut the pipe to create new ends near the site of the twist. The pipe could then be untwisted locally and resealed to remove the kink and reform the pipe. It is in this fashion that cells deal with strains in DNA. They possess enzymes, called **DNA topoisomerases**, which make cuts in DNA molecules – either single- or double-stranded breaks in the phosphodiester backbone – to allow local twisting and untwisting of the DNA chain. The topoisomerase enzyme then reseals the DNA backbone to reform the DNA molecule.

1.6   The Eukaryotic Nucleosome

Each cell within our body contains a huge amount of DNA. The different chromosomes of the human genome contain approximately 3.2×10^9 base pairs of DNA. Since we are diploid organisms, having two sets of each chromosome, the total amount of DNA in most of our cells totals 6.4×10^9 base pairs. At 0.33 nm per base pair (Figure 1.7), this corresponds to an overall length of approximately 2.1 m. How can this fit into a nucleus measuring just 5–10 μm across? The answer is that the DNA is highly compacted. It is associated with a number of proteins that results in the wrapping of DNA into **nucleosomes**. During interphase, the genetic material (together with its associated proteins) is relatively uncoiled and dispersed throughout the nucleus as **chromatin**. When mitosis begins, the chromatin condenses greatly, and during prophase it is compressed into recognizable chromosomes. This condensation represents a contraction in length of some 10 000-fold.

The genetic material when isolated from bacteria and viruses consists of strands of DNA or RNA almost devoid of proteins. In eukaryotes, however, a substantial amount of protein is associated with the DNA to form chromatin.

Electron microscopic observations have revealed that chromatin fibres are composed of linear arrays of spherical particles. The particles occur regularly along the axis of a chromatin strand and resemble beads on a string. These particles, initially referred to as ν-bodies (Olins and Olins, 1974), are now called nucleosomes.

The digestion of chromatin with certain nucleases, such as micrococcal nuclease, yields DNA fragments that are approximately 200 bp in length, or multiples thereof (Wingert and Von Hippel, 1968). If the digestion of chromatin DNA were random, then a wide range of fragment sizes would be produced. This therefore demonstrates that the DNA of chromatin consists of repeating units that are protected from enzymatic cleavage. The DNA between the units is attacked and cleaved by the nuclease, and multiples occur where two or more units are joined together.

The proteins associated with DNA in chromatin are divided into basic, positively charged **histones** and less positively charged **non-histones**. Of the proteins associated with DNA, the histones play the most essential structural role. Histones contain large amounts of the positively charged amino acids lysine and arginine (see Appendix 1), making it possible for them to bind through electrostatic interactions to the negatively charged phosphate groups of the DNA nucleotides. There are five different types of histone protein – H1, H2A, H2B, H3 and H4. A nucleosome core particle consists of two copies each of histones H2A, H2B, H3 and H4 to form a histone octamer around which ~150 base pairs of DNA are wrapped in a left-handed superhelix, which completes about 1.7 turns per nucleosome (Figure 1.14). Recent insights into the structure of the nucleosome came in 1997 when Richmond and colleagues were able to solve the X-ray crystal structure of the nucleosome–DNA complex at high resolution (Luger *et al.*, 1997). At this resolution, most atoms of each histone are visible, and the precise path the DNA helix as it encircles the histone octamer can be traced (Figure 1.15). The high-resolution structure also revealed that the amino-terminal ends of some of the histones protrude from the octamer and project away from the nucleosome. The significance of the amino-terminal tails is that they have the potential to interact with adjacent nucleosomes to create nucleosome–nucleosome contacts. We will discuss the significance of the histone tails further when we look at the regulation of gene expression.

Extensive investigation of the structure of nucleosomes has provided the basis for predicting how the chromatin fibre within the nucleus is formed, and how it coils up into the mitotic chromosome. This model is illustrated in Figure 1.16. The 2 nm DNA double helix is initially coiled into a nucleosome core particle that is about 10 nm in diameter. Approximately 200 base pairs of DNA link each core particle to form the 'beads on a string' seen in electron

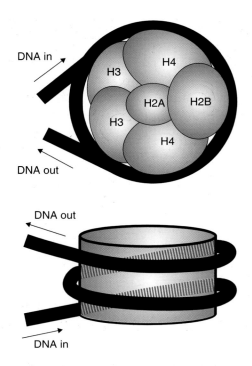

Figure 1.14. Wrapping of DNA around the nucleosome core. The nucleosome is composed of two molecules each of histones H4, H3, H2A and H2B. In the representation shown here only a monomer of H2A and H2B can be observed, the other monomers being located at the back of the octamer. Almost 150 bp of DNA wrap around the octamer core, forming approximately two turns

microscopy images. Histone H1, which is not part of the core octamer, may be located at the site where DNA enters and leaves the nucleosome and possibly functions to seal the DNA around the nucleosome.

The formation of nucleosomes represents the first level of packing, whereby the DNA is reduced to about one-third of its original length. In the nucleus, however, chromatin does not exist in this extended form. Instead, the 10 nm chromatin fibre is further packed into a thicker 30 nm fibre, which was originally called a solenoid. It is not clear whether the transition between the 10 nm fibre and the 30 nm fibre represents a physiological event or whether it merely occurs *in vitro* as a consequence of altering the salt concentration. The 30 nm fibre does, however, consist of numerous nucleosomes packed closely together, but the precise orientation and details of the structure are not clear. It has recently been suggested that the 30 nm fibre might adopt a compact helical zig-zag pattern with about four nucleosomes per 10 nm (Beard and Schlick,

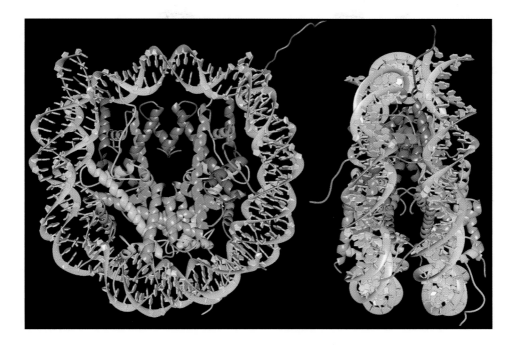

Figure 1.15. The structure of the nucleosome–DNA complex as solved by X-ray crystallography at high resolution. In both the top and side views of the complex, the strands of the DNA (shown in green and orange) wrap around the histone octamer. The protein core is almost exclusively internal with the exception of the amino-terminal histone tails that extend out from the complex. This figure was kindly provided by Tim Richmond (ETH Zurich) and is reprinted by permission from *Nature* (Luger *et al.*, 1997) copyright (1997) Macmillan Publishers Ltd

2001). The formation of the 30 nm fibre creates a second level of packaging, in which the overall length of the DNA is reduced some two fold.

In the transition to the mitotic chromosome, still another level of packing occurs. The 30 nm fibre forms a series of looped domains that further condense the structure of the chromatin fibre. The fibres are then coiled into the chromosome arms that constitute a **chromatid**, which is part of the metaphase chromosome.

In the overall transition from fully extended DNA helix to the extremely condensed status of the mitotic chromosome, a packaging ratio of about 500:1 must be achieved. The model presented above only accounts for a ratio of about 50:1. The remainder of the packing arises from the coiling and folding of the 30 nm fibre. The tight packing of the DNA into a chromosome presents an enormous challenge to both the replication of DNA and to its **transcription**. We will discuss these issues in the sections below.

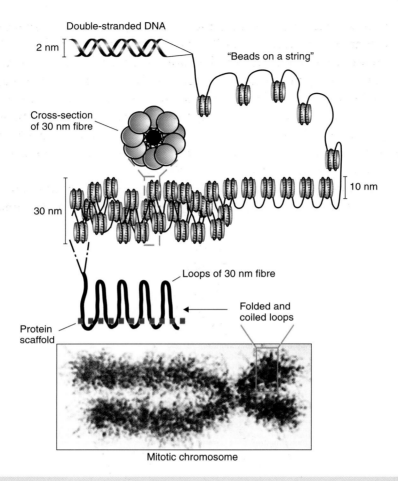

Double-stranded DNA

2 nm

"Beads on a string"

Cross-section
of 30 nm fibre

10 nm

30 nm

Loops of 30 nm fibre

Folded and
coiled loops

Protein
scaffold

Mitotic chromosome

Figure 1.16. Packaging of DNA into the eukaryotic chromosome. The 2 nm DNA double helix is wrapped into nucleosomes as illustrated in Figure 1.15. At some point histone H1 enters the complex, possibly at the DNA entry/exit point on the nucleosome. The nucleosomes can form extended 10 nm fibres, which are long arrays of ordered nucleosomes. The nucleosomes can further condense to form a 30 nm fibre. This may be a 'zig-zag' array of nucleosomes. The 30 nm fibre is then wrapped onto a protein scaffold, which can be additionally folded and coiled to form the mitotic chromosome that is observed under the electron microscope

1.7   The Replication of DNA

Watson and Crick finished their famous 1953 paper with the statement 'It has not escaped our notice that the specific pairing we have postulated immediately suggests a possible copying mechanism for the genetic material'. It was apparent to them from the arrangement and nature of the bases that each strand of the DNA double helix could serve as a template for the synthesis of a complementary strand (Figure 1.17). They proposed that if the double helix were unwound, each nucleotide on the parental strands would have an affinity for its complementary nucleotide (Watson and Crick, 1953b). Thus, each of the parental DNA strands could act as a template for the synthesis of a new DNA strand that would result in the production of two identical double-stranded DNA duplexes. In this model, each of the newly synthesized DNA molecules consists of one parental and one newly synthesized DNA strand; hence, the process is known as **semi-conservative replication**.

Two other modes of replication that also rely on the parental strands serving as templates for new DNA synthesis were also considered. In **conservative**

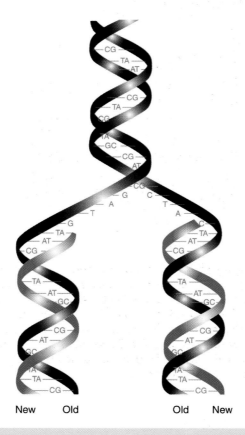

New   Old   Old   New

Figure 1.17. The replication of double-stranded DNA as suggested by Watson and Crick. The separation of the DNA strands of the parent helix (black) means that each strand can act as a template of the synthesis of a new strand of DNA (blue). This is termed semi-conservative replication, as the daughter DNA molecules contain one old strand and one new strand

replication, the parental helix would be opened to reveal the base sequence, and a newly synthesized helix would be produced. The two newly synthesized DNA strands would come together to form a helix and the parental helix would reform. In the second alternative mode, called **dispersive replication**, the parental DNA from one strand could be dispersed into either of the newly synthesized strands, which would then be a mixture of new and old DNA.

Matthew Meselson and Franklin Stahl provided the experimental evidence to suggest that semi-conservative replication was used by bacteria to produce new DNA molecules (Meselson and Stahl, 1958). The results of their classic experiment are depicted in Figure 1.18. They grew *E. coli* cells for many generations in a medium in which the sole source of nitrogen was a heavy – but

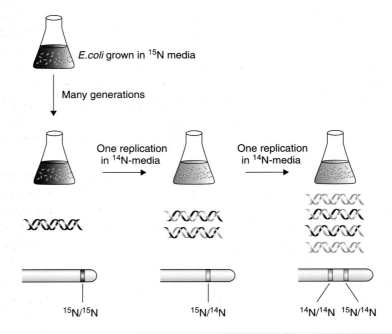

Figure 1.18. The Meselson–Stahl experiment. Bacterial cells were grown for many generations in a medium containing the heavy isotope of nitrogen, ^{15}N. The DNA isolated from such bacteria was isolated and centrifuged through a density gradient. The mobility of the DNA band was noted. The bacteria were then transferred to media containing the normal isotope of nitrogen (^{14}N). After one generation, all the DNA isolated from the bacteria ran through the density gradient with a mobility different to that seen previously. After two generations, two DNA species were observed. One had the same mobility as that seen after one generation, and the other ran with the same mobility as DNA isolated from bacteria grown on unlabelled media. In subsequent generations, the intensity of the latter band increased and the band of intermediate density decreased. These data are most readily explained by DNA replication occurring through a semi-conservative mechanism

non-radioactive – derivative of ammonium chloride ($^{15}NH_4Cl$). This meant that the DNA of these *E. coli* cells, into which the ^{15}N was incorporated as part of the bases, was heavier than the DNA from cells grown on normal ammonium chloride ($^{14}NH_4Cl$). The normal and heavy forms of DNA could be distinguished after centrifugation through a density gradient of caesium chloride. The heavier DNA will travel further under such conditions than the lighter normal DNA. This provided a means of monitoring the levels of each nitrogen isotope within the bacterial replicated DNA as it was replicated. Meselson and Stahl took their bacteria that contained heavy DNA and transferred them into media containing normal ammonium chloride and then isolated DNA after each successive generation. After a single round of DNA replication on the normal media, they found that the bacterial DNA now possessed an intermediate density between the expected positions for wholly ^{15}N-containing and wholly ^{14}N-containing DNA. This result is consistent with semi-conservative replication, but not with conservative replication, where two distinct bands would have been expected. After two cell divisions, Meselson and Stahl did observe the presence of two bands of different densities – the first corresponding to the position of a mixed $^{14}N/^{15}N$ DNA species and the second, of equal intensity, corresponding to the expected position of unlabelled $^{14}N/^{14}N$ DNA. Subsequent rounds of DNA replication resulted in an increased proportion of the $^{14}N/^{14}N$ DNA band.

The experiments described above suggest that conservative replication does not occur, but could not rule out dispersive replication. Meselson and Stahl, however, dismissed this mode of replication by the analysis of individual single strands of replicated DNA. They heated their DNA (Figure 1.11) and found that individual strands possessed either the density of ^{15}N- or ^{14}N-containing single-stranded DNA, but not that of an intermediate density – which would have been expected if a newly synthesized DNA strand contained a mixture of the parental and newly synthesized sequences. Semi-conservative replication has also been shown to occur in eukaryotes (Taylor, Woods and Hughes, 1957), and is now widely accepted as the universal mechanism for DNA replication.

1.8 DNA Polymerases

Both bacteria and eukaryotic cells possess multiple DNA polymerase enzymes. *E. coli* contain three such enzymes (Table 1.2). The first to be isolated, now called DNA polymerase I, is primarily involved in DNA repair rather than DNA replication. DNA polymerase II is also involved in DNA repair, while DNA polymerase III is the main replicating enzyme (Kornberg and Baker, 1992). DNA polymerase III is a multisubunit enzyme that functions as a dimer of these

Table 1.2. The properties of DNA polymerases I, II and III from *E. coli*

	Pol I	Pol II	Pol III
Main function	DNA repair and primer removal	DNA repair	DNA replication
Size, kDa	109	120	140
Structural genes	*polA*	*polB*	*DnaE, N, Q, X, Z*
Molecules per cell	400		10–20
5′ to 3′ polymerization	+	+	+
3′ to 5′ exonuclease	+	+	+
5′ to 3′ exonuclease	+	−	−

multiple subunits. Many DNA polymerase enzymes have been studied at the structural level, which, combined with extensive biochemical analysis, has led to the elucidation of their molecular mechanism of action.

DNA polymerase enzymes can possess three distinct enzymatic activities.

- All DNA polymerases direct the synthesis of DNA fragments in a 5′ to 3′ direction.

- Many DNA polymerases also possess 3′ to 5′ exonuclease activity. This enables the enzyme to remove nucleotides at the 3′- end of the newly synthesized chain. This is often considered a 'proof-reading activity', which can be used to excise incorrectly incorporated nucleotides before their replacement with the correct bases.

- Some DNA polymerases also have a 5′ to 3′ exonuclease activity. This allows the enzyme to remove sequences that have already been synthesized and, as we will see below, is involved in joining up discontinuous DNA fragments produced on the lagging strand during DNA replication.

DNA polymerases are not able to initiate DNA synthesis on their own. They have an absolute requirement for a free 3′-end onto which the enzyme can add new nucleotides. This means that two primers are required to initiate DNA replication in each direction from the origin – one complementary to each DNA strand to be replicated. As stated above, DNA synthesis only proceeds in a 5′ to 3′ direction. That is, new nucleotides are added on to the 3′ end of an existing polynucleotide chain. This leads to a problem. Both strands of the parent DNA molecule must be replicated, but only one of the strands is orientated in a 5′ to 3′ direction from the origin of replication. This strand, called the

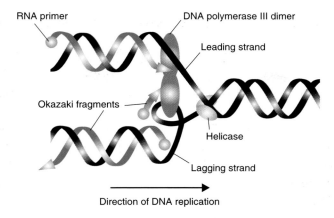

RNA primer

DNA polymerase III dimer

Leading strand

Okazaki fragments

Helicase

Lagging strand

Direction of DNA replication

Figure 1.19. DNA replication by DNA polymerase III. The strands of the parent DNA molecule (black) are unwound by a helicase at the replication fork. The leading strand and the lagging strand are then fed into a dimer of DNA polymerase III. DNA replication on the leading strand occurs in a 5′ to 3′ direction by extending a single RNA primer. Many primers are required to produce the lagging strand in short sections (Okazaki fragments), which are later joined using DNA ligase

leading strand, can be copied in a continuous manner – DNA replication can occur without stopping as the DNA polymerase is working in its favoured orientation. Replication of the other strand, the **lagging strand,** would appear to have to occur in the 3′ to 5′ direction, which is not possible. The lagging strand is replicated in a discontinuous fashion (Figure 1.19). For discontinuous strand synthesis, a series of short DNA segments are produced in a 5′ to 3′ direction (Okazaki *et al.*, 1968) that are then ligated together to produce the intact newly synthesized strand.

An enzyme that we will return to in later chapters is the RNA-dependent DNA polymerase called **reverse transcriptase.** This enzyme is found in certain viruses (called retroviruses) that contain RNA, rather than DNA, as their genetic material. Following infection of the host cell, the viral RNA serves as a template for the synthesis of a complementary DNA (**cDNA**) molecule. The DNA may then be incorporated into the host's genome where, if the DNA is expressed, the retroviral RNA genome will be produced. The ability to use the reverse transcriptase enzyme *in vitro* to produce DNA from RNA will be discussed in Chapter 5.

1.9   The Replication Process

How does DNA replication start and finish? As we have already seen, the DNA molecules inside cells are big and the need to tightly control the replication of

DNA sequences is obvious. We can consider the replication process in three phases – initiation, elongation and termination.

- *Initiation of replication.* The initiation of DNA replication does not occur at random sites around the genome, but rather it is initiated at specific points called **origins of replication**. Once DNA synthesis has been initiated, two replication forks, extending in either direction from the origin of replication, proceed to allow the full replication of the genome. Bacteria, such as *E. coli*, have a single origin of replication (called *Ori*C). Eukaryotic cells, on the other hand, have multiple origins of replication which are different from *Ori*C – the yeast *Saccharomyces cerevisiae* has been estimated to have about 300 replication origins, whilst human cells utilize over 20 000 origins during the replication of the genome. *Ori*C is a 245 bp DNA sequence that acts as the binding site for a number of proteins (namely DnaA, B and C). The binding of these proteins promotes the melting (opening) of the DNA helix, a process that is essential so that DNA replicating enzymes can read the base sequence. Replication origins in both prokaryotes and eukaryotes probably serve the same overall function, but the replication origins of prokaryotic cells will not substitute for their eukaryotic counterparts and *vice versa*. Once the helix has been unwound, the DNA polymerase can access the base sequences by 'reading' the base pair hydrogen bonds such that a complementary DNA chain may be synthesized. However, as we noted above, the polymerase can only function if a free 3′ hydroxyl group is present. This hydroxyl group is provided by an RNA primer (which is complementary to the DNA) that is 5–15 nucleotides long. The synthesis of the primer is directed by a form of RNA polymerase (called primase) that does not require a free 3′- end to initiate synthesis. DNA replication then proceeds simultaneously on both strands. Other proteins are also required for DNA replication to occur. The sections of single-stranded DNA produced during replication are stabilized through the binding of single-stranded binding proteins (SSBs), and the DNA is unwound into the polymerase complex with the help of DNA helicases (Figure 1.19). Additionally, topoisomerase enzymes (e.g. DNA gyrase) are required to relieve tension in the helix that results as a consequence of the unwinding process.

- *Elongation.* Given that DNA polymerase III can only replicate in a 5′ to 3′ direction, how can both DNA strands be simultaneously replicated? A mechanism for leading and lagging strand DNA synthesis is shown in Figure 1.19. In this model, DNA polymerase III, which is a dimer, is able to replicate the leading strand in a continuous fashion. The lagging strand, however, forms a loop so that nucleotide polymerization can occur on both template strands in a 5′ to 3′ direction. Looping will invert the orientation of the template with respect to the enzyme but not the direction of actual synthesis on a lagging strand. After the synthesis of approximately 1000–2000 base pairs, the monomer of the enzyme on a lagging strand encounters a completed Okazaki fragment, at which point it releases the lagging strand. A new loop is then formed with the lagging strand and the process is repeated. The 5′ to 3′ exonuclease activity of DNA polymerases may aid the complete formation of Okazaki fragments. DNA ligase is then required to join the phosphodiester backbone of the Okazaki fragments to form a complete newly synthesized lagging strand. We will discuss the mechanism of action of DNA ligase and its use in genetic engineering in Chapter 2.

- *Termination.* Just as important as the correct initiation of DNA synthesis is its correct termination. For bacteria, such as *E. coli*, which contain a circular genome, if replication initiates bidirectionally from a single point, then termination should occur at a point halfway round the genome. Termination is not a passive process. It occurs at defined DNA sequences (called terminator sequences) that act as binding sites for a protein called Tus. Tus binding to the terminator sequences is highly asymmetric, which allows replication forks travelling in one direction to pass, but blocks DNA replication in the opposite direction (Kamada *et al.*, 1996). Little is known about the termination of DNA replication in eukaryotes.

In eukaryotic cells, genome replication must be coordinated with the cell cycle so that two copies of the entire genome are available when the cell divides. The cell cycle is a four-stage process that is based upon microscopic observations of dividing cells (Figure 1.20). These observations showed that dividing cells pass through repeated cycles of metaphase, when nuclear and cell division occurs, and interphase, where few changes can be detected using a microscope. DNA replication occurs during interphase. The four stages of the cell cycle are the following.

- M-phase (mitosis) – the period when the nucleus and the cell divides.

- G1-phase (gap 1) – a growth phase where transcription, translation and other general cellular activities occur.

- S-phase (synthesis) – where DNA synthesis occurs and the genome is replicated.

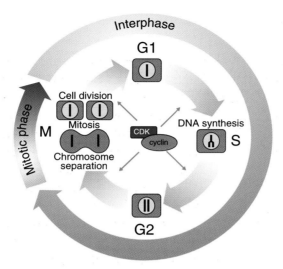

Figure 1.20. The eukaryotic cell cycle is split into cell division (mitosis) and the period between divisions (interphase). After a division is complete, the cell grows during the first gap phase (G1). DNA synthesis then occurs (S) before a second gap phase (G2). Finally, the duplicated chromosomes separate and cell division occurs (M). The decisions to proceed with particular phases of the cycle are controlled by proteins – cyclins and cyclin-dependent kinases (CDKs). These are termed checkpoint controls

- G2-phase (Gap 2) – the second interval period. Another growth phase when final preparations for cell division take place.

A human cell in culture takes about 20 hours to progress through one complete cell cycle. Of this, over 9 h will be spent in G1, while S-phase takes about 8 h to complete, and G2 lasts about 2 h. The actual time the cell spends dividing (i.e. in M-phase) is only about 45 min. Clearly, it is important that the S- and M-phases are coordinated so that the genome is completely replicated once, but only once, before cell division occurs. The periods immediately before entry into S- and M-phases are the key **cell-cycle checkpoints**. Proteins, cyclins and cyclin-dependent kinases (CDKs) act to control entry into each phase of the cell cycle. Cyclins are a diverse family of proteins that bind and activate some members of the CDK family. Most cyclins display dramatic changes in concentration during the cell cycle, which help to generate the oscillations in CDK activity that form the foundation of the cell-cycle control. Other checkpoints also exist, primarily within S-phase, to ensure that, for example, DNA damage is repaired before replication can be completed. Lee Hartwell, Paul Nurse and Tim Hunt were awarded the 2001 Nobel Prize in Physiology or Medicine for their discoveries of key regulators of the eukaryotic cell cycle.

1.10 Recombination

DNA should never be considered as a static molecule. It is constantly changing, and one mechanism by which this change is brought about is recombination. Recombination is the large-scale rearrangement of a DNA molecule. This type of rearrangement occurs as a consequence of two DNA molecules sharing either extensive regions of similar sequence (**homologous recombination**) or very short regions of homology (**site-specific recombination**). Cells containing a diploid set of chromosomes have plenty of opportunities to find a homologous partner for recombination to occur. Recombination in a bacterial system was first demonstrated independently by Alfred Hershey and Max Delbrück in 1947. They studied the infection of *E. coli* with bacteriophages. If an *E. coli* cell was infected at the same time with two genetically different T2 bacteriophages, the resulting phage population included recombinant phage types as well as the original parental phage types (Hershey, 1947).

Homologous recombination occurs between two DNA molecules that have essentially the same sequence. The genetic information is swapped and mixed up between the two DNA versions such that the recombined DNA molecules are a mixture of the starting ones. Several mechanisms have been proposed to explain the molecular basis of these events. The key to understanding the molecular processes involved in recombination was first articulated in 1964 by Robin Holliday (Holliday, 1964). His model for recombination required that a nick (cleavage at the phosphodiester bond between two nucleotides) occurred at the same site in two homologous DNA molecules (called the donor and recipient DNA). A strand exchange reaction then occurred, followed by sealing of the nicks using DNA ligase. The exchange of donor and recipient DNA strands leads to the formation of a **Holliday junction** – a structure in which the crossed DNA strands connect the donor and recipient DNA molecules. The Holliday junction can undergo branch migration, leading to more recombination between strands from different DNA molecules. Finally, resolution of the junction gives rise to the recombined DNA molecules. Here, however, we will concentrate on the modified mechanism proposed by Meselson and Radding (1975) since the event initiating recombination can be brought about by a single nick in one of the DNA strands rather than earlier models, which relied on two DNA nicks before recombination could occur. The basic model is shown in Figure 1.21(a). The formation of the Holliday junction and its resolution, together with some of the *E. coli* proteins known to be involved in the process, are outlined below.

- The nick is created by the RecBCD endonuclease, which cleaves DNA strands at sequences called chi (χ) sites (5′-GCTGGTGG-3′).

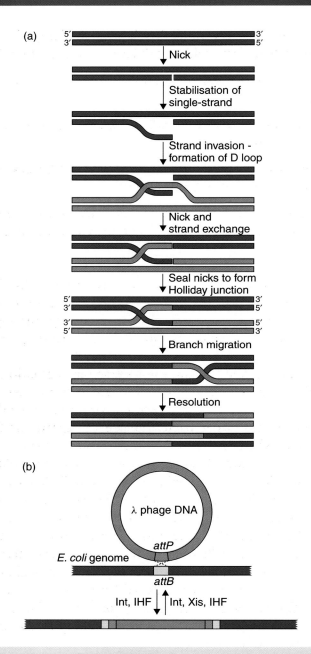

Figure 1.21. Homologous and site-specific recombination. (a) The Meselson–Radding model for homologous recombination. (b) The site-specific integration of λ DNA into the *E. coli* genome. See text for details

- Strand invasion is catalysed by the RecA protein, along with single-strand DNA binding protein (SSB), which stabilizes the free single DNA strand.

- The 'D-loop' that is formed after strand invasion is cut, again by the RecBCD endonuclease.

- Strand exchange occurs and the ends are sealed using DNA ligase. This forms the Holliday junction.

- The formation of the Holliday junction is followed by branch migration, catalysed by the proteins RuvA and RuvB.

- The resolution of the Holliday junction requires the RuvC protein, which nicks two of the DNA strands. DNA ligase then seals the strands after they have been cut.

At the end of these processes, the cell will have two molecules that have exchanged DNA. The same process outlined above also takes place in eukaryotes, using similar kinds of enzymatic activity. Recently, the high-resolution structure of the Holliday junction stabilized by the RuvA protein has been solved (Hargreaves *et al.*, 1998).

Unlike homologous recombination, site-specific recombination occurs when a particular DNA, that is not at all homologous to another DNA, has a sequence that acts as a target for the recombination of that other DNA. In prokaryotes, this is the case when bacteriophage λ integrates its DNA into the *E. coli* chromosome, where it will reside in the lysogenic state. We will look at the life cycle of λ in more detail in Chapter 3. This recombination event occurs between sites in the bacteriophage DNA (*attP*) that have a sequence homology with sites in the bacterial DNA (*attB*), even though the two DNA molecules are completely different in all other sequences. This is shown in Figure 1.21(b). The enzyme that catalyses this event is called integrase (Int), along with a host protein called integration host factor (IHF). When the bacteriophage DNA is excised from the *E. coli* genome to begin the formation of new phage particles, a phage protein called Xis is required. This site-specific recombination still requires the base pairing of a short homologous region (the *attP* with the *attB* sites), but does not involve any of the proteins of the homologous recombination pathways.

1.11 Genes and Genomes

The genetic information contained within the DNA base sequence directs the production of the proteins and enzymes that build the cell. The nucleic acids are arranged into units called **genes**, and the whole collection of genes constitutes the **genome**. Most genes are used as templates to produce proteins. The complement of proteins within a particular cell type is distinctive – a hair follicle will produce keratin and a pancreatic β-cell will produce insulin – but the protein content of a cell can also change dramatically depending upon, for example, the availability of nutrients. This means that in certain cell types, certain genes will be expressed and others will not, and the cell must have the ability to activate certain genes in response to external signals. This raises a number of issues. How is the cell able to discern within its genome what is a gene and what is not, and how is the cell able to turn a particular gene on while at the same time not affecting the expression of other genes?

1.12   Genes within a Genome

What is a gene? Surprisingly, there is not a universally accepted answer to this question. In its broadest terms a gene may be considered as a unit of heredity composed of nucleic acid. In classical genetics, a gene is described as a discrete part of a chromosome that determines a particular characteristic. Here, we will use the term gene to describe an open reading frame (ORF – the region of the gene that will be transcribed and translated into a protein sequence) together with its transcriptional control elements (promoter and terminator). Thus a gene is not just the coding sequence itself, but contains the information required for the expression of that coding sequence. Most protein coding ORFs have the same overall format (Figure 1.22). They start with a particular triplet of DNA bases (ATG) and end at stop signal, again a triplet of DNA bases (either TGA, TTA or TAG). The sequence in between will either directly code for the protein (prokaryotes) or be split into a series of **exons** and **introns** (eukaryotes). We will discuss exons and intron later, but with the advent of fully sequenced genomes (Chapter 9), a gene is usually defined as a DNA sequence that has at least 100 codons between the start and the stop signals. This sequence alone, is however insufficient to direct protein production. Each gene requires other DNA sequences surrounding it that act as binding sites for proteins such as the RNA polymerase and for those involved in transcriptional and translational initiation and termination. When combined, the coding portions of the gene and these promoter and terminator sequences form the functional gene, sometimes referred to as a 'transcriptional unit'.

How many genes are present within a genome, how many genes are essential to the cell and how many are expressed at one time? The advent of completely sequenced genomes (Chapter 9) has allowed us to address, at least in part, some of these questions.

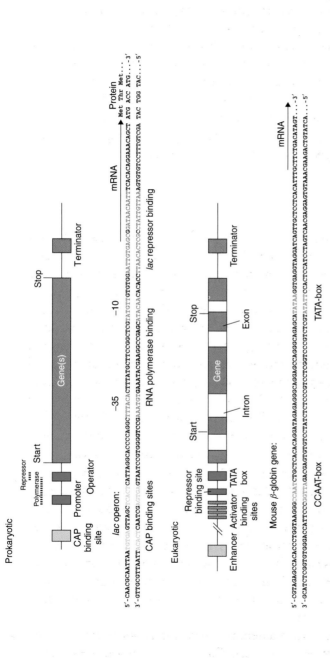

Figure 1.22. The architecture of a typical prokaryotic and a typical eukaryotic protein coding gene. In prokaryotes, families of genes required, for instance, to produce all the enzymes of a pathway in prokaryotes are often transcribed together as a polycistronic message. The operon contains the binding sites for the RNA polymerase (promoter) and enhancer elements, such as the cyclic AMP activator protein (CAP). The binding of these proteins triggers transcription. Additionally, the operon may contain the binding sites for repressor proteins, which, when bound to DNA, occlude polymerase binding and thus stop transcription. The transcript ends at terminator sequences at the 3′-end of the operon. Below is shown the sequence of the control region of the *lac* operon. In eukaryotes, single genes are produced and regulated individually. They contain activator and repressor binding sites at the 5′-end of the gene. They also may contain enhancer elements several thousand base pairs upstream of the gene which are required for its full expression. The activator promotes the assembly of the RNA polymerase II holoenzyme complex at the TATA box, and transcription begins a short distance to the 3′ side of this. Again, termination occurs at specific sites at the 3′-end of the gene. Perhaps the major difference between eukaryotes and prokaryotes is the presence of introns that split the coding sequences (exons). The introns are removed by splicing before the transcript is translated. The DNA sequence of the promoter region of the mouse β-globin gene is also shown

- *How many genes are contained within a genome?* The *E. coli* genome (4.6 Mb in length) contains about 4200 genes. The average length of a gene in *E. coli* is ~950 bp, and the average separation between genes is 118 bp, so the majority of the genome codes for RNA or protein. The yeast *Saccharomyces cerevisiae*, a single-cell eukaryote, has a larger genome size (13.5 Mb) and some 6000 genes. Humans, on the other hand have a genome that is almost 250-fold larger that of yeast (at 3300 Mb) but may have as few as 30 000 genes – representing an increase of only fivefold.

- *How many genes are essential to the organism?* To determine whether a gene is essential for the organism, the function of the gene must be impaired in some way. This is usually achieved by **gene knockouts**, where the gene is removed from the genome and its effects on cell viability determined. Analyses such as these suggest that about half the *E. coli* and yeast genes are required for viability. Many apparently non-essential genes may play specialist roles under conditions not examined in this type of experiment (nutrient starvation for example). The apparently low percentage of essential genes, however, may reflect a functional degeneracy among certain sets of genes.

- *How many genes are expressed at one time?* The genes that are expressed in a particular cell define that cell. Some of the genes expressed in a liver cell must be different to those expressed in a skin cell. Genome-wide analysis of the expression of all genes within fully sequenced organisms such as yeast suggests that about 90% of genes are expressed at any one time, but 80% of these are expressed with very low abundance levels – in the order of 0.1–2 transcripts per cell (Causton *et al.*, 2001). The level of expression of individual genes will vary widely. Highly abundant proteins are produced from highly expressed genes, whilst other proteins that may be present at a much lower level (e.g. one copy per cell) are often produced from genes that are expressed at a very low level.

The intermediary between DNA and its encoded protein is RNA. The process of converting DNA sequences into proteins occurs in two steps. The DNA is first **transcribed** into RNA which is subsequently **translated** into the polypeptide sequence of the protein. As we have already seen (Figure 1.5), RNA differs from DNA in that it contains a ribose sugar rather than a deoxyribose sugar. Additionally, uracil replaces thymine as the complementary base to adenine in RNA. Transcription is similar to DNA replication except that only one of the two DNA strands is copied. There are two distinct classes of RNA that are produced by transcription, messenger RNA (mRNA) and structural RNAs (ribosomal RNA (rRNA) and transfer RNA (tRNA)). mRNA, although by far the least abundant form of RNA – in most cells representing a few percentage points of the total RNA – is the carrier of the genetic information.

1.13 Transcription

Transcription is the process by which an RNA copy of one of the strands in the DNA double helix is made. The antisense strand of the DNA directs the synthesis of a complementary RNA molecule. The RNA molecule produced is therefore identical to the sense strand of the DNA – except that it contains U instead of T. There are fundamental differences in the ways in which genes are transcribed in prokaryotes and eukaryotes. Here, it is important to understand the processes involved in each case. Many of the experiments we will look at in later chapters involve the use of eukaryotic cells, but the bacterium *E. coli* still plays a vital role in almost all genetic engineering experiments.

Transcription begins at specific DNA sequences called **promoters**. Like DNA replication, transcription occurs in three phases – initiation, elongation and termination. Initiation of transcription usually occurs to the 3′ side of the promoter, and termination occurs at specific sites downstream of the coding sequence of the gene. At first glance, the overall architecture of a typical prokaryotic gene and a typical eukaryotic gene may appear to be similar (Figure 1.22). However, the controlling region for eukaryotic genes will not function in a prokaryotic cell, and *vice versa*.

Most protein coding genes in prokaryotes are transcriptionally active by default. That is to say, in the absence of other factors, the RNA polymerase can recognize the promoter of a gene, bind to it and produce RNA. Transcriptional control is brought to bear on the gene by repressor proteins that bind to DNA sequences adjacent to the RNA polymerase binding site. DNA binding by the repressor either occludes RNA polymerase binding and/or prevents a bound polymerase from transcribing. The eukaryotic RNA polymerase involved in the production of protein coding genes (pol II) is unable to recognize promoter sequences on its own. Therefore, eukaryotic genes are transcriptionally inactive in the absence of other factors. In both prokaryotes and eukaryotes, transcription is a highly regulated process. Proper timing and levels of gene expression are essential to almost all cellular processes.

1.13.1 Transcription in Prokaryotes

François Jacob and Jacques Monod were the first to elucidate a transcriptionally regulated system (Jacob and Monod, 1961). They worked on the lactose metabolism system in *E. coli*. Lactose is a disaccharide of galactose and glucose. When the bacterium is in an environment that contains lactose as the sugar source it expresses the following structural genes.

- *β*-galactosidase – the enzyme hydrolyses the bond between the two sugars, glucose and galactose. It is coded for by the gene *lac*Z.

- Lactose permease – the enzyme spans the cell membrane and brings lactose into the cell from the outside environment. The membrane is otherwise essentially impermeable to lactose. It is coded for by the gene *lac*Y.

- Thiogalactoside transacetylase – the function of this enzyme is not known. It is coded for by the gene *lac*A.

These three enzymes are located adjacent to each other in the *E. coli* genome (Figure 1.23) and a region of DNA that is responsible for their transcriptional regulation is located just to the 5′ side of these structural genes. This assortment of genes and their regulatory regions is called the **lac operon**. In order to metabolize lactose, the structural genes must be expressed and the protein products produced. One could imagine, therefore, a simple **genetic switch** in which the *lac* structural genes are transcriptionally inert in the absence of lactose and on in its presence. There are, however, other factors that influence the transcriptional activity of the *lac* structural genes. The favoured sugar source for *E. coli* is glucose, since it does not have to be modified to enter the respiratory pathway. So, if both glucose and lactose are available, the bacterium turns off lactose metabolism in favour of glucose metabolism. The *lac* operon has the following DNA sequence elements.

(a) Operator (*lac*O) – the binding site for repressor.

(b) Promoter (*lac*P) – the binding site for RNA polymerase.

(c) Repressor (*lac*I) gene – which encodes the Lac repressor protein. This protein binds to DNA at the operator and blocks transcription of the structural genes by RNA polymerase bound at the promoter.

(d) Pi the promoter for *lac*I.

(e) CAP binding site for cAMP/CAP complex.

In the absence of lactose, RNA polymerase binds to the promoter but is prevented from transcribing the structural genes by the *Lac* repressor which is bound to the operator site (Lee and Goldfarb, 1991). Since no RNA can be made, no protein is produced. The repressor itself is produced from the *lac*I

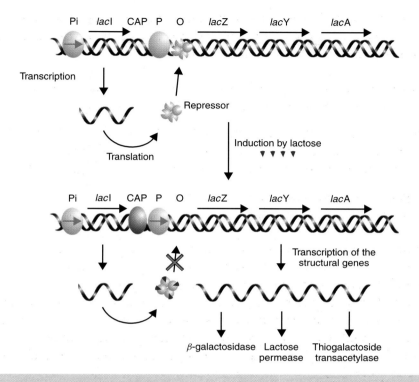

Figure 1.23. The activation of the lac operon in *E. coli*. In the absence of lactose, the *lac* repressor binds to the operator (O) and prevents RNA polymerase bound to the promoter (P) from transcribing the structural genes. When lactose enters the cell, it binds to the *lac* repressor and causes a conformational change that inhibits its ability to bind DNA. Consequently, the polymerase bound at P will transcribe the *lac* structural genes (*lacZ*, *lacA* and *lacY*). Full activation of the *lac* genes only occurs in the absence of glucose when the catabolite activator protein (CAP) binds to the operon and aids the binding of RNA polymerase to P

gene by an RNA polymerase bound to another promoter (Pi). When lactose is present within the cell, it acts as an inducer of the operon. It binds to the *lac* repressor, and induces a conformational change that results in the repressor dissociating from DNA. Now the RNA polymerase bound to the promoter is free to move along the DNA and RNA can be made from the three genes. Lactose can now be metabolized.

A pertinent question to ask here is how the *lac* genetic switch actually gets turned on initially. As stated above, the *lac* structural genes, including the permease required to get lactose into the cell, are switched off in the absence of lactose. It therefore seems impossible to achieve activation, since activation is required to get lactose into the cell. This is, however, an over-simplification. Each of the *lac* structural genes is transcribed at a low, basal, level (approximately five copies per cell) in the absence of lactose. This means that when the bacterium encounters a source of lactose it is able to transport a few molecules into the cell so that full induction of the *lac* structural genes can occur. When fully induced, approximately 5000 copies of each protein product are present in the cell.

Regulatory control of the *lac* operon is more complicated since it needs to be turned off (repressed) if glucose is present, no matter whether lactose is present or not. The repression of the *lac* operon by glucose is mediated through the CAP site. When levels of glucose (a catabolite) in the cell are high, the production of cyclic AMP (cAMP) is inhibited, so when glucose levels drop, more cAMP forms. cAMP binds to a protein called CAP (catabolite activator protein), which then binds DNA at the CAP site in the *lac* operon. This activates transcription of the *lac* structural genes by increasing the affinity of the promoter for RNA polymerase. This phenomenon is called **catabolite repression**, since glucose inhibits the transcription of the *lac* operon by not allowing its full activation.

Like many operons in prokaryotes, the three *lac* structural genes are transcribed as a single mRNA. This **polycistronic** message is translated as it is produced since there is no physical separation of the DNA from the cytoplasm. Each of the ORFs contains its own ribosome binding site (see below) within the encoded mRNA and is translated independently of the other *lac* genes.

We have seen above how the transcript is initiated, but how does transcription end? Once RNA polymerase has started transcription, the enzyme moves along a DNA template, synthesising RNA, until it meets a **terminator** sequence. Here, the polymerase stops adding nucleotides to the growing RNA chain, releases the completed product and dissociates from the DNA. Termination can be brought about in one of two ways.

- *Rho-dependent termination.* The protein rho binds to the newly synthesized RNA chain and appears to cause termination when the polymerase pauses at certain DNA sequences.

- *Rho-independent termination.* Intrinsic termination sequences may form hair-pin loops in the RNA that promote polymerase dissociation. This type of termination is independent of additional factors and is the most common form of termination in *E. coli*.

1.13.2 Transcription in Eukaryotes

The process of activating gene expression in eukaryotes is far more complex than in prokaryotes. There are several differences between eukaryotes and prokaryotes that impinge upon gene activation and RNA processing.

- Eukaryotes possess multiple RNA polymerase enzymes.

- The wrapping of DNA into nucleosomes represents a barrier to transcription.

- The physical separation of the nucleus (where transcription occurs) and the cytoplasm (where translation occurs) in eukaryotes means that mechanisms must exist to protect the mRNAs from degradation in the cytoplasm before they have been translated.

- Eukaryotic genes are **monocistronic**, with each gene being produced as a separate transcript from its own promoter.

- The genes of eukaryotes are not continuous and are split into coding regions and non-coding regions.

We will touch upon each of these issues as we consider how a typical eukaryotic gene may be transcribed, and the subsequent fate of the RNA that is produced.

Eukaryotes contain three different RNA polymerases, each responsible for the transcription for a certain class of genes (Table 1.3). Each of the polymerases is a large complex of proteins that are involved in, amongst other things, regulating polymerase activity. Activation of RNA polymerase II genes (those coding for proteins) is brought about by transcriptional activator proteins binding to DNA sequences within the promoter of a gene (Figure 1.24). In the absence of these activators, the genes are not transcribed. The role of the activator has been the focus of much speculation. However, it appears to act simply as a way of recruiting various protein complexes to a particular promoter. Control of gene expression is therefore targeted through the activator. Activators have at least two separate functions. First, they must be able to bind to DNA so that they can find their target gene within the mass of genomic DNA. Second, they participate in protein–protein contacts necessary to recruit the RNA polymerase and other complexes to the gene through an activation domain. Either of these functions can be modulated to regulate gene expression in response to specific signals.

Table 1.3. Properties of eukaryotic RNA polymerases. In older papers, these are sometimes referred to as polymerases A, B and C

RNA polymerase	Location	Products
I	Nucleolus	28S, 18S and 5.8S rRNA
II	Nucleus	mRNA, and some snRNA
III	Nucleus	tRNA, 5S rRNA, and some snRNA

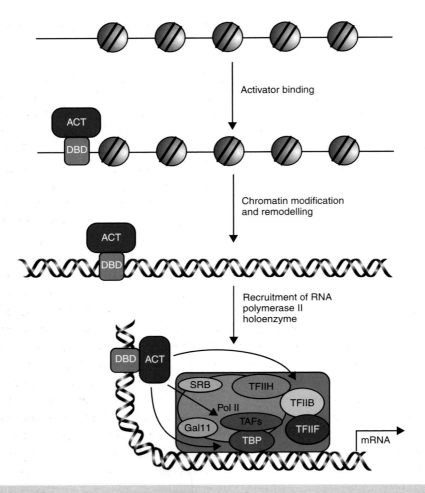

Figure 1.24. The multiple roles of transcriptional activators in switching on gene expression in eukaryotes. The binding of the activator protein through its DNA binding domain (DBD) can occur either in a nucleosome-free region, as shown, or on the DNA already bound within a nucleosome. The activation domain of the activator (ACT) recruits chromatin modifying and chromatin remodelling complexes to the gene. The activator then functions to recruit an RNA polymerase II holoenzyme complex to the gene so that transcriptional initiation can occur. The polymerase holoenzyme may exist as one or more discrete sub-complexes within the cell

An additional problem for the transcription of eukaryotic genes is that the majority of the DNA is wrapped into nucleosomes, which act as physical barriers to the assembly of the transcriptional complexes. Many activators are able to bind to their cognate DNA recognition sequences whether or not these are contained within nucleosomal DNA (Taylor *et al.*, 1991), while others require that the nucleosomes are removed before they can bind DNA. After

DNA binding, the activator recruits one, or more, of the following complexes to the gene.

- *Chromatin-modifying complexes*. Histone proteins form the core of the nucleosome. However, as can be seen in Figure 1.15, the amino-terminal ends of the histones protrude from the complex. These 'tails' can be modified, by acteylation, phosphorylation or methylation of particular amino acids (Figure 1.25(a)), and the modifications play a significant role in the ability of nearby genes to be transcribed. Although these tails are not needed to maintain the structural integrity of the nucleosome, they do have roles in higher-order chromatin structure and in interactions with non-histone chromosomal proteins. Much attention has been focused recently on the role of histone acteylation in the process of transcription. It has been known for many years that increased levels of histone acetylation at a gene or chromosomal region are associated with transcriptional activity, whereas under-acetylation of histones is observed in non-transcriptionally active regions (Allfrey, Falkner and Mirsky, 1964), but the significance of this observation was not fully realized until it was discovered that transcriptional activators recruit chromatin modifying complexes to the promoters of genes (Bhaumik and Green, 2001; Larschan and Winston, 2001). The acetylation of histones is performed by enzymes called histone acetyltransferases (HATs) while the reverse reaction is catalysed by enzymes call histone deacetylases (H-DACs). The acetylation of lysine residues results in the elimination of positive charge from the protein (Figure 1.25(b)). It has been speculated that in under-acetylated histones (transcriptionally inactive) the histone tails wrap around the DNA with their positive charge being attracted to the negatively charged sugar–phosphate backbone. Acetylated histone tails, on the other hand, may interact with the DNA more loosely so that the histone–DNA complex is less rigid and may be more easily dissociated to allow transcription to occur.

- *Chromatin remodelling complexes*. These complexes, such as the yeast SWI/SNF proteins (pronounced 'switch–sniff', based on the nomenclature used in the original genetic screens that isolated the genes), use energy derived from ATP to move nucleosomes. Their precise mechanism of action is unclear, but they may function as ATP driven motors that move along DNA and disrupt protein–DNA interactions as they move (Pazin and Kadonaga, 1997).

- *RNA polymerase II holoenzyme*. RNA polymerases were first isolated in 1960 (Stevens, 1960; Weiss, 1960). Multiple forms of the enzyme in

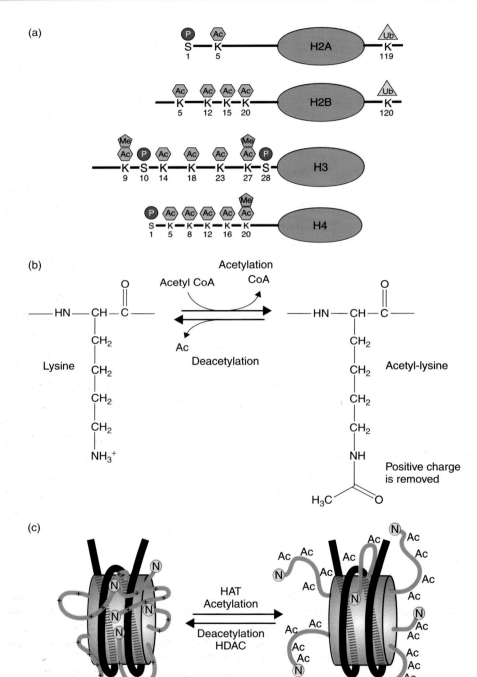

eukaryotic cells were identified by their sensitivity to the bicyclic octapeptide α-amanitin – RNA polymerase I is not affected by this compound; RNA polymerase II is rapidly inhibited at low concentrations of α-amanitin; RNA polymerase III from mammalian cells is inhibited by high levels of α-amanitin. Each of the polymerases is a large (~500 kDa) multi-protein complex typically containing 8–14 individual subunits. Here we will concentrate on RNA polymerase II (pol II) since it is responsible for the transcription of protein coding genes. Pol II is an enzyme comprising 12 subunits encoded, in yeast, by the RPB1 to RBP12 genes. The crystal structure for 10 of the subunits of RNA pol II has been solved recently (Cramer *et al.*, 2000; Cramer, Bushnell and Kornberg, 2001) and is shown in Figure 1.26; it gives tantalizing insights into the structural mechanism of RNA production. The largest subunit Rpb1p contains a highly conserved carboxy-terminal repeat domain (CTD) with the consensus sequence Tyr–Ser–Pro–Thr–Ser–Pro–Ser. The yeast CTD includes 26 or 27 repeats and the human CTD has 52 repeats. This sequence can be extensively phosphorylated, which is an important step in converting RNA pol II from a form involved in promoter recognition to a form involved in transcriptional elongation (Hampsey, 1998). We will see later that, besides its role in transcription, the CTD is also critical for co-transcriptional RNA processing events such as capping, polyadenylation and splicing. As stated previously, pol II is incapable of binding DNA on its own. It requires other proteins – called general transcription factors – to achieve this function. There are six principal general transcription factors (TFIIA, IIB, IID, IIE, IIF and IIH) that are essential for gene-specific DNA binding and the formation of a polymerase complex capable of transcriptional initiation. Again, most of the general transcription factors are not single proteins, but are multi-protein complexes – for example TFIID is composed of 13 polypeptides – the TATA-box binding protein (TBP) and 12 TATA-box binding

Figure 1.25. Modifications of the histone tails. (a) The amino-terminal 'tails' of the histone proteins may be modified by phosphorylation (P), acetylation (Ac) or methylation (Me) at the serine (S) or lysine (K) residues indicated. The numbers refer to the amino acid positions within each protein. Additionally, histones H2A and H2B can be ubiquitinated (Ub) at their carboxy-terminal ends. The expression of genes contained within or close to nucleosomes containing these modified histones can be drastically altered. Reprinted from *FEMS Microbiology Review*, Vol 23, Pérez-Martín, Chromatin and transcription in *Saccharomyces cerevisiae*, pp. 503–523, Copyright (1999), with permission from Elsevier. (b), (c) The acetylation of the histone tails alters their charge, and results in a 'loosening' of the histone–DNA complex. This allows transcription to occur more readily

(a)

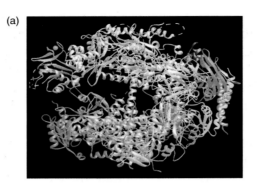

(b)

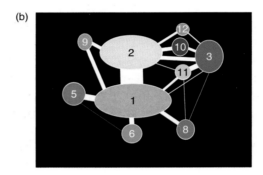

(c)

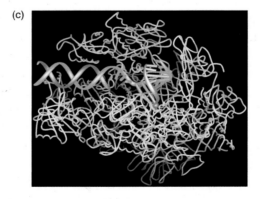

Figure 1.26. Architecture of yeast RNA polymerase II. (a) The backbone model for 10 subunits of RNA polymerase as determined by X-ray crystallography. The red sphere in the large cleft at the centre of the molecule represents the magnesium ion at the active site. A diagrammatic arrangement of the subunits of the protein present in the structure is shown in panel (b) and the protein complexed with both double-stranded DNA (blue and green) and RNA (red) is shown in (c). These figures were kindly provided by Patrick Cramer (University of Munich) and are reprinted with permission from *Science* (Cramer *et al.*, 2000; Gnatt *et al.*, 2001). Copyright American Association for the Advancement of Science

associated factors (TAFs) (Lee and Young, 2000). Pol II exists in the cell as a large complex of proteins, termed the holoenzyme. The holoenzyme is composed of SRB (suppressor of RNA polymerase B) proteins that interact with the CTD, Med (mediator) proteins and some of the general transcription factors. The precise composition of the general transcription factors involved with the holoenzyme is still unclear. The SRB and Med proteins appear to act as a regulatory 'glue' that stabilizes interactions between RNA pol II and general transcription factors and may also confer responsiveness to transcriptional activators (Koleske and Young, 1994; Malik and Roeder, 2000). The mediator complex is assumed to enter the initiation complex with RNA pol II and is released at the end of initiation or early in RNA chain elongation (Svejstrup *et al.*, 1997). Transcriptional activators interact with a number of the components of the Pol II holoenzyme, including TBP, TAFs, TFIIB and some of the SRB and Med proteins. Whatever interactions occur within the cell, the end result appears to be simply the recruitment of the pol II complex to a promoter that has been freed from nucleosomes. Once the polymerase is bound to the promoter, transcription will be initiated.

The interactions between activators and the proteins described above lead to a model for gene activation in which the activator must first bind to DNA. The activator then recruits chromatin modifying and remodelling complexes to remove nucleosomes from the promoter, and finally pol II is brought to the promoter so that transcription can begin. The entire gene need not be devoid of nucleosomes for full transcription to occur. If the promoter is accessible to RNA polymerase and transcription factors, the presence of nucleosomes will not inhibit the elongation of the message. The structure of the nucleosome is such that the positive charges (lysines and arginines) of the histones are placed adjacent to the negative DNA helix. As the polymerase advances, it displaces DNA from the nucleosome and forms a closed loop, while the torsion ahead of the RNA polymerase generates supercoiling of the DNA. This displaces the histone octamer, which keeps contact with the DNA behind the RNA polymerase. In this way, the nucleosome need never lose contact with DNA as the RNA polymerase passes by (Studitsky, Clark and Felgerfeld, 1995). So once the RNA polymerase is transcribing the gene, nucleosomes do not stop it. The important event for transcription-level gene expression is to get the RNA polymerase bound and functioning at the promoter.

The molecular processes involved in transcriptional termination in eukaryotes are relatively poorly defined. Some transcripts terminate over 1000 bp downstream of the 3′ end of the mature mRNA, and appear to end at termination regions rather than at specific sites. The transcript is cleaved from the extending polymerase before the polymerase itself terminates transcription

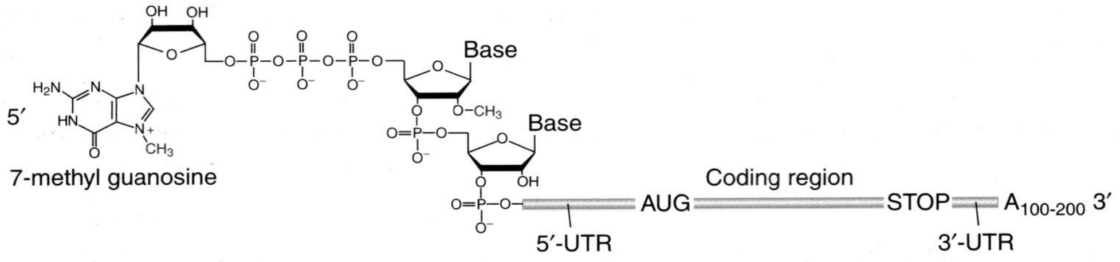

Figure 1.27. The structure of a mature mRNA molecule in eukaryotes. After transcription, a 7-methyl guanosine cap is added to the 5'-end of the message, and the ribose sugar of the first, and sometimes second, nucleotide is methylated at the 2'-position. The 3'-end of the transcript is polyadenylated with the addition of 100–200 A residues

(Figure 1.27). The function of the RNA polymerase complex is not finished once the transcript has been cleaved. The polymerase also functions to direct processing of the transcript, such as splicing and polyadenylation (McCracken *et al.*, 1997; Hirose and Manley, 1998). Indeed, recent evidence suggests the enzymes and cellular machinery involved in transcription, RNA processing and translation may be extensively coupled to form 'gene expression factories' that maximize the efficiency and specificity of each stage of gene expression (Maniatis and Reed, 2002).

1.14 RNA Processing

Perhaps the most obvious difference between the genes of prokaryotes and eukaryotes is that genes of the later are split into **exons** (coding regions) and **introns** (non-coding regions). In the mid-1970s, Philip Sharp and Richard Roberts independently hybridized a messenger RNA to the DNA from which it had been transcribed, and viewed the resulting hybrids using electron microscopy. They found that RNA hybridization occurred at discontinuous sequences within the DNA. That is, the binding of mRNA to the DNA sequence from which it was derived would result in the formation of loops in the DNA, corresponding to sequences in the DNA that were not present in the mRNA molecule (Chow *et al.*, 1977; Berget, Moore and Sharp, 1977). Walter Gilbert gave names to these various regions. The parts of the gene that are represented in the mRNA, and are therefore part of the expressed region, he called exons. The parts of the gene that are the intervening sequences he called introns. The mechanism by which the precursor RNA (pre-mRNA) is converted into the final mRNA – in other words, how introns are removed – has been the subject of intense scrutiny since their discovery. Before we address this issue, we need to see what happens to an mRNA molecule as it is produced.

Once a gene has been transcribed, the mRNA produced is extensively modified. Almost as soon as mRNA synthesis is initiated, the 5′-end of the message is capped by the addition of a 7-methyl guanosine residue. The cap is added to the 5′-end of the mRNA *via* a 5′–5′ condensation of guanosine to the mRNA to form a structure consisting of 3′–G–5′ppp5′–N-3′p. After cap formation, a methyl group is added to the guanine residue and to the first and/or second adjacent nucleotide (Figure 1.27). The function of the cap is not entirely clear. It has been suggested that the cap, and proteins that bind to it, direct ribosome binding and correct translational initiation. In addition to the cap, the 3′-end of the mRNA is cleaved from the extending chain and then polyadenylated.

RNA transcript cleavage and polyadenylation are directed by a polyA signal within the RNA (Figure 1.28). The core polyA signal for vertebrate pre-mRNAs consists of two recognition elements flanking a cleavage–polyadenylation site. Typically, an almost invariant 5′-AAUAAA-3′ hexamer is found 20–50 nucleotides upstream of a more variable element rich in U or GU residues. Cleavage of the newly formed transcript occurs between these two elements and is coupled to the addition of approximately 200 adenosines to the 3′-end of the 5′ cleavage product. Two protein factors required for this process are the cleavage and polyadenylation specificity factor (CPSF), which binds to the AAUAAA motif, and the cleavage stimulation factor (CStF), which binds the downstream GU-rich element. As the polyA signal is extruded from the polymerase, it is recognized by a subset of cleavage and polyadenylation factors that have already been recruited to the RNA polymerase CTD. Then assembly continues as more factors join the complex. Ultimately cleavage occurs, followed by polyadenylation in which a stretch of ~200 A residues is added to the 3′-end of the message. Polyadenylation only occurs to mRNAs, but not all mRNAs are polyadenylated; e.g. the mRNA species that encode histone proteins are not polyadenylated. The process of polyadenylation is carried out by an enzyme called polyA polymerase (PAP), and not only confers stability on the transcript, but is also required for the movement of the mature mRNA out of the nucleus into the cytoplasm, where translation occurs. As we will see in later chapters, the addition of polyA to mRNAs has important practical consequences. The hybridization of polyT sequences to only mRNA, and not other RNAs, is a vital tool to the genetic engineer.

1.14.1 RNA Splicing

There are several different mechanisms of RNA splicing, which function on different RNA species. Here, however, we will concentrate on the mechanism

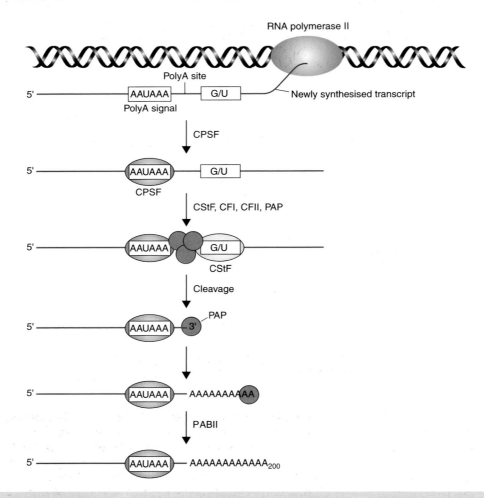

Figure 1.28. Cleavage and polyadenylation of the mRNA transcript. The polyA site within the newly formed transcript is shown, together with the signal AAUAAA and the downstream U/GU region. A complex of cleavage and polyadenylation specificity factor (CPSF), cleavage factors I and II (CFI and II) and cleavage stimulation factor (CStF) bind to these sequences. PolyA polymerase (PAP) also joins the complex at this stage. RNA cleavage occurs, and CPSF remains bound to the 5′-end of the cleaved message. The 3′-end is rapidly degraded. PAP begins the synthesis of the polyA tail, resulting in the addition of the first 10 A residues. Finally, polyA binding protein II (PABII) joins the reaction, stimulating the synthesis of polyA and extending the tail to about 200 A residues

of mRNA splicing since understanding it, and its consequences, are most applicable to the genetic engineer. The capped and polyadenylated mRNA is **spliced** to remove the introns and fuse the exons together into a single unit that can be translated. The need to produce precisely spliced products is obvious. Any slippage in the fusing of exons together would have disastrous

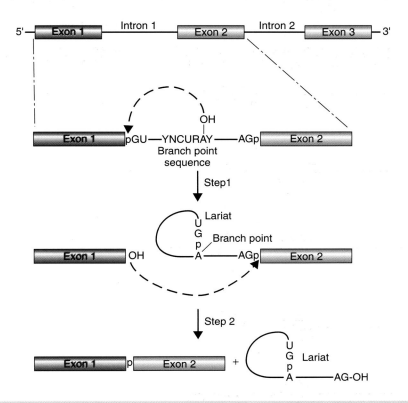

Figure 1.29. The splicing of exons to form a mature mRNA. Nuclear pre-mRNA splicing takes place in two distinct mechanistic steps. Step 1 involves a nucleophilic attack by the 2′-OH group of the branch point A residue on the phosphodiester bond of the 5′ exon–intron boundary, displacing the 5′ exon (shown in blue) as the leaving group and giving the lariat structure shown. The second step is a nucleophilic attack of the 3′-end of the 5′ exon on the phosphodiester bond of the 3′ intron–exon boundary, displacing the intron (as a lariat form) and sealing the two exons together (the blue and red)

consequences for the protein that is to be eventually made. How are the boundaries between an intron and an exon marked, and by what molecular mechanism does splicing actually occur? Perhaps surprisingly, there is relatively little RNA sequence conservation at the exon–intron boundary. Most introns begin with the dinucleotide sequence 5′-GU-3′ and end with the dinucleotide sequence 5′-AG-3′ (Figure 1.29). There are, however, other less conserved sequences present within the intron that act as binding sites for complexes that are essential for splicing. These complexes, collectively called the **spliceosome**, contain both RNA and protein components, and are found exclusively in the nucleus. The RNA molecules found within the spliceosome are small (100–200 nucleotides) and are complexed with proteins to form small nuclear ribonucleoproteins (snRNPs, or snurps). The RNA molecules within these complexes are rich in uridine residues, and consequently the snRNPs have been designated by the names U1, U2, U4, U5 and U6. A model of the overall splicing process catalysed by the spliceosome is shown in Figure 1.29. The mechanism of splicing for exon coupling and intron extrusion depends on two transesterification reactions, which result in the formation of a **lariat** form of the intron and the fused exons. The branching associated with the A residue within the intron in the lariat form contains two phosphodiester linkages to the 2′ and 3′ of the A. The snRNPs play roles both in the catalytic process and in structurally maintaining the two exons in proximity with each other as the splicing reaction proceeds. Readers interested in the mechanism of splicing are directed to specific reviews on the topic (Sharp, 1994).

1.14.2 Alternative Splicing

Since each exon in a eukaryotic gene encodes a portion of a protein, you can imagine that it is possible, by altering how the pre-mRNA is spliced, to produce different versions of the mRNA and ultimately, different proteins. Alternative splicing is a widely occurring phenomenon, with recent estimates suggesting that at least 30% of all human genes are subject to this type of processing (Sorek and Amitai, 2001). How do splice variants differ from the original sequence from which they are derived? The physiological activity of proteins produced from splice variants may be the same, opposite or completely different and unrelated. Alternative splicing therefore greatly increases the number of different protein activities that can be generated from a defined set of genes within the genome.

Perhaps one of the most dramatic examples of alternative splicing is found in the genes determining sex in the fruit fly *Drosophila melanogaster*. There are three genes that are involved in the sex determination process: *sxl* (sex lethal), *tra* (transformer) and *dsx* (double sex) (Chabot, 1996). Each of these genes produces a pre-mRNA that has two possible splicing patterns, depending upon whether the fly is male or female. Figure 1.30 shows these three genes and their splicing patterns. For males, the inclusion of two exons (exon3 in *sxl* and exon2 in *tra*) produces mRNA molecules that have termination (stop) codons and results in the formation of inactive proteins. The only active male product is the protein translated from the *dsx* gene. This protein inactivates all female-specific genes. The female, on the other hand, produces mRNAs without the stop codon containing exons. In the case of *sxl* and *tra*, the protein products have a positive effect on the splicing patterns observed, controlling the choice of introns removed in the spliceosome reaction.

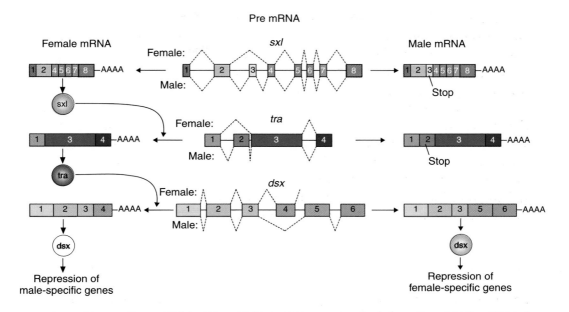

Figure 1.30. Alternative splicing during sex determination in *Drosophila*. In the centre of the diagram are the pre-mRNAs for three genes involved in the sex determination process (*sxl*, *tra* and *dsx*), with the splicing pattern for the female (top) and the male (bottom). The product mRNAs are shown to either side (female on the left, male on the right). In the female, the protein products of the *sxl* and *tra* mRNA control splicing site selection. These proteins are not produced in the male due to the inclusion of exons containing stop codons. Males, however, produce the *dsx* protein, which represses the transcription of other female-specific genes. The female version of the *dsx* protein represses the transcription of male-specific genes

1.15 Translation

Translation is the process whereby the structural RNAs, and their associated proteins, decode the linear sequence of information contained within the mRNA to produce linear chains of amino acids, called **polypeptides**, that make up proteins. The nucleotide sequence of the mRNA is read as a series of triplets (a **codon**). Each codon specifies the insertion of a single amino acid into the growing peptide chain. Translation begins at an initiation codon, AUG, and ends at one of three different termination codons, UAA, UGA or UAG (see Appendix 1). Each triplet is read in turn from the initiation codon onwards (Figure 1.31). There are 20 different amino acids found within proteins. The triplets of the genetic code, however, provide up to 64 possible different codons (Appendix 1). Thus, there is degeneracy in the genetic code, with most amino acids being encoded by more than one codon. For example, methionine is

Sense strand   Antisense strand

```
DNA:    5'-AGCCTCCTGAAAGATGAAGCTACTGTCTTCTATCGAACAAGCATGCGATATTTGCTAATTTGAAGTCA...-3'
        3'-TCGGAGGACTTTCTACTTGCATGACAGAAGATAGCTTGTTCGTACGCTATAAACGATTAAACTTCAGT...-5'
mRNA:   5'-AGCCUCCUGAAAGAUGAAGCUACUGUCUUCUAUCGAACAAGCAUGCGAUAUUUGCUAAUUUGAAGUCA...-3'
Protein:              MetLysLeuLeuSerSerIleGluGlnAlaCysAspIleCysStop
```

Point mutation:

```
DNA:    5'-AGCCTCCTGAAAGATGAAGCTACTGTCTGCTATCGAACAAGCATGCGATATTTGCTAATTTGAAGTCA...-3'
        3'-TCGGAGGACTTTCTACTTGCATGACAGACGATAGCTTGTTCGTACGCTATAAACGATTAAACTTCAGT...-5'
mRNA:   5'-AGCCUCCUGAAAGAUGAAGCUACUGUCUGCUAUCGAACAAGCAUGCGAUAUUUGCUAAUUUGUUCAGT...-3'
Protein:              MetLysLeuLeuSerAlaIleGluGlnAlaCysAspIleCysStop
```

Insertion:

```
DNA:    5'-AGCCTCCTGAAAGATGAAGCTACTGTCTTGCTATCGAACAAGCATGCGATATTTGCTAATTTGAAGTCA...-3'
        3'-TCGGAGGACTTTCTACTTGCATGACAGAACGATAGCTTGTTCGTACGCTATAAACGATTAAACTTCAGT...-5'
mRNA:   5'-AGCCUCCUGAAAGAUGAAGCUACUGUCUUGCUAUCGAACAAGCAUGCGAUAUUUGCUAAUUUGAAGUCA...-3'
Protein:              MetLysLeuLeuSerCysTyrArgThrSerMetArgTyrLeuLeuIleStop
```

Figure 1.31. The transcription and translation of a DNA sequence results in the forma-tion of the encoded protein. Mutations in the DNA sequence can have dramatic effects on the protein produced. Shown are the effects of a point mutation on, and the insertion of a single nucleotide into, the coding region of a gene. The resulting changes in the sequence of the resulting protein are shown

encoded by a single codon (AUG), while leucine and arginine are encoded by six codons each. The genetic code is almost universal. The vast majority of genes from all organisms obey the same code; however, some mitochondrial encoded genes deviate from the code. For example, human and yeast mitochondria use the UGA termination codon to insert a tryptophan amino acid into an extending polypeptide chain.

Alterations in the DNA sequence can drastically affect the resulting protein sequence. In the example shown in Figure 1.31, mutation of the DNA sequence altering a sense strand T residue to a G results in the change of a serine codon (UCU) to an alanine codon (GCU). The insertion, or deletion, of nucleotides has an even more dramatic effect on the encoded protein. In the example shown, the insertion of a GC base pair alters the reading frame of the transcript so that the resulting protein is completely different from the original sequence after the point of insertion. Mutations like this are referred to as **frame-shift** mutations. Not all DNA mutations will result in changes to the protein sequence. For example, if the serine codon discussed above had been changed from UCU to UCG then this new codon would still encode serine. Mutations such as these are termed **silent,** as they do not alter the protein sequence.

The tRNAs act as intermediaries in the translation process. They transport a specific amino acid to the mRNA triplet it encodes. Enzymes called aminoacyl-tRNA synthetases couple the amino acid encoded by a particular codon to the tRNA that contains the appropriate **anticodon**. The codon–anticodon pairing therefore directs the addition of the correct amino acid to the growing polypeptide chain. The mRNA is translated upon **ribosomes**. These bipartite structures consist of a large and a small subunit. Each subunit contains structural RNAs (the rRNAs) complexed with the proteins that perform the various chemical reactions involved in translation. The mechanism of translation can, like DNA replication and transcription, be split into three steps – initiation, elongation and termination. In bacteria, initiation begins when the ribosome assembles on the purine-rich **Shine–Dalgarno sequence** (5′-AGGAGGU-3′) 4–8 bases upstream of the initiator codon (Shine and Dalgarno, 1975). This sequence is complementary to the 3′-end of the 16S rRNA (5′-ACCUCCU-3′) of the 30S ribosome subunit, and positions the ribosome to initiate translation. In eukaryotes, ribosome assembly occurs at the **Kozak sequence** (optimal consensus 5′-ACCAUGG-3′) that surrounds the initiator codon (underlined) (Kozak, 1986). The ribosome then moves along the mRNA in a 5′ to 3′ direction (Figure 1.32). Translation starts at the AUG initiation codon and continues along the mRNA until the ribosome dissociates when it reaches the termination codon. No tRNAs exist for the terminator codons, and instead RF (release factor) protein binds to dissociate the ribosome. Under optimal conditions about 15–20 amino acids can be polymerized per second (Crowlesmith and Gamon, 1982). However, the actual rate of translation may be considerably higher than this since more than one ribosome may be bound to single mRNA at one time, forming a **polysome**. The number of ribosomes bound to a particular mRNA will influence the rate of its translation: the higher the number of ribosomes bound, then the higher will be the number of protein chains produced. The **codon usage** of an individual mRNA can also significantly influence the rate of elongation. The tRNA molecules for certain codons are more abundant than others, leading to a discrepancy in codon usage for highly expressed genes.

Translation has proven to be particularly important to the genetic engineer since many antibiotics target translation as their mechanism of inhibiting bacterial growth. The following antibiotics function by the inhibition of translation:

- *chloramphenicol* – inhibits peptidyl transferase on the 50S ribosomal subunit.

- *erythromycin* – inhibits translocation by 50S ribosomal subunit.

- *fusidic acid* – inhibits translocation by preventing the dissociation of an elongation factor from the ribosome.

- *puromycin* – an aminoacyl-tRNA analogue that causes premature chain termination.

- *streptomycin* – causes mRNA misreading and inhibits chain initiation

- *tetracycline* – inhibits binding of aminoacyl-tRNA to ribosomal A-site.

The protein chain contains information that directs post-translational processes and cellular compartmentalization in eukaryotes. All protein translation begins with a methionine residue at the amino-terminal end of the newly synthesized polypeptide (coded by the AUG codon). Many proteins, in both prokaryotes and eukaryotes, have this residue removed by an enzyme called **methionine aminopeptidase** such that the final protein sequence begins with the amino acid encoded by the second codon (Bradshaw, Briday and Walter, 1998). Soluble cytosolic proteins are simply released from the ribosome after polypeptide synthesis is complete, and are already in the correct location to undertake

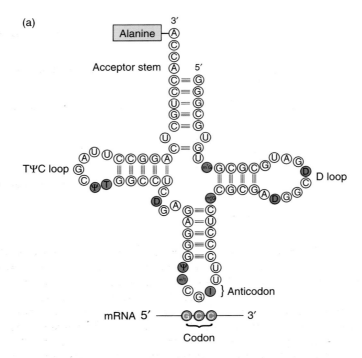

Figure 1.32. (a) Cloverleaf structure of yeast alanine tRNA showing the location of the unusual bases (shaded green). tRNA molecules can contain the following unusual bases: pseudouridine (Ψ), inosine (I), dihydrouridine (D), ribothymidine (T), methylguanosine (m¹G), dimethylguanosine (m₂²G) and methylinosine (m¹I). The tRNA anticodon base pairs with the appropriate codons in mRNA

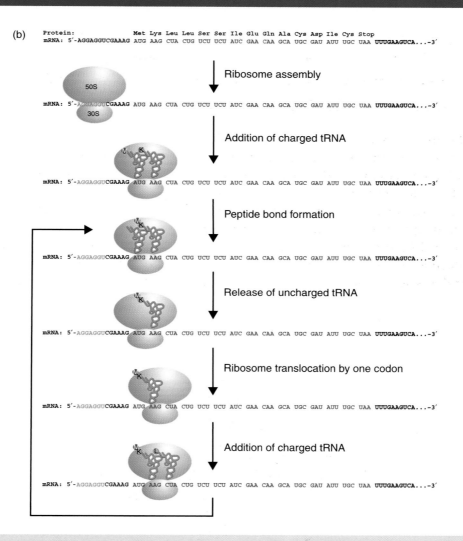

Figure 1.32. (b) Translation of an mRNA molecule on a ribosome in bacteria. The ribosome assembles on the mRNA just upstream of the initiation codon (AUG) at the Shine–Dalgarno sequence (shown in red). The 30S ribosomal subunit binds directly to this sequence and then recruits the 50S subunit. The assembled 70S ribosome is a large complex of proteins and rRNA that is of a physical size to cover approximately 35 bases of mRNA. A special initiator tRNA (fMet-tRNA$_f$) binds to the initiator codon (in the P site) through base pairing of the codon with the anticodon present in the tRNA. The next aminoacyl-tRNA (charged with its appropriate amino acid) binds to the next codon (in the A site) prior to peptide bond formation. The formation of a peptide bond involves the transfer of the amino acid(s) attached to the tRNA in the P site to the aminoacyl-tRNA in the A site. The uncharged tRNA then leaves the complex and the ribosome moves to the next codon. Subsequent amino acids are added in a similar fashion

their specific function. Many eukaryotic proteins are, however, destined for a particular compartment of the cell. The polypeptide itself encodes the information required for its final destination. Eukaryotic proteins that are destined to accumulate within the nucleus contain a nuclear localization signal (NLS), consisting of a short stretch of predominately basic amino acids (e.g. PKKKRLV), which direct the protein into the nucleus (Goldfarb *et al.*, 1986). Proteins that are destined to be exported from the cell, or to be incorporated in cellular membranes, are transported into the endoplasmic reticulum (ER) during translation. Such proteins containing a stretch of predominantly hydrophobic amino acids at their amino terminus, called the **signal sequence,** which directs the polypeptide–ribosome complex to the ER, where the newly synthesized amino acid chain is inserted through the ER membrane. Inside the ER, the signal is cleaved off as the growing peptide chain is secreted into the ER. The protein is then transported by the Golgi apparatus to its final destination or, if it contains an ER retention signal (the amino acids KDEL) at its extreme carboxy-terminal end, it is retained within the ER itself (Munro and Pelham, 1987).

2 Basic techniques in gene analysis

> ## Key concepts
>
> - The coincident discovery of restriction enzymes, bacterial transformation, and agarose gel electrophoresis formed the basis of the explosion of molecular biology in the 1970s
> - Cutting DNA at defined sites and joining foreign DNA molecules together became possible
> - Cloning – the study of single genes in isolation – now became feasible
> - High-resolution analysis of DNA using gels
> - Blotting techniques to detect homologous nucleotide sequences
> - Rapid purification of DNA

One of the main problems in gene analysis is the relatively uniform nature of the DNA molecule itself. As we have already seen, the DNA that makes up the human genome is extremely long, with some 6.4×10^9 base pairs of DNA in most cells, and is composed of only four different nucleotides. This size and relative lack of complexity makes isolating and studying single DNA fragments or genes appear a daunting task. We know that the base sequence of DNA is vital for encoding genes, but to investigate the function of a single gene it is essential that the gene be studied in isolation, freed from the rest of its native genome. Ideally, we would want to cut a single gene out of a genome. Until the 1970s, however, there were no methods available for cutting DNA at specific sequences. The ability to fragment DNA at specific sites became the cornerstone of molecular biology. The discovery of enzymes able to cleave DNA at specific

Analysis of Genes and Genomes   Richard J. Reece
© 2004 John Wiley & Sons, Ltd   ISBNs: 0-470-84379-9 (HB); 0-470-84380-2 (PB)

sequences, **restriction enzymes**, led to the award of the 1978 Nobel Prize to Werner Arber, Daniel Nathans and Hamilton O. Smith.

When thinking about how we might like to isolate a particular piece of DNA, there are several important factors to take into account. We need to isolate the piece of DNA that we are interested in, and **clone** it so that it can be replicated and amplified in the absence of other human genes. How might this be achieved? The most favoured method of studying the function of a gene is to clone it into the molecular biologists' favourite bacterium *Escherichia coli*. *E. coli* has several advantages that make it amenable for cloning – the cells grow quickly, the genetics are well characterized and strains have been engineered so that they are relatively harmless to ourselves. There are, however, many problems are associated with inserting foreign DNA sequences into bacteria. In general, DNA sequences will only be replicated if they contain a **replicon**. The genomes of bacteria and viruses usually contain a single replicon. So, simply putting a piece of foreign DNA into a bacteria cell will not result in the replication of that DNA. Indeed the most likely fate of foreign DNA in bacteria is degradation. As far back as the 1950s, it was noted that if a bacteriophage was prepared from a particular strain of *E. coli* cells (say strain C) then it would very efficiently infect cultures of the same *E. coli* strain. The bacteriophage would, however, infect other *E. coli* strains (strain K) at very low efficiency. Interestingly, the same bacteriophage prepared from an *E. coli* K strain would efficiently infect both *E. coli* K and C strains. The bacteriophage produced in *E. coli* C was in someway being restricted from entering *E. coli* K strains. Werner Arber noted that DNA from *E. coli* C prepared bacteriophages was rapidly degraded upon entering an *E. coli* K strain (Arber, 1965). Degradation occurs because the bacteria contain restriction–modification systems specifically designed to protect them from foreign DNA sequences.

2.1   Restriction Enzymes

Bacterial restriction–modification systems have two components – a restriction endonuclease and a DNA methylase. The restriction enzyme (or **restriction endonuclease**) cleaves DNA at specific sequences. The term 'endonuclease' applies to sequence specific nucleases that break nucleic acid chains somewhere within the DNA, rather than at the ends of the molecule. The first restriction enzyme to be isolated was that from *E. coli* K laboratory strains in 1968 (Meselson and Yuan, 1968). This enzyme was able to cleave DNA, but the precise site of DNA cleavage remained unclear. The enzyme displayed a number of complex activities that made it difficult to study (Table 2.1).

Table 2.1. Properties of restriction endonucleases. Adapted from Dryden, Murray and Rao (2001)

Property	Type I	Type II	Type III
Restriction and modification	Single multifunctional enzyme	Separate nuclease and methylase	Separate enzymes sharing a common subunit
Nuclease subunit structure	Heterotrimer	Homodimer	Heterodimer
Cofactors	ATP, Mg^{2+}, SAM	Mg^{2+}	Mg^{2+} (SAM)
DNA cleavage requirements	Two recognition sites in any orientation	Single recognition site	Two recognition sites in a head-to-head orientation
Site of DNA cleavage	Random, approx 1000 bp away from recognition site	At or near recognition site	24–26 bp to the 3′-side of the recognition site
Enzymatic turnover	No	Yes	Yes
DNA translocation	Yes	No	No
Site of methylation	At recognition site	At recognition site	At recognition site

In 1970, Hamilton Smith and his co-workers isolated a restriction enzyme activity from the bacterium *Haemophilus influenzae* strain Rd and showed that it was able to cleave DNA at specific sites. The enzyme, called HindII, recognizes a six-base-pair double-stranded DNA sequence of 5′-G–T–pyrimidine–purine–A–C-3′:

$$5' - GT(T/C) \, | \, (G/A)AC - 3'$$
$$3' - CA(A/G) \, | \, (C/T)TG - 5'$$

and cleaves DNA on both strands in the centre of the sequence (indicated by the dotted line). Smith found that his enzyme was unable to cleave *Haemophilus influenzae* genomic DNA, but it cleaved the bacteriophage T7 genome (39 937 bp in length) in over 40 places, to give a highly specific fragmentation pattern. The restriction enzyme HindII enzyme always recognizes the sequence above and always cuts directly in the centre of this sequence. This sequence is known as the **recognition site** for HindII. Wherever this particular sequence of six base pairs occurs unmodified in a DNA molecule, HindII will cleave the DNA.

In fact, Smith was very fortunate in his experiments. *Haemophilus influenzae* contain two different restriction enzymes, HindII and HindIII, which were probably both present in his partially purified restriction enzyme activity. HindIII recognizes and cleaves the sequence 5'-AAGCTT-3', but this sequence does not occur within the bacteriophage T7 genome, and thus would not interfere with the digestion patterns he obtained.

Shortly after Smith's discovery, another restriction enzyme, EcoRI, was isolated and characterized from *Escherichia coli* strain RY13 (Hedgpeth, Goodman and Boyer, 1972). This enzyme was found to have the recognition site 5'-GAATTC-3'. More than 900 restriction enzymes have now been isolated from over 230 species of bacteria. Restriction enzymes have names that reflect their origin – the first letter of the name comes from the genus and the second and third letters from the species of bacteria from which they were isolated. The letters were, by convention, written in italics, but recently this has changed so that they are now written in plain font (Roberts *et al.*, 2003). The numbers following the nuclease name indicate the order in which the enzyme was isolated from the bacterial strain. The number is written as a Roman numeral. For example, the first restriction enzyme to be isolated from the bacterium *Providencia stuartii* was named PstI, and the second to be isolated from *Bacillus stearothermophilus* strain ET was named BstEII.

A pertinent question to ask here is why the DNA of an organism producing a restriction enzyme is not itself attacked by the restriction enzyme. In Smith's experiments described above, he was able to observe cleavage of T7 genomic DNA by HindII, but genomic DNA from *Haemophilus influenzae*, from which the enzyme is derived, was not cleaved, even though it contains many recognition sites for HindII. The answer lies in the modification part of the restriction–modification system. *Haemophilus influenzae*, like all organisms that produce restriction enzymes, also produces enzymes that modify its own DNA so that it will not be cleaved by the restriction enzyme. For example, *E. coli* RY13 strains, which produce EcoRI, also make EcoRI methylase. Like many restriction enzymes, the genes encoding both the nuclease and the modifying enzymes are adjacent to each other in the genome. EcoRI methylase attaches methyl groups to the N6 position (Figure 1.5) of the second adenine residues in the EcoRI recognition site, and this covalent modification renders the site uncleavable by EcoRI (Figure 2.1). Indeed, a methylation of just one of the DNA strands in the EcoRI recognition sequence is sufficient to prevent cleavage by the restriction enzyme. The resistance of the **hemimethylated** DNA site to cleavage protects the bacteria DNA from degradation immediately after

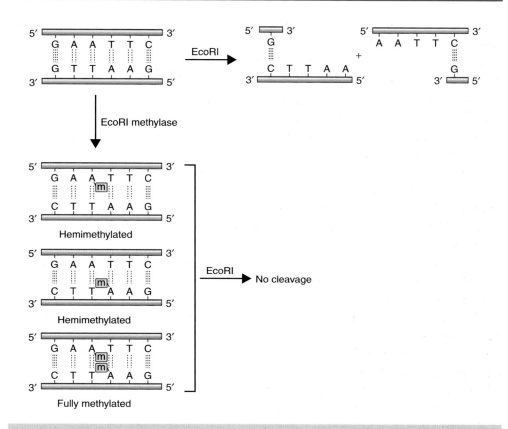

Figure 2.1. Protection of DNA cleavage by modification. The restriction enzyme EcoRI recognizes and cleaves the sequence 5'-GAATTC-3'. *E. coli* R strains, from which this enzyme is derived, protect their own DNA from fragmentation by also producing a specific methylase. EcoRI methylase takes a methyl group from S-adenosylmethionine and places it on the N6 of the second adenine in the recognition sequence. EcoRI restriction enzyme is unable to cleave the methylated DNA

semiconservative DNA replication of the fully methylated sequence, until EcoRI methylase can once again restore the fully methylated state. Methylation of the adenine bases in the EcoRI recognition site does not affect the overall DNA structure, but inhibits cleavage by the enzyme (Jen-Jacobson *et al.*, 1996). DNA methylase enzymes (sometimes called DNA methyltransferases) operate using an ingenious mechanism to add methyl groups to specific bases within double-stranded DNA. A number of these enzymes have been studied structurally (Figure 2.2). Binding of the methylase to DNA results in the flipping out of the base that is to be methylated, and stabilization of the orphaned base left in the helix by other parts of the protein. Base flipping involves rotation of backbone bonds in double-stranded DNA to expose an out-of-stack nucleotide, which

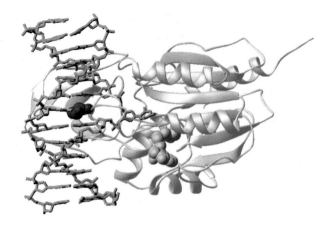

Figure 2.2. The structure of the *Hha*I methylase bound to DNA. The enzyme (shown in white) inserts an amino acid (dark blue) into the double helix of its recognition site (5'-GCGC-3') and flips the first C residue (green) completely out of the helical axis (Blumenthal and Cheng, 2001). This residue is methylated using S-adenosylmethionine as a donor (shown in gold). The methylated base presumably flips back into the helix when the enzyme dissociates. Reproduced from Blumenthal and Cheng (2001) by permission of Xiaodong Cheng (Emory University)

can then be a substrate for the enzyme-catalysed chemical reaction (Cheng and Roberts, 2001).

So, the organism is able to protect its own DNA while degrading foreign DNA that it not correctly methylated. As we will see later, methylase enzymes are important tools for the genetic engineer who wants to protect certain DNA sequences from cleavage by restriction enzymes *in vitro*.

2.1.1   Types of Restriction–Modification System

Restriction–modification systems are divided into three types on the basis of enzyme complexity, cofactor requirements and the position of DNA cleavage, although new systems are being discovered that do not fit readily into this classification. The properties of each type are summarized in Table 2.1. The restriction enzyme *Eco*K, the first to be isolated in 1968, is a type I endonuclease that recognizes the DNA sequence 5'-AACN$_6$GTGC-3', where N can be any nucleotide, but cleaves DNA in an apparently random fashion well away from the recognition site. Type I restriction enzymes require ATP, S-adenosylmethionine (SAM) and magnesium ions (Mg^{2+}) for activity. The enzymes are composed of three subunits, a specificity subunit that determines the DNA recognition site, a modification subunit and a restriction subunit. In the presence of SAM, the enzyme binds to its recognition site irrespective of its methylation state. If the recognition site is methylated, then ATP hydrolysis stimulates the dissociation of the enzyme from DNA. If the recognition site is methylated on one strand only (hemimethylated), then ATP stimulates the methylation of the other strand, with SAM being a methyl donor. If the site is unmethylated, then DNA cleavage occurs. Cleavage occurs several kilobases away from the recognition site, with the intervening DNA sequences looping out from the enzyme. The relatively random nature of the DNA cleavage event makes enzymes of this type not particularly useful for genetic engineering experiments.

The HindII, HindIII and EcoRI restriction enzymes are all examples of type II enzymes. They recognize DNA sequences that are rotationally symmetrical – the sequence in the 5′ to 3′ direction on one DNA strand is the same as the sequence in the 5′ to 3′ direction on the other strand. Such DNA sequences are often referred to as being **palindromic**. The palindromic binding site may either be continuous (e.g. KpnI recognizes the sequence 5′-GGTACC-3′) or interrupted (e.g. BstEII recognizes the sequence 5′-GGTNACC-3′, where N can be any nucleotide). Type II restriction enzymes require only magnesium ions for activity and usually cleave DNA either within the recognition site or very close to it. Different type II restriction enzymes will recognize and cleave different sites, but one restriction enzyme will always cut a particular base sequence the same way, no matter what the origin of the DNA. Different restriction enzymes isolated from different sources often recognize the same DNA sequence, although they may cleave the DNA differently. Such enzymes are called **isoschizomers** of each other. For example, the restriction enzymes SmaI and XmaI both recognize the same DNA sequence (5′-CCCGGG-3′) and are therefore isoschizomers, but SmaI cleaves in the middle of the site between the central C and G nucleotides, while XmaI cleaves after the first C nucleotide on each DNA strand.

The type III restriction systems act as complexes of two different subunits: one subunit (M) is responsible for DNA sequence recognition and modification, and the other subunit (R) is responsible for nuclease action. DNA cleavage requires magnesium ions, ATP, and is stimulated by SAM. The recognition sites are approximately symmetric, and cleavage occurs by nicking one DNA strand at a measured distance to one side of the recognition sequence. The two recognition sites in opposite orientation are necessary to break the DNA duplex.

2.1.2 Other Modification Systems

Most *E. coli* strains used in the laboratory for DNA isolation contain three site-specific DNA methylation systems (Figure 2.3). DNA modification by these systems can lead to the protection of potential restriction enzyme recognition sites from cleavage. Consequently, knowledge of the sensitivity of restriction endonucleases to DNA methylation conferred by these systems in *E. coli* is of great importance to the genetic engineer.

(a) *Dam methylation.* The methylase specified by the *Dam* gene methylates the N6-position (see Figure 1.5) of the adenine residue in the sequence 5'-GATC-3'. The methylation of adenine at this point does not affect its ability to base pair with thymine. Methylation by *Dam* has been implicated in a variety of cellular functions, including DNA replication. *Dam* methylase is a relatively slow enzyme, and as a consequence newly synthesized DNA will be hemimethylated – the old DNA strand will be methylated, while the newly synthesized strand will not (Campbell and

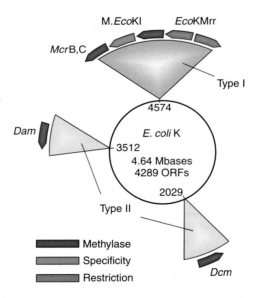

Figure 2.3. Restriction – modification systems found in laboratory *E. coli* K strains. The genomic location of restriction – modification systems in *E. coli* K12. The circle represents the genome of *E. coli*, with the numbers on the inside indicating genomic location. *E. coli* contains two type II modification systems (*Dam* and *Dcm*) and two type I restriction – modification systems (*Mcr* and *EcoKI*). The various methylase, restriction and specificity subunits are colour coded as indicated

Kleckner, 1988). Later, however, both strands will become methylated. This is important since the methylation state of the *E. coli* origin of replication (*ori*C) controls whether replication will proceed or not. If the DNA on both strands of *ori*C is methylated, then this site can serve as an origin of replication and DNA synthesis will proceed. If, however, the DNA at *ori*C is hemimethylated, *ori*C is not active and DNA replication will not start. The slow remethylation of *ori*C after one round of replication delays another round until the first is complete (von Freiesleben *et al.*, 2000).

(b) *Dcm methylation.* The methylase specified by the *Dcm* gene methylates the C5-position of the internal cytosine residue in the sequence 5′-CCWGG-3′ – where W is either an A or T residue. The precise role of *Dcm* methylation in the cell remains obscure since *E. coli* deleted for the *dcm* gene have no obvious phenotype (Palmer and Marinus, 1994).

(c) *Mcr system.* In the 1980s it became apparent that DNA from various bacterial and eukaryotic sources could only be cloned at very low efficiency in certain *E. coli* strains, the problem being that the incoming DNA was being restricted by the host. The phenomenon is caused by methylcytosine in DNA, and is called modified cytosine restriction (*Mcr*). One *Mcr* system is encoded by two genes, *Mcr*B and *Mcr*C, from within the *E. coli* genome, close to the site of the *Eco*KI restriction–modification system (Figure 2.3). The proteins encoded by these genes recognize the DNA sequence 5′-R–mC–N_{40-80}–R–mC-3′, where R is either A or G, N is any nucleotide and mC is methylated cytosine. Cleavage occurs at multiple sites in both strands between the methylated cytosine residues, and requires GTP to translocate DNA between the two 5′-RmC-3′ binding sites (Panne, Raleigh and Bickle, 1999; Stewart *et al.*, 2000; Panne *et al.*, 2001).

The *Dam*- and *Dcm*-dependent methylation may interfere with cleavage by restriction endonucleases with recognition sites partially or completely overlapping such methylation sites. For example, MboI (recognition sequence 5′-GATC-3′) does not cut DNA methylated by the *Dam* methylase, while its isoschizomer Bsp143I is insensitive to *Dam* methylation. Similarly, *Eco*RII (5′CCWGG-3′- where w = A or T) does not cleave *Dcm* methylated DNA, meanwhile its isoschizomer MvaI does.

The precise nature of cutting by the type II restriction enzymes makes them most applicable for use in genetic engineering. The other systems are important, however, since they must be removed from host cells into which foreign DNA is to be inserted if the DNA is to be cut at certain sites.

2.1.3   How Do Type II Restriction Enzymes Work?

Despite little, if any, sequence homology at the amino acid level, all of the type II restriction enzymes that have been studied at the structural level have a highly conversed catalytic core composed of a five-stranded β-sheet flanked by two α-helices (Figure 2.4) (Kovall and Matthews, 1999). Restriction endonucleases appear to function by 'scanning' the length of a DNA molecule by binding to it in a non-specific fashion (Pingoud and Jeltsch, 2001). Once it encounters its particular recognition sequence, the restriction enzyme undergoes a large conformational change, which activates the catalytic sites. The enzyme will then make one cut in each of the two sugar–phosphate backbones of the DNA double helix to generate a 3′ hydroxyl and a 5′ phosphate (Figure 2.4). The positions of these two cuts, both in relation to each other, and to the recognition site itself, are determined by the individual restriction enzyme. Once the cuts have been made, the DNA molecule will break into fragments. Not all restriction enzymes cut symmetrically and leave blunt ends like HindII described above. Many cleave the DNA backbones in positions that are not directly opposite each other and consequently leave overhanging ends at either the 5′- or the 3′-end of a DNA molecule (see Figure 2.5).

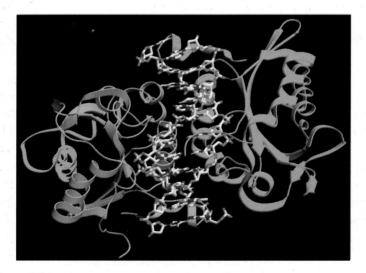

Figure 2.4. The structure of the restriction enzyme BamHI bound to DNA. The enzyme recognizes double-stranded DNA of the sequence 5′-GGATCC-3′, and cleaves the phosphodiester backbone between the two G residues. This results in formation of two DNA fragments that have 5′ overhanging ends – called cohesive or sticky ends. The protein is a dimer of identical subunits (coloured in green and cyan)

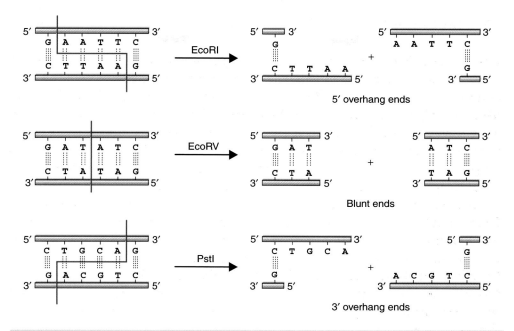

Figure 2.5. DNA cleavage by type II restriction enzymes. Different restriction enzymes recognize different DNA sequences and can cleave to give either overhanging ends or blunt ends. The recognition site of each enzyme is shown, together with the cleavage site, indicated by the blue line. After cleavage, the weak hydrogen bonds holding the overhanging ends will fall apart, and two new DNA ends will be formed

Take, for example, the restriction enzyme EcoRI. The recognition sequence for this enzyme is 5′-GAATTC-3′. When the enzyme encounters this sequence, it cleaves each backbone between the G and the closest A base residues (Mertz and Davis, 1972). Once the cuts have been made, the resulting fragments are held together only by the relatively weak hydrogen bonds that hold the four complementary bases to each other. The weakness of these bonds allows the DNA fragments to separate from one each other. Each resulting fragment has a protruding 5′-end composed of unpaired bases. Other enzymes, for example PstI, create cuts in the DNA backbone that result in protruding 3′-ends. The protruding ends – both 3′ and 5′ – are sometimes called 'sticky' or 'cohesive' ends because they will bond with complementary sequences of bases. In other words, if an unpaired length of bases (5′-AATT-3′) encounters another unpaired length with the sequence (3′-TTAA-5′), they will bond to each other – they are 'sticky' for each other. Although compatible sticky-ended DNA fragments can associate with each other through complementary base pairing, they will not form a continuous sugar–phosphate DNA backbone. The nicks in the DNA backbone of paired sticky ends can, however, be repaired using DNA ligase enzymes (see below). The cellular origin, or even the species origin, of the sticky ends does not affect their stickiness. Any pair of complementary sequences will tend to bond, even if one of the sequences comes from a length of human DNA, and the other comes from a length of bacterial DNA.

How often do restriction enzyme recognition sites occur within a particular DNA sequence? Most of the restriction enzymes that we have talked about so far recognize DNA sequences that are six base pairs in length. Some other restriction enzymes, however, recognize either longer (seven or eight base pairs) or shorter (four or five base pairs) DNA sequences. The length of the recognition site plays a role in determining the frequency of DNA cleavage of a particular DNA molecule. We would expect that any particular restriction enzyme that recognizes a 4 bp target sequence would cut DNA once every 4^4 (i.e. 256) bp in a random DNA sequence, assuming all bases occur equally frequently. Any particular 6 bp target would be expected to occur once every 4^6 (4096) bp, and an 8 bp target every 4^8 (65 536) bp. So, the treatment of genomic DNA with, say EcoRI, should result in the formation of a series of DNA fragments of approximate size 4 kb. The actual frequency and DNA cleavage by any particular restriction enzyme is obviously dependent upon the frequency at which the bases that make up its recognition site actually occur within that DNA. As we saw in Chapter 1, Chargaff noted that the amount of $C + G$ residues in different organisms was different to the amount of $A + T$. Such differences – often expressed as the **GC content** of a genome – represent the percentage of nucleotides within the genome that are either G or C. Restriction enzyme cleavage sites that contain a high proportion of G and C residues will cut DNA relatively infrequently if the overall GC content of the genome is low. Additionally, higher-eukaryotic genomes tend to contain fewer 5′-CG-3′ dinucleotide pairs (also called CpG) than would be predicted by chance (Bird, 1980). Consequently, restriction enzymes that cleave recognition sequences bearing this motif will cut DNA from higher eukaryotes less frequently than might be predicted.

2.2   Joining DNA Molecules

The discovery of restriction enzymes meant that the cleavage of DNA molecules at specific places along their length now became possible. The next problem that faces a genetic engineer is to join two DNA molecules together. The solution to the problem of DNA joining came from work carried out in the late 1960s. Several laboratories simultaneously discovered **DNA ligase**, an enzyme that catalyses the formation of a phosphodiester bond between two DNA chains (Weiss and Richardson, 1967; Zimmerman *et al.*, 1967). DNA ligase enzymes require a free hydroxyl group at the 3′-end of one DNA chain and a phosphate group at the 5′-end of the other. The formation of a phosphodiester bond between these groups requires energy. In *E. coli* and other bacteria NAD$^+$ serves this role, whereas in animal cells and bacteriophage ATP drives the reaction. DNA ligases are only able to join DNA molecules that are part of a double helix – they are unable to join two molecules of single-stranded DNA. As we saw in Chapter 1, the role of DNA ligases is to seal nicks in the backbone of double-stranded DNA after replication. This joining process is essential for the normal synthesis of DNA and for repairing damaged DNA, and has been exploited by genetic engineers to join DNA chains to form recombinant DNA molecules.

The mechanism of action of DNA ligases has been studied in detail (Timson, Singleton and Wigley, 2000). ATP, or NAD$^+$, reacts with the ligase enzyme to form a covalent enzyme–AMP complex in which the AMP is linked to an ε-amino group of a lysine residue in the active site of the enzyme in the form of a phosphoamide bond (Figure 2.6). The AMP moiety activates the phosphate group at the 5′-end of the DNA that is to be joined. The final step is a nucleophilic attack by the 3′-hydroxyl group on this activated phosphorus atom. A phosphodiester bond is formed and AMP is released. This sequence of reactions is driven by the hydrolysis of the pyrophosphate that was released in the formation of the enzyme–adenylate complex. Thus, two high-energy phosphate bonds are spent in forming a phosphodiester bond in the DNA backbone if ATP is the energy source.

The temperature optimum for the ligation of nicked DNA is 37 °C, but at this temperature the hydrogen bonding between the sticky ends of restriction enzyme cleaved DNA is unstable. For example, EcoRI generated DNA ends associate through only four AT base pairs, and these are not sufficient to resist thermal disruption at such a high temperature. The optimum temperature for ligating the sticky ends of restriction enzyme generated DNA fragments is therefore a compromise between the rate of enzyme action and the association of the DNA fragments. The temperature optimum for such reactions has been found to be in the range of 4–20 °C (Sgaramella and Ehrlich, 1978). The DNA ligase from *E. coli* will only ligate blunt-ended DNA fragments if they are at a very high concentration. The high concentration is thought to promote association of the DNA fragments so that the ligation reaction may occur. The ligase enzyme from bacteriophage T4 will ligate blunt-ended DNA fragments, although at low efficiency with respect to the efficiency of ligation of sticky-ended DNA fragments (Ferretti and Sgaramella, 1981). A general scheme for the cutting and rejoining of DNA molecules is shown in Figure 2.7.

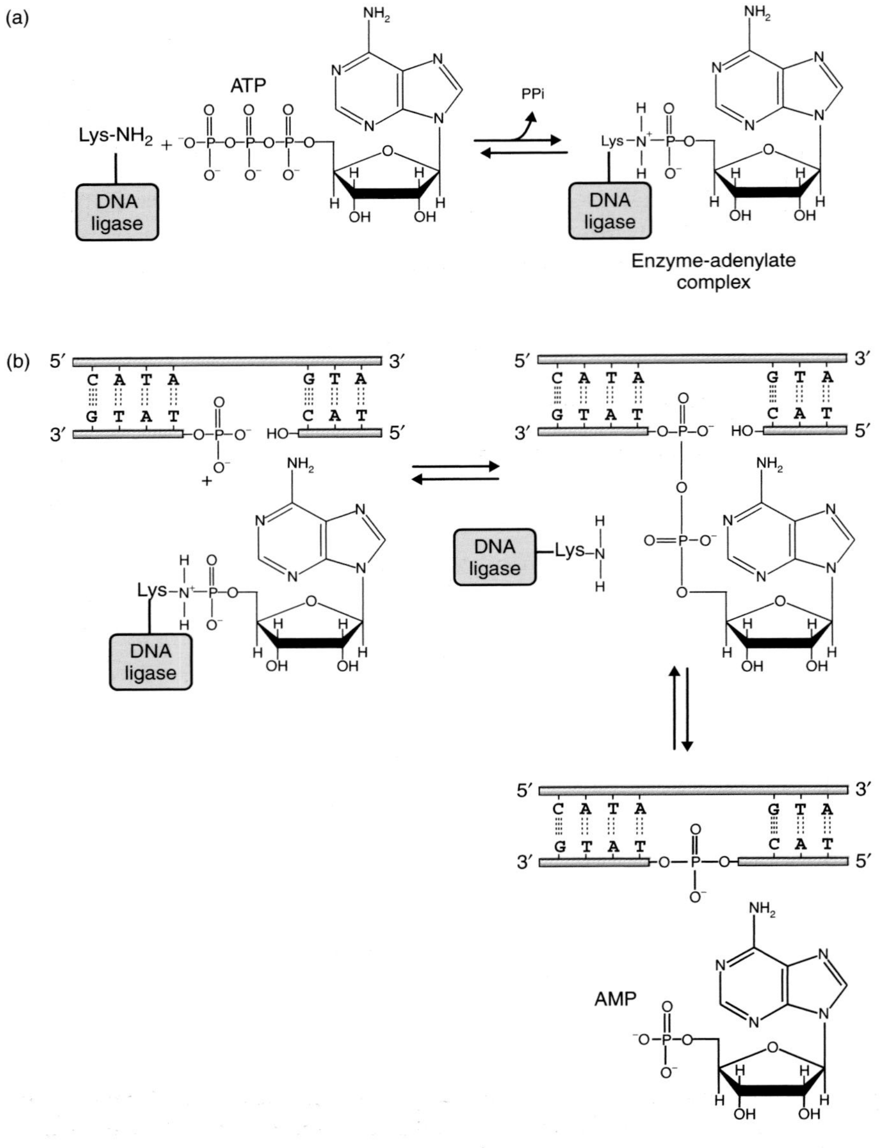

(a)

(b)

Figure 2.6. The mechanism of DNA joining by DNA ligase. See the text for details. This figure is adapted from Doherty *et al.* (1996)

2.3 The Basics of Cloning

The ability to break and rejoin DNA molecules almost at will led to the first experiments in DNA cloning in 1972 (Jackson, Symons and Berg, 1972).

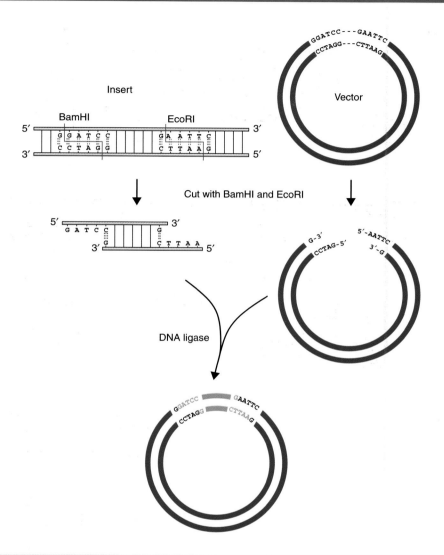

Figure 2.7. Breaking and joining DNA using restriction enzymes and DNA ligase. Linear DNA (insert) and a closed-circular plasmid DNA (vector) each contain the recognition site for BamHI and EcoRI. Mixing the DNA fragments with compatible ends together in the presence of DNA ligase can result in the formation of vector–insert hybrid DNA molecules

For the first time it was possible to extract a fragment of DNA from one source and insert, or **clone**, it into the DNA from another source. Perhaps the most common type of cloning experiment involves the insertion of a foreign piece of DNA into a suitable vector so that the foreign DNA may be propagated in *E. coli*. In Chapter 3 we will discuss the various different types of vector that are available, but at this stage we could consider the vector as a closed-circular double-stranded plasmid DNA molecule. If we wish to insert foreign DNA sequences into this vector, we need to cut it to produce a linear DNA onto which we can attach other DNA sequences using DNA ligase.

Let us first consider the insertion of DNA into the vector using two different restriction enzymes (Figure 2.7). Treatment of both a vector and insert DNA sequences with the restriction enzymes will generate a number of DNA fragments. In the vector the recognition sites for the restriction enzymes are located close to each other. As we will see in Chapter 3, this is very common in engineered plasmids. Cutting such a vector with BamHI and EcoRI will yield two fragments – a large one, comprising the majority of the vector, and a small one, representing the DNA between the restriction enzyme recognition sites. In the presence of DNA ligase, neither of these fragments is able to ligate to itself because the DNA ends are not compatible with each other. Digestion of the linear insert DNA sequence with BamHI and EcoRI results in the generation of three DNA fragments. Only one of the fragments contains a BamHI- and EcoRI-compatible end; the others represent the DNA at either end of the fragment. Mixing the vector DNA and insert DNA that are compatible with each other will result in the formation of hydrogen bonds between the two DNA molecules. If the ends were not compatible, this hydrogen bonding would not occur. The addition of DNA ligase to the hydrogen bonded intermediate will result in the sealing of the DNA backbone and the formation of a vector–insert hybrid DNA molecule. If one of the other insert DNA fragments becomes hydrogen bonded to the vector, say *via* its BamHI-compatible end, then ligation will not result in the formation of a closed-circular vector. As we will see in the next section, such DNA molecules are not replicated when they are transformed into bacteria.

This type of cloning scheme works because the vector DNA that has been cut with the two restriction enzymes contains non-complementary DNA ends. If the vector had been cut using a single restriction enzyme, or with two restriction enzymes that left the same sticky ends, then the vector could easily recircularize in the presence of DNA ligase. Treating the vector with a phosphatase enzyme after it has been cut with the restriction enzymes, however, can prevent this. Phosphatases catalyse the removal of 5′ phosphate groups from nucleic acids and nucleotide triphosphates. Since phosphatase treated DNA fragments lack the 5′ phosphate required by DNA ligase, such treatment will inhibit vector self-ligation and will promote the formation of vector–insert DNA hybrids. Such a cloning scheme is shown diagrammatically in Figure 2.8. When the vector has been treated with phosphatase, DNA ligase

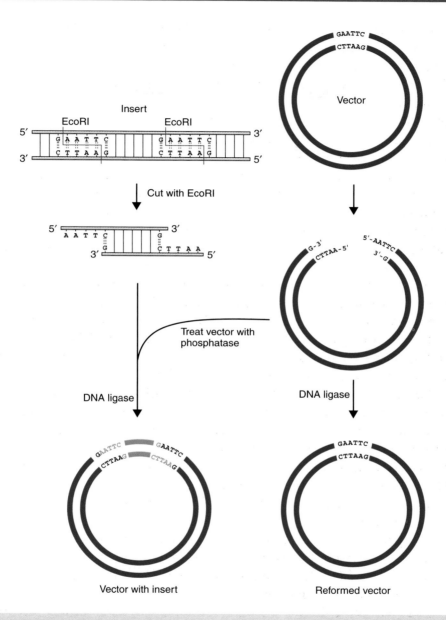

Figure 2.8. The basics of cloning into a plasmid vector containing a single unique restriction enzyme recognition site. The vector contains a single EcoRI recognition site, while the insert has two. Cutting both with the enzyme generates the fragments shown. Adding ligase to the cut vector will probably result in its reformation. This can be prevented by treating the cut vector with phosphatase to remove the 5′ phosphate residues from the ends of the DNA. The cut insert can provide the missing phosphate that DNA ligase requires, so mixing the vector and insert will result in the formation of hybrid DNA molecules

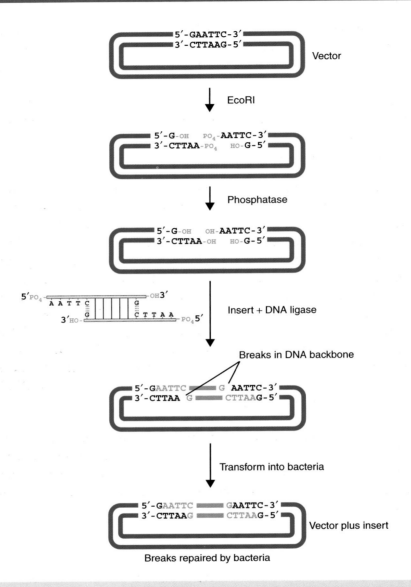

Figure 2.9. The ligation of vector DNA that has been treated with phosphatase to a compatible insert. The vector is cut with the restriction enzyme EcoRI and then treated with phosphatase to remove the phosphates from the free 5'-ends of the cut DNA. The ligation of a compatible insert into this vector will result in the ligation of only the 5'-ends of the insert with the vector. The 3'-end of the insert (a hydroxyl group) and the 5'-end of the vector (also a hydroxyl group) will be unable to ligate. Transformation of the vector–insert hybrid into bacteria, however, will result in the repair of the broken DNA strands to form the complete vector plus insert plasmid

will be able to seal the nicks in the DNA phosphodiester backbone on one strand only (Figure 2.9). However, once these molecules are transformed into bacteria, the break on the other strand is repaired using the bacterial DNA repair systems.

In cloning DNA fragments, there are of course many occasions when restriction enzyme recognition sites either do not occur or do not occur in the correct place within a fragment you are try to clone. This problem can be overcome in a number of ways.

- *Clone into a blunt-ended restriction site.* If the restriction enzyme in the vector leaves blunt ends (like EcoRV shown in Figure 2.5) then any other blunt-ended DNA fragment can be ligated into the cut vector. A number of restriction enzymes give rise to blunt ends after cutting DNA. Other sites can be made blunt by either cleaving off the overhanging ends with a nuclease (e.g. mung bean nuclease) or by 'filling in' the overhanging ends using a DNA polymerase. Such fill-in reactions are often performed with the Klenow fragment of DNA polymerase I in the presence of the appropriate deoxynucleotides. For example, filling in the ends of EcoRI cut DNA (Figure 2.5) would require both dATP and dTTP, and filling the ends of BamHI cut DNA would require all four deoxynucleotide triphosphates. The major drawback to blunt-end cloning is the inefficiency of DNA ligase at carrying out these reactions.

- *Using oligonucleotide linkers.* If you want to join DNA fragments together that have, say, EcoRI and PstI ends respectively, you can synthesize a small synthetic DNA molecule to link the two ends together. DNA ligase will then be able to efficiently seal the two non-compatible sticky ends by using the linker as a bridge between the two.
  Single-stranded linker:

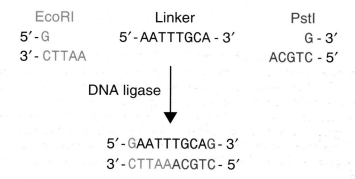

Double-stranded linker:

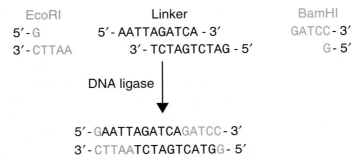

- Mutagenesis to create new restriction sites. The sequence of the DNA may be altered to create or destroy restriction sites. We will discuss this further in Chapter 7.

Now that it was possible to construct hybrid DNA molecules, the next problem was to try to get these hybrid DNA molecules into living cells so that the DNA could be replicated and the genes for which they code could be expressed.

2.4 Bacterial Transformation

Before 1970, there had been many attempts to transform *E. coli* cells with foreign DNA. In general, however, little progress could be made. Going back to the experiments of Griffith and Avery, MacLeod and McCarty (Chapter 1), we know that transformation of some bacteria will occur with naked DNA, but it is a rare event that occurs at low frequency. Additionally, as we have already seen, bacteriophages can efficiently infect various strains of *E. coli*, but if the same experiment is performed with naked DNA, the efficiency of transformation is very low. There are several reasons for this.

(a) *Getting naked DNA into cells is not a trivial problem.* DNA is highly charged and will not easily pass through the membranes that surround the bacterium. In the early 1970s, however, methods were devised to make *E. coli* cells competent for the uptake of naked DNA – such methods are discussed below.

(b) *What is the fate of the foreign DNA once it enters the cell?* For the foreign DNA to be maintained and replicated with the bacterium, it must either be integrated into the bacterial chromosome, so that it will be subsequently propagated as part of the bacterial genome, or be independently replicated. The exact mechanism whereby integration occurs is not clear and it is usually a rare event. If the foreign DNA fails to be integrated, it will probably be lost during growth of the bacterial cells. The reason for this is straightforward; in order to be replicated DNA molecules must contain an origin of replication. Fragments of DNA lacking an origin of replication – even if they survive the bacterial restriction systems – will be diluted out of the host cells after cell division and will eventually be lost. Even if a foreign DNA molecule contains an origin of replication, this may not function in the bacterial cells into which the DNA has been transformed. If fragments of DNA are not able to independently replicate, the obvious solution is to attach them to a suitable replicon. Such replicons are known as **vectors** or **cloning vehicles**. Small plasmids and bacteriophages are the most convenient vectors since they are replicons in their own right, maintenance does not necessarily require integration into the host genome and their DNA can be readily isolated in an intact form. The different plasmids and bacteriophages that are used as vectors are described detail in Chapter 3.

(c) *Monitoring the transformation process.* Assuming that you are able to get foreign DNA into a bacterial cell and have it stably maintained, how can you distinguish the transformed cells from those that have not been transformed? Even if the foreign DNA encodes a gene product, the differences between prokaryotic and eukaryotic gene expression (Chapter 1) mean that it would be unlikely that foreign genes would be efficiently transcribed and translated in the transformed cells. The solution to this problem is to insert the foreign DNA into a cloning vector that also contains a selectable marker that will be expressed in the transformed cells. These markers, usually antibiotic resistance genes, will be discussed in more detail in Chapter 3.

In 1970, it was found that treating *E. coli* cells with calcium chloride ($CaCl_2$) allowed them to take up naked bacteriophage DNA (Mandel and Higa, 1970). This chemical transformation treatment was also subsequently shown to allow plasmids to enter bacterial cells, at varying levels of efficiency. Increased transformation efficiencies have been observed using high voltage electric pulses in a process called **electroporation**, and using a **gene gun**. The process of transformation results in the insertion of a DNA molecule into the host cell. All commonly used plasmid and bacteriophage vectors used to clone foreign DNA fragments allow for the insertion of a single vector molecule into the host cell. This single molecule may be amplified many times within the host, but all of the resulting molecules are identical. A consequence of this is that if a mixed population of DNA fragments is ligated into a common vector and transformed into, say, *E. coli*, then the resulting bacterial colonies will each contain one, and only one, type of recombinant DNA molecule. The mixed population of DNA fragments is segregated into its individual components during the transformation and cell growth processes. This is particularly important in the isolation of single recombinant DNA species from complex DNA libraries (Chapter 5).

2.4.1   Chemical Transformation

Chemical transformation of *E. coli* cells is a simple process. Essentially, the cells are grown to mid-log phase, harvested by centrifugation and resuspended in a solution of calcium chloride. The foreign DNA – often contained within a plasmid – and the now competent cells are then incubated on ice and subsequently subjected to a brief (30 s) heat shock at 37–45 °C. Nutrient medium is then added to the cells and they are allowed to grow for a single generation to allow the phenotypic properties conferred by the plasmid (e.g. antibiotic resistance) to be expressed. Finally, the cells are plated out onto a selective medium such that only cells that have taken up the foreign DNA will grow. The role of calcium chloride in this process is not clear. It is thought to affect the bacterial cell wall, and may also be responsible for binding DNA to the cell surface. The actual uptake of DNA is thought to be stimulated by the brief heat shock (Figure 2.10).

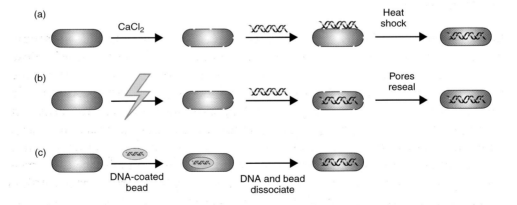

Figure 2.10. Three methods for the transformation of cells. a) *Chemical transformation.* Treatment of cells with calcium ions can make cells competent for the uptake of DNA. The DNA may adhere to the surface of the cell and uptake is mediated by a pulsed heat-shock. b) *Electroporation.* Cells are treated with an electrical pulse, which mediates the formation of pores. DNA can enter the cell before the pores spontaneously reseal. c) *The gene gun.* DNA molecules bound to a bead are fired at cells and are able to enter the cytoplasm. Here, the bead and the DNA dissociate

Since the transformation of *E. coli* is an essential step in many cloning experiments, the process should be as efficient as possible. The efficiency of transformation is governed by a number of host-specific and other factors, but the molecular processes by which transformation occurs are not well understood, and conditions by which efficient transformation can take place are determined empirically. Transformation efficiencies are usually increased if

- the bacterial cells to be transformed are derived from strains that are deficient in restriction systems to reduce the likelihood of degrading the foreign DNA,

- certain exonucleases (e.g. *rec*BC) are mutated in the *E. coli* host cell and

- the competent *E. coli* cells are treated not just with calcium ions, but also with a variety of other divalent cations (e.g. rubidium and manganese) (Hanahan, 1983).

This chemical transformation procedure is applicable to most *E. coli* K12 strains, with typical transformation efficiencies of 10^7–10^9 transformants per microgram of DNA added being achieved, depending on the particular strain of *E. coli* being employed (Liu and Rashidbaigi, 1990).

2.4.2 Electroporation

Electroporation is the use of an electric field pulse to induce microscopic pores within a biological membrane. These pores, called 'electropores', allow molecules, ions and water to pass from one side of the membrane to the other. If a suitable electric field pulse is applied, then the **electroporated** cells can recover, with the electropores resealing spontaneously, and the cells can continue to grow. Pore formation is extremely rapid (approximately 1 μs), while pore resealing is much slower, and is measured in the order of minutes. The use of electroporation to transform both bacterial and higher cells became very popular throughout the 1980s. The mechanism by which electroporation occurs is not well understood and hence, like chemical transformation, the development of protocols for particular applications has usually been achieved empirically by adjusting electric pulse parameters (amplitude, duration, number and inter-pulse interval) (Ho and Mittal, 1996; Canatella *et al.*, 2001).

Two main factors seem to influence the formation of electropores – the types of cell that are used, and the amplitude and duration of the electric pulse that is applied to them. Certain cell types respond well to this type of treatment while others are more refractory – in general, however, most cells can take up DNA when they are electroporated with varying degrees of efficiency. The pulse amplitude and duration are critical if electropores are to be induced in a particular cell. The product of the pulse amplitude and duration has to be above a lower limit threshold before pores will form, beyond which the number of pores and the pore diameter increase with the product of amplitude and duration. An upper limit threshold is eventually reached, at high amplitudes and durations, when the pore diameter and total pore area are too large for the cell to repair. The result is irreversible damage to the cell. During the electroporation pulse, the electric field causes electrical current to flow through the cells that are to be transformed. Buffers and bacterial growth media contain ionic species (e.g. Na^+) at concentrations high enough to cause high electric currents to flow. These currents can lead to dramatic heating of the cells that can result in cell death. Heating effects are consequently minimized by using a relatively high-amplitude, short-duration pulse or by using two very short-duration pulses (Sukharev *et al.*, 1992). Additionally, the cells to be electroporated are extensively washed in distilled water to remove any traces of salt that could 'spark' when the pulse is applied to them.

2.4.3 Gene Gun

The gene gun is a device that literally fires DNA into target cells (Johnston and Tang, 1994). The DNA to be transformed into the cells is coated onto microscopic beads made of either gold or tungsten. The coated beads are then attached to the end of a plastic bullet and loaded into the firing chamber of the gene gun. An explosive force fires the bullet down the barrel of the gun towards the target cells that lie just beyond the end of the barrel. When the bullet reaches the end of the barrel it is caught and stopped, but the DNA-coated beads continue on towards the target cells. Some of the beads pass through the cell wall and into the cytoplasm of the target cells. Here, the bead and the DNA dissociate and the cells become transformed. The gene gun is particularly useful for transforming cells that are difficult to transform by other methods, e.g. plant cells. It is also gaining in use as a method for transferring DNA constructs into whole animals. For example, a vaccine has been developed against foot and mouth disease, a highly virulent viral infection of farm animals. The vaccine is composed of several viral genes that when expressed in the pig will give the animal resistance to infection by the natural virus (Benvenisti *et al.*, 2001).

2.5 Gel Electrophoresis

The progress of the first experiments on cutting and joining DNA molecules were monitored using velocity sedimentation in sucrose gradients. This type of technique, which relies on separation based on size alone after extensive centrifugation through a tube containing high levels of sucrose, requires relatively large amounts of DNA, and is unable to distinguish small changes in the size of a DNA molecule. Separation techniques that needed less material and gave a high degree of separation were required to effectively monitor genetic engineering experiments. As we have already seen, DNA is a highly charged molecule. The phosphates that form the sugar–phosphate backbone of each DNA strand provide a high degree of negative charge. A small DNA fragment will have less negative charge than a large DNA fragment since it contains fewer phosphates. The overall charge per unit length for both a small and a large DNA molecules is, however, identical. So, if an electric current is applied to a sample of small and large DNA fragments in free solution, they will both move to the positive electrode (anode) at the same rate, assuming that friction is negligible in free solution. Therefore, a mechanism by which DNA molecules could be separated would be to increase the amount of friction so that small DNA molecules would move to the anode faster by virtue of having less friction than larger DNA molecules. Running the DNA fragments through a gel can provide the necessary friction to separate DNA fragments of different sizes.

2.5.1 Polyacrylamide Gels

Polyacrylamide gel electrophoresis (PAGE) had been introduced in 1959 as a method for separating proteins (Raymond and Weintraub, 1959). The technique is, however, equally applicable to the separation of nucleic acids. The pore size of this kind of gel may be varied, by altering the percentage polyacrylamide used to construct the gel (from 3 to 30 per cent), for separating molecules of different sizes. PAGE is a powerful technique in the analysis of DNA molecules, and is able to very effectively separate DNA molecules that differ in size by as little as a single base pair. This high level of resolution makes PAGE ideal for the analysis of DNA sequence. The technique is, however, limited to relatively small DNA molecules (less than 1000 bp in length). Large DNA molecules are unable to enter the pores of the polyacrylamide and are consequently not separated by the gel. Since most vectors that are commonly used for cloning genes are bigger than can be resolved by PAGE, an alternative technique was required.

2.5.2 Agarose Gels

Agarose is a naturally occurring colloid that is extracted from seaweed. It is a linear polysaccharide made up of the basic repeat unit agarobiose, which comprises alternating units of galactose and 3,6-anhydrogalactose. Agarose gels are formed by suspending dry agarose, at concentrations ranging between 1 and 3 per cent, in aqueous buffer, then boiling the mixture until a clear solution forms. This is poured into a suitable gel former containing a comb to form wells, and allowed to cool to room temperature to form a rigid gel (Figure 2.11). After the gel has set, the comb is removed and samples of DNA can be loaded into the resultant wells. The gel is then subjected to a constant electric field (in the range of 10 V/cm gel) and the DNA will migrate toward the positive electrode (anode).

The resolution of an agarose gel is inferior to that obtained by PAGE. The bands formed in an agarose gels are relatively fuzzy because the pore size cannot be accurately controlled. A 1 per cent agarose gel contains a wide variety of pore sizes, while a 2 per cent gel on average contains smaller pores but these are still widely variable. Once DNA fragments have been

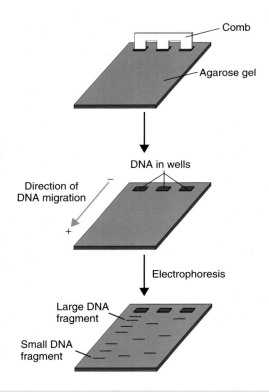

Figure 2.11. Agarose gel electrophoresis. Molten agarose is poured into a former containing a comb. After the gel has set, the comb is removed to form wells into which DNA samples may be loaded. An electric current is the applied to move the negatively charged DNA through the gel towards the positive electrode. Smaller DNA molecules generally run through the gel with a higher mobility than larger DNA fragments

separated through an agarose gel, they must be stained so that the DNA can be visualized. The most common method of staining involves soaking the gel in a solution of ethidium bromide. Ethidium bromide is a flat planar molecule that is able to intercalate between the stacked base pairs of DNA (Figure 2.12). The binding of ethidium bromide to DNA results in distortion of the double-helical structure and localized unwinding of the helix. Ethidium bromide will bind very efficiently to double-stranded DNA, but less so to single-stranded DNA and RNA because of the relative lack of base stacking. Soaking a DNA containing gel in ethidium bromide will result in concentration of the chemical within the DNA. Illumination of the soaked gel with light in the ultraviolet range (260–300 nm) results in fluorescence of ethidium bromide, and the DNA shows up on the gel as a band of fluorescence.

As we have already seen, the charge per unit length of DNA is constant, and in a gel you would expect friction to increase in direct proportion to the length of DNA – a 2000 bp fragment should experience twice as much friction as a 1000 bp fragment. Therefore there should be a direct and inverse relationship between the mass of DNA and its migration rate through a gel.

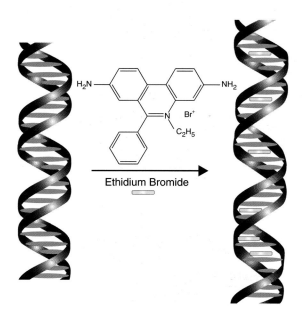

Figure 2.12. The binding of ethidium bromide to DNA. Ethidium bromide is a flat, planar molecule that is able to intercalate in between the stacked bases of double-stranded DNA. The binding of ethidium bromide distorts the double helix and increases its overall length. DNA to which ethidium bromide is bound fluoresces when viewed under ultraviolet light

Excellent separation of DNA molecules in the range of 200–15 000 bp is achieved using agarose gels. Two main factors govern the speed at which a DNA fragment will migrate through an agarose gel when a constant electric current is applied – its molecular mass (or length) and its shape. In general small DNA fragments will migrate faster through an agarose gel than large DNA fragments (Figure 2.13(a)). However, if we look carefully at the way in which DNA fragments of a known length run through an agarose gel, we can see that there is not a direct inverse relationship between DNA fragment size and distance migrated (Figure 2.13).

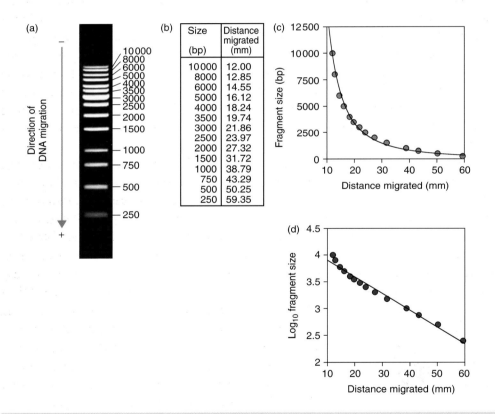

Figure 2.13. The migration of DNA fragments through an agarose gel. (a) An agarose gel showing the separation of DNA fragments of known size. (b) The size of a DNA fragment and the distance migrated through the gel. (c) A plot of fragment size against distance migrated taken from the data shown in (b). This indicates that the relationship between fragment size and distance migrated is not linear. (d) A plot of the log of fragment size against distance migrated. This indicates that there is a direct, inverse, relationship between distance migrated and the log of the size of a DNA fragment

If we measure the distance that DNA molecules of a known size migrate through an agarose gel, and plot the data as size of the DNA fragment against distance migrated, we see that the migration distance is inversely proportional to the log of size (Figure 2.13(d)). This effect is most readily observed by inspection of the distances that particular DNA fragments migrate from the well after electrophoresis. DNA fragments of 500 and 1000 bp are separated widely on the gel, while DNA fragments of 8000 and 10 000 bp run close to each other on the gel (Figure 2.13(a)). A consequence of the inverse log relationship is that any DNA fragment greater than about 30 kbp will migrate in approximately the same location as much larger fragments on an agarose gel. That is to say, large DNA fragments will not be resolved from one another, and so agarose gels do not effectively separate large DNA molecules. A DNA fragment of 30 kbp runs on an agarose gel in the same place as a DNA fragment of 60 kbp. Both of these DNA fragments are able to enter the pores of the gel, and do pass through the gel driven by the electric current, but separation is not achieved. Separation of DNA fragments of this size cannot be achieved by either running the gel longer, or by lowering the concentration of agarose within the gel. To understand this phenomenon, we need to think how DNA fragments actually travel through the pores of a gel.

DNA is a long, thin highly charged polymer. To travel through the pores of a gel, the DNA will tend to take the path of least resistance and travel end-first through the gel pore (Figure 2.14). Several different theoretical models have been put forward to explain the movement of DNA through gels (Slater, Mayer and Drouin, 1996). Perhaps the simplest way to think about highly flexible DNA molecules travelling through a gel is to imagine them snaking (or reptating) their way through the pores of the gel matrix (Figure 2.15). Small DNA fragments will be able to pass through the gel pores more easily than longer DNA fragments due to the sieving effect of the pores. Large DNA molecules travelling in this fashion, however, may become entangled or knotted within the gel and will thus be retarded beyond the level expected based on size alone. DNA fragments above 30 000 bp in length appear to knot sufficiently to inhibit reptation through the gel pores. Larger fragments suffer the same fate, and consequently DNA fragments above 30 kbp run in the same place on an agarose gel.

The second major factor influencing migration through a gel is the topology or structure of a particular DNA fragment. For instance, plasmid DNA isolated from *E. coli* cells is invariably negatively supercoiled closed-circular molecules. These are relatively compact structures that run quickly through agarose gels (Figure 2.16). If one strand of the plasmid double helix becomes broken (nicked) then the supercoiling within the plasmid will be lost, and the more open structure of the relaxed plasmid will migrate more slowly through an

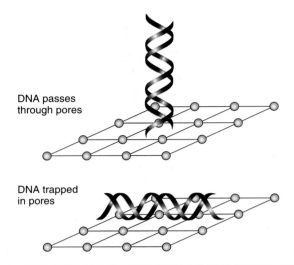

Figure 2.14. DNA is thought to travel through the pores of a gel in an end-on fashion. If we think of an agarose gel as a meshed network of pores, then we can imagine DNA can more readily pass through the pores if it travels end-on rather than side-on. This end-on movement is sometimes referred to as snaking or reptation

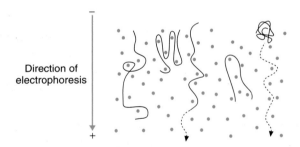

Figure 2.15. DNA snaking through the pores of a gel. DNA molecules moving through the pores of a gel may become trapped in a variety of ways. Larger DNA molecules are more likely to be trapped due to their length than smaller ones. This may be the reason that DNA molecules larger than about 30 kbp all run in about the same place in a conventional agarose gel

agarose gel. If the same plasmid is treated with a restriction enzyme that cleaves it once, then this linearized DNA will run with a mobility intermediate between those of the supercoiled and the nicked molecules. Therefore, DNA molecules that all contain precisely the same number of base pairs can run in several different locations on an agarose gel depending upon the topology of the DNA.

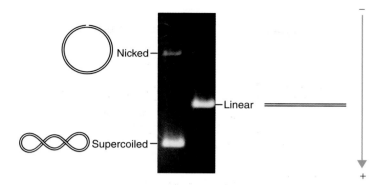

Figure 2.16. The effect of topology on the mobility of DNA fragments. Three DNA fragments, each containing precisely the same number and sequence of base pairs, can run in different places on an agarose gel. Supercoiled DNA is highly compacted and runs rapidly through the gel. If just one of the DNA strands of supercoiled DNA becomes nicked – i.e. a single break in one strand of the sugar–phosphate backbone – then the molecule adopts an open structure with a low mobility. Linear DNA, in which the DNA backbone is broken in both strands, runs with an intermediate mobility on an agarose gel

2.5.3 Pulsed-field Gel Electrophoresis

As we have seen above, all DNA fragments above about 30 kbp run with the same mobility regardless of their size. This is seen in the gel as a large diffuse band. To overcome the size limitation of resolution in an agarose gel, Schwartz and Cantor introduced pulsed-field gel electrophoresis (PFGE) as a method for resolving extremely large DNA molecules (Schwartz and Cantor, 1984). Rather than subjecting DNA fragments in a gel to a continuous static electric field, they altered the direction of the electric current to alter the path of DNA molecules as they travel through the gel (Figure 2.17). Using this technique, the upper size limit of DNA separation in agarose gels was raised from 30–50 kbp to well over 10 Mbp (10 000 kbp).

If the DNA is forced to change direction during electrophoresis, different sized fragments within the diffuse unresolved DNA band begin to separate from each other; perhaps the changes in direction inhibits, or reduces, knot formation. With each reorientation of the electric field relative to the gel, the smaller-sized DNA fragments will begin moving in the new direction more quickly than the larger DNA fragments. Thus, the larger DNA lags behind, providing a separation from the smaller DNA. The original pulsed-field systems used the uneven electric fields generated from static electrodes. As a consequence, the DNA did not run in straight lanes, making interpretation of gels difficult. Ideally, the DNA should separate in straight lanes to simplify

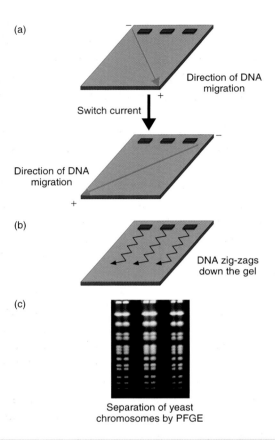

(a)

Direction of DNA migration

Switch current

Direction of DNA migration

(b)

DNA zig-zags down the gel

(c)

Separation of yeast chromosomes by PFGE

Figure 2.17. Pulsed-field gel electrophoresis (PFGE). (a) The switching of the electric current during PFGE. Current is applied across the gel for a defined period – the pulse time – which is often in the range of 0.5–2 min. After this time the direction of the current is switched. (b) The repetitive switching of the current means that the DNA will zig-zag down the gel. The original PFGE technique used two non-homogenous electric fields to change the direction of DNA migration during electrophoresis (Schwartz and Cantor, 1984). The zig-zagging motion allowed the separation of large DNA molecules, but the non-homogenous electric field resulted in bowed DNA banding patterns. (c) Using homogenous electric fields, straight-line separation patterns are obtained, like the separation of whole yeast chromosomes shown here

lane-to-lane comparisons. The simplest approach to obtaining straight lanes is termed field inversion gel electrophoresis (FIGE), which uses parallel electrodes to assure an homogeneous electric field. FIGE works by periodically inverting the polarity of the electrodes during electrophoresis. Because FIGE subjects the DNA to a 180° reorientation, the DNA spends a certain amount of time moving backwards. The 180° reorientation angle of FIGE results in a separation range most useful under 2000 kbp. Furthermore, FIGE has mobility inversions, in which larger DNA can move ahead of smaller DNA during electrophoresis. The use of homogenous electric fields in conjunction with PFGE also results in the DNA running in a straight line. For example, contour-clamped homogeneous electric field (CHEF) electrophoresis reorients the DNA at smaller oblique angle, generally between 96 and 120°. This causes DNA to always move forward in a zigzag pattern down the gel, but the DNA does not move from its lane and straight-line patterns are obtained (Figure 2.18).

Several parameters act together during PFGE to affect the effective separation range of a particular gel. These include the type and concentration of agarose used, the buffer composition, the buffer temperature, the electric field strength, the reorientation angle etc. However, the **pulse time** is primarily responsible for changes in the effective separation range. The pulse time is the duration of the each of the alternating electric fields. Shorter pulse times lead to separation of shorter DNA molecules because the smaller DNA fragments will begin to move more quickly upon reorientation than the larger fragments. Similarly, longer pulse times lead to separation of larger DNA molecules (Figure 2.19).

The advent of PFGE meant that single DNA molecules representing whole chromosomes could be separated with ease using gels. Large DNA molecules such as intact chromosomes are easily sheared and also difficult to pipette due to their high viscosity. The need for intact material during electrophoresis is obvious. Attempts to separate partially degraded or truncated material will result in the smearing of bands, and consequently gels become difficult to interpret. The solution to this problem is to first embed unbroken cells (e.g.

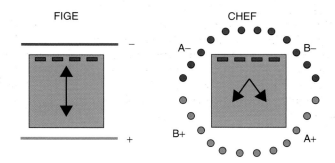

Figure 2.18. The two main types of PFGE. Field inversion gel electrophoresis (FIGE) is where the current is switched back and forth at a 180° angle to the direction of overall DNA movement. Contour-clamped homogenous electric field (CHEF) electrophoresis systems employ multiple electrodes such that the precise position of the electric field can be accurately varied

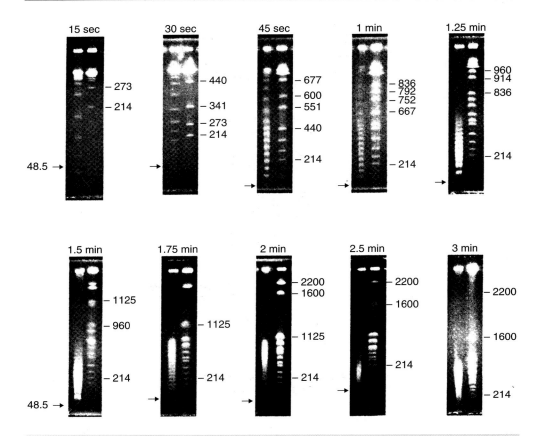

Figure 2.19. The effect of pulse time on the separation of DNA fragments of different sizes. The chromosomes of the yeast *Saccharomyces cerevisiae* and a set of known molecular size DNA markers were subjected to PFGE under otherwise identical conditions using different pulse times. Notice that increasing the pulse time gives rise to better separation of large DNA fragments. The sizes of the markers are shown in kbp. Reproduced from Wrestler *et al.* (1996) by permission of Bruce Birren (Whitehead Institute)

bacteria or yeast) in agarose plugs and then treat the plugs with enzymes to digest away the cell wall and proteins, thus leaving the naked DNA undamaged in the agarose. The plugs then are cut to size, treated with restriction enzymes if necessary and then loaded into the well of an agarose gel (Figure 2.20).

2.6 Nucleic Acid Blotting

When we look at DNA fragments on an ethidium bromide stained gel, all the DNA molecules appear to be identical (Figure 2.13(a)). We know, however,

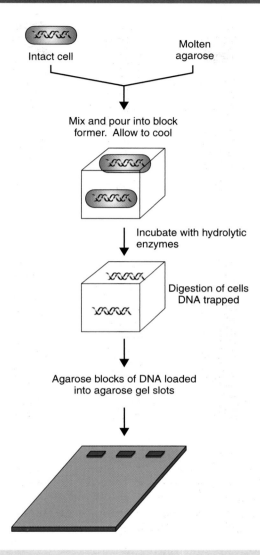

Figure 2.20. Preparation of high-molecular-weight DNA for analysis by PFGE. Intact cells are mixed with molten, but cool, agarose and poured into a block former. Once set, the agarose blocks can be treated with hydrolytic enzymes to break open the cell walls and release the chromosomal DNA. If required, the DNA can be digested with restriction enzymes whilst also in the agarose block. The treated block is then loaded into the well of an agarose gel before being subjected to PFGE

that the sequence of the DNA in each band on the gel is different to other bands – even bands that are identical in length may have a very different DNA sequence. It is, of course, the DNA sequence itself that plays a vital role in the function of a molecule. In the mid-1970s methods were developed to distinguish bands on gels that contain a particular DNA sequence. These methods rely on the hybridization of nucleic acid sequences in order to detect the presence of complementary sequences. The original method of blotting was developed by Ed Southern in 1975 for detecting DNA fragments in an agarose gel that were complementary to a given nucleic acid sequence (Southern, 1975).

2.6.1 Southern Blotting

In the procedure, referred to as **Southern blotting**, the DNA fragments separated on an agarose gel are transferred and immobilized onto a membrane. After the gel has been run, the DNA fragments are denatured (i.e. the strands are separated) using alkali. The single-stranded nature of the DNA on the membrane is important to allow complementary DNA sequences to be able to bind to the DNA fragments attached to the membrane. After the DNA in the gel has been denatured, the DNA fragments are transferred to a membrane. The transfer process can be mediated either electrophoretically or through capillary attraction by placing the gel and membrane in contact with each other and allowing buffer to flow through the gel onto the membrane – the DNA fragments move with the buffer and become trapped on the membrane. Initially, nitrocellulose membranes were used, but these were fragile and easily broken. Nylon membranes are commonly used today. After transfer, the DNA fragments need to be fixed to the membrane so that they cannot detach. A number of methods of fixing are available including baking at 80 °C and ultraviolet cross-linking. UV cross-linking is based on the formation of cross-links between a small fraction of the T residues in the DNA and the positively charged and amino groups on the surface of the nylon membrane. Following fixation, the membrane is placed in a solution of labelled (often radioactive) single-stranded nucleic acid – either single-stranded DNA or RNA. The labelled nucleic acid (or **probe**) is allowed to hybridize to its complementary partner sequence on the membrane. The interaction between the single-stranded probe and its complementary sequence will result in the binding of the probe to the membrane through non-covalent hydrogen bonding that normally holds the DNA double helix together. The membrane is then washed extensively to remove non-specifically bound probe, and specific interactions are detected by exposing the membrane to X-ray film (Figure 2.21).

Southern blotting is used to detect DNA sequences that are either identical or similar to the sequence of the probe. The hybridization of the probe to the DNA sequences trapped on the membrane is the critical component of success of these experiments. As we have seen previously, single-stranded DNA will bind with high affinity to its complementary partner sequence. It will also

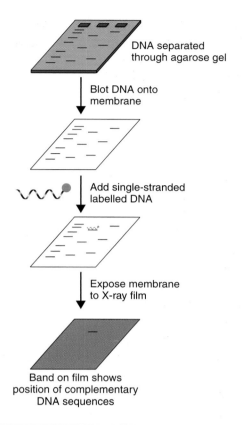

DNA separated through agarose gel

Blot DNA onto membrane

Add single-stranded labelled DNA

Expose membrane to X-ray film

Band on film shows position of complementary DNA sequences

Figure 2.21. Southern blotting. DNA fragments separated on an agarose gel are transferred onto a membrane. The DNA is made single stranded before the addition of a labelled probe. The binding of the probe to the membrane – through base pair hydrogen bonding – can be detected by exposing the membrane to X-ray film. The radio-labelled probe will form a band on the film at a position corresponding to the complementary sequence on the membrane

bind to similar, but non-identical, DNA sequences with a reduced affinity. This differential binding affinity to different DNA sequences can be used to identify DNA molecules that are not identical but are merely related to the probe. This sort of analysis is achieved by altering the temperature (or salt concentration) at which the probe is washed off the filters after it has bound (Figure 2.22). Washing the membrane at high temperature (high stringency) will result in thermal disruption of all but the most tightly bound sequences. Consequently, the bands that show up on the X-ray will film will be either identical or highly related to the probe. Washing the membrane to lower temperatures will reduced the overall level of stringency, and sequences that are less related to the probe

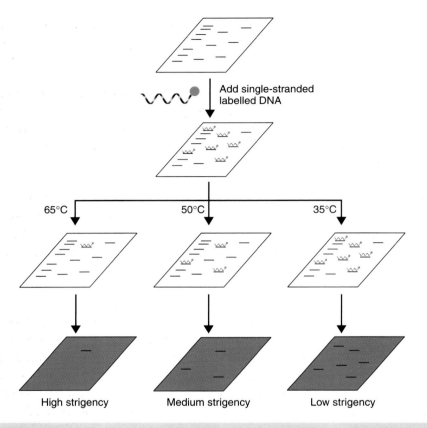

Add single-stranded labelled DNA

65°C   50°C   35°C

High strigency   Medium strigency   Low strigency

Figure 2.22. Washing Southern blot membranes at different temperatures results in different stringencies. Washing the membrane at high temperature will remove all but the most tightly bound DNA molecules – those most similar to the sequence of the probe. Lower-temperature washes can reveal sequences on the membrane that are similar, but not identical, to the sequence of the probe. The actual temperature at which washes are performed will depend on the length of the probe and the likelihood of it finding an exactly matching sequence on the membrane

will give positive signals on the X-ray film. This approach is immensely useful when you do not precisely know the sequence of the gene you wish to identify. This situation often arises when knowledge of the protein sequence is known, but the DNA sequence encoding that protein is unknown. We shall discuss this issue further in Chapter 6.

2.6.2   The Compass Points of Blotting

Since Southern's first description of blotting and hybridization, a number of variants have been described. These all involve the transfer of material from a gel to a membrane, and subsequent specific detection of particular molecules. Such techniques have been assigned names reflecting the points of a compass rather than a description.

- *Northern blotting* – the detection of RNA sequences using probes (Thomas, 1980). RNA is separated on an agarose gel and transferred to a membrane, much as described above. Specific RNA molecules are then detected by hybridization using labelled single-stranded DNA or RNA sequences that are complementary to particular RNAs.

- *Western blotting* – the detection of specific proteins using antibodies (Burnette, 1981). Proteins are separated through a polyacrylamide gel containing the detergent SDS to keep them in an unfolded (**denatured**) state. The proteins are transferred from the gel onto a membrane in much the same way as described above for Southern blotting. Particular proteins are then detected using antibodies. The specific interaction between the antibody and its antigen occurs on the membrane, and the position of the bound antibody is detected. Initially radio-labelled antibodies were used, but these have been largely superseded by antibody 'sandwiches'. The sandwiches work through the binding of one unlabelled antibody (the primary antibody) to the antigen on the membrane. A second, labelled, antibody (the secondary antibody) is then used to detect the presence of the first antibody. This has several advantages; firstly, the multivalent nature of antibody binding means that a substantial increase in sensitivity is achieved, and secondly, a single secondary antibody can be used to detect a number of different primary antibodies. For example, primary monoclonal antibodies are often raised in mice. A secondary antibody raised, say, in the rabbit against mouse immunoglobulins will be capable of interacting with a number of different mouse derived primary antibodies (Figure 2.23).

- *South-Western blotting* – the identification of proteins that bind to particular DNA sequences (Singh *et al.*, 1988). Proteins are either separated on a gel or blotted directly onto membranes. The membrane is then challenged with a labelled double-stranded DNA oligonucleotide. If a protein is able to bind the DNA, it will be detected by the presence of the label.

2.7 DNA Purification

Many of the techniques we have discussed in this chapter are heavily reliant upon the availability of relatively large quantities of purified DNA. The isolation of nuclear material from cells is a relatively straightforward process. Lysis of

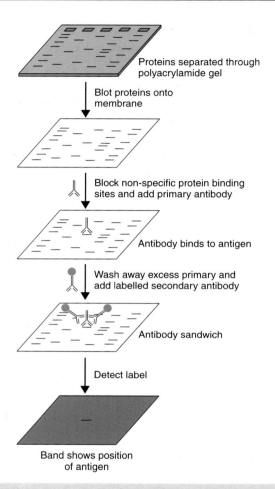

Proteins separated through
polyacrylamide gel

Blot proteins onto
membrane

Block non-specific protein binding
sites and add primary antibody

Antibody binds to antigen

Wash away excess primary and
add labelled secondary antibody

Antibody sandwich

Detect label

Band shows position
of antigen

Figure 2.23. Western blotting using antibodies to detect specific proteins. Proteins are separated through a polyacrylamide gel, usually containing SDS to maintain them in a denatured state, prior to blotting onto a nitrocellulose membrane. Non-specific protein binding sites are blocked on the membrane – using solubilized milk powder – before the primary antibody is added. The primary antibody will specifically bind to the antigen to which it was raised. A labelled secondary antibody is then added to detect the location of the primary antibody. The antibody sandwich results in an amplification of the signal. The secondary antibody is often labelled with an enzyme whose activity, in the presence of appropriate substrates, results in either a colour change on the membrane or the emission of light that can be detected using X-ray film

the cell wall will result in the nuclear material spilling out from the broken cells. This material can be harvested, but the DNA will be contaminated with both RNA and with proteins. Methods are therefore required to remove these contaminants to yield a purified DNA fraction.

Total genomic DNA isolation is often carried out according to the following procedure.

- The host cell wall is lysed. The method used for the lysis procedure depends upon the nature of the host cell itself. For instance, bacterial cells are often treated with the enzyme lysozyme to weaken their cell wall before being lysed with detergents. Yeast cells, on the other hand, are treated with zymolase to disrupt the integrity of their cell wall before lysis proceeds – often by grinding the cells using glass beads to break them open.

- The cellular debris will be removed by centrifugation.

- The remaining soluble material will then be vortexed in the presence of phenol to remove proteins by denaturing then. Chloroform is often used in conjunction with phenol (as a phenol/chloroform solution) since it is also a protein denaturant, but it also stabilizes the rather unstable boundary between the aqueous phase and a pure phenol layer.

- The nucleic acid remaining in the aqueous layer can then be precipitated using ethanol. Upon addition of ethanol the DNA will form a stringy white precipitate that can be collected by a centrifugation.

- RNA can be removed from the preparation by treatment with RNase.

For many applications, the DNA must be purified further. **Isopycnic centrifugation** can be used to separate DNA from contaminants, and also to isolate individual types of DNA molecule. The DNA sample is mixed with a substance (caesium chloride, CsCl) that is capable of forming a self-generating uniform density gradient during the centrifugation process. DNA molecules will sediment only to the position in the centrifuge tube at which the gradient density is equal to its own density, and they will remain there to form a band. After the DNA band has formed, it is removed from the centrifuge tube and the CsCl removed by precipitating the DNA with alcohol. A major advantage of this technique is the ability to separate different species of DNA molecules (Figure 2.24). The inclusion of ethidium bromide in the centrifugation mix results in its binding to DNA. The topological constraints of supercoiled DNA mean that closed-circular DNA molecules (e.g. plasmids) are able to bind only limited amounts of ethidium bromide. Nicked DNA circles, and linear DNA fragments, are able to bind much more ethidium bromide and are extensively unwound in its presence (Figure 2.12). This results in nicked and linear DNA being less dense than their supercoiled equivalent, and consequently being separated in a different location on a CsCl/EtBr gradient. The main drawback of isopycnic centrifugation is the long hours of centrifugation required to form

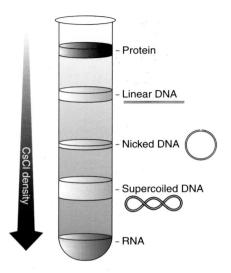

Figure 2.24. Isopycnic centrifugation of DNA molecules through a caesium chloride–ethidium bromide gradient. The DNA sample is mixed with caesium chloride and ethidium bromide and centrifuged at high speed for approximately 48 h. Each component bands at it own density, which, for the DNA, is influenced by the extent to which the molecule can be unwound by the ethidium bromide

the gradients, with isopycnic DNA banding taking 36–48 h to form in a CsCl gradient. The technique has lost favour recently with the introduction of commercially available 'kits' that allow rapid and inexpensive DNA isolation with minimal exposure to harmful reagents. The basis of these 'kits' is discussed below.

The purification of plasmid DNAs (see Chapter 3) is vitally important for cloning and the *in vitro* manipulation of genes. An obvious problem is distinguishing the relatively small amount of plasmid DNA contained within a cell from the relatively large amount of chromosomal DNA that will also be present. Plasmid DNA is most commonly isolated from bacterial cells. Again, cell lysis is an important stage in the process. Insufficient cell lysis will result in low plasmid yields, while cells that have been lysed too much suffer similar problems. The most popular methods of extracting DNA from cells are based on the original procedure of Birnboim and Doly (1979). This method, termed the **alkaline lysis procedure**, takes advantage of the fact that at alkali pH (between 12.0 and 12.5) linear DNA, but not closed-circular DNA, will become denatured. These procedures inevitably result in shearing of the circular *E. coli* genome to generate large linear DNA fragments. Most methods to purify plasmid DNA based on this observation require small cultures of

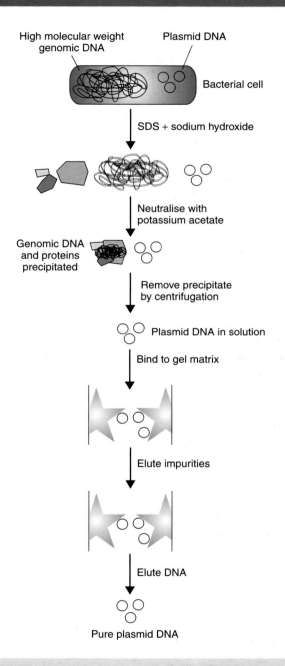

High molecular weight genomic DNA

Plasmid DNA

Bacterial cell

SDS + sodium hydroxide

Neutralise with potassium acetate

Genomic DNA and proteins precipitated

Remove precipitate by centrifugation

Plasmid DNA in solution

Bind to gel matrix

Elute impurities

Elute DNA

Pure plasmid DNA

Figure 2.25. Plasmid DNA purification. Bacterial cells harbouring plasmid DNA are broken open using SDS. The presence of sodium hydroxide ensures the denaturation of genomic DNA. Upon neutralization, the genomic DNA aggregates and forms a precipitate together with the SDS−protein complexes. These can be removed by centrifugation. The plasmid DNA is then bound to a matrix to which contaminants are unable to adhere. The pure plasmid DNA can then be eluted from the matrix

bacterial cells (1–5 mL of bacterial culture) and can yield up to 10 µg of purified plasmid DNA in a typical **mini-prep** procedure. The basic method for plasmid purification is as follows (Figure 2.25).

- The bacterial cells are treated with the detergent sodium dodecyl sulphate (SDS) in the presence of sodium hydroxide to ensure an alkaline pH. The SDS disrupts the cell membranes and denatures proteins. During this process, the chromosomal DNA becomes denatured, but plasmid DNA molecules will remain in solution.

- The mixture is then neutralized using acidic potassium acetate (pH 5.2). The chromosomal DNA renatures upon neutralization, but aggregates to form an insoluble network. The high concentration of potassium acetate also causes precipitation of the protein–SDS complexes and of the high-molecular-weight RNA. Again, plasmid DNA molecules will remain in solution.

- The precipitate is then removed by centrifugation and the plasmid DNA can be further purified from the supernatant.

- The plasmid DNA in solution is then purified from any contaminants (metabolites, RNA, fragments of genomic DNA etc.) by binding the DNA to either silica or a positively charged (anionic) ion exchange matrix. For many small-scale mini-prep procedures, the plasmid DNA is bound to a silica matrix in buffer containing high levels of salt. Under these conditions, other contaminants will not bind to the matrix and can be washed away. Subsequently, the DNA can be eluted from the matrix using a low-salt buffer (or water). Larger-scale plasmid preparations procedures will often bind the plasmid DNA to a positively charged ion-exchange resin, such as DEAE (diethylaminoethanol), in buffer containing, for example, 1 M NaCl. This level of salt allows the DNA to bind to the matrix, but not the contaminants. After washing, the salt concentration is raised to 1.6 M to elute the DNA. Procedures such as this require that the DNA is precipitated (with ethanol or isopropanol) and resuspended in a low-salt buffer before it can be used in many molecular biology protocols.

3 Vectors

Key concepts

♦ Vectors are autonomously replicating DNA molecules that can be used to carry foreign DNA fragments

♦ DNA fragments cloned into vectors can be isolated in quantities sufficient for most laboratory manipulations

♦ Vectors are based on naturally occurring DNA sequences that have been modified and combined to serve particular functions

♦ The choice of vector depends chiefly upon the size of the DNA molecules that must be inserted into it

♦ Vectors have been developed for a variety of specific purposes, including the production of single-stranded DNA, the high-level expression of protein encoding genes, and the production of RNA

The vast majority of molecular cloning experiments utilize the bacterium *Escherichia coli* for the propagation of cloned DNA fragments. Even if the final destination of a cloned DNA fragment is a eukaryotic cell, DNA constructs are invariably produced in *E. coli* prior to being shuttled into their ultimate host. Cloning is possible in other organisms, but the advantages of *E. coli* have led to its widespread acceptance as the genetic engineering organism of choice. *Escherichia coli*, named for the German physician Theodor Escherich (1857–1911), is a gram-negative, rod shaped bacterium propelled by long, rapidly rotating flagella (Figure 3.1). It is part of the normal flora of the human mouth and gut, helping to protect the intestinal tract from bacterial infection, aiding digestion and producing small amounts of vitamins B_{12} and K. The bacterium, which is also found in soil and water, is widely used in laboratory research and is probably the most thoroughly studied life form. As a laboratory organism, *E. coli* has a number of distinct advantages.

Analysis of Genes and Genomes   Richard J. Reece
© 2004 John Wiley & Sons, Ltd   ISBNs: 0-470-84379-9 (HB); 0-470-84380-2 (PB)

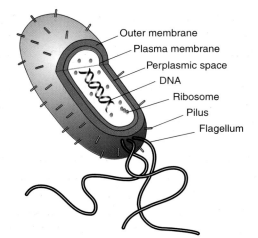

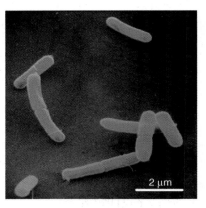

Figure 3.1. Structure of the bacterium *Escherichia coli*. A cut-away model of the bacterium showing some of the cellular layers and components, and *E. coli* cells viewed using an electron microscope

- It is easy to grow in simple, inexpensive growth medium.

- The organism has a rapid doubling time of about 20–30 minutes during log-phase growth.

- Its genetics are well understood.

- Laboratory strains of *E. coli* are generally safe and contain mutations that do not allow them to escape the laboratory environment.

- It has a fully mapped and sequenced genome.

- Extra-chromosomal copies of DNA (plasmids and bacteriophage DNA) can be exploited to carry foreign DNA fragments.

E. coli cells generally reproduce asexually but, in order to increase diversity and share the gene pool, they have mechanisms for the transfer of genetic material from one bacterium to another. As far back as the mid-1940s it was known that bacterial cells were able to exchange genetic material with each other in a semi-sexual manner. The experiments of Lederberg and Tatum clearly demonstrated the transfer of genetic information through **bacterial conjugation** (Lederberg and Tatum, 1946). The ability to perform this transfer is conferred by a set of genes called F (for fertility). These genes can exist on a circular piece of DNA that replicates independently from the bacterial chromosome, or they can be integrated into the chromosome. A bacterium containing these genes (often referred to as the 'male' bacterium) uses a **pilus** to attach to a neighbouring bacterium. The two cells then are drawn together, and DNA is transferred from one bacterium to another. There are three manifestations of the fertility factor.

- F – also called the **F episome**. This is a large circular double-stranded DNA molecule (99 159 bp) that carries only the fertility genes. It is maintained in the bacterium as an extra-chromosomal plasmid.

- Hfr – the F element has become integrated into the *E. coli* genome. When conjugation occurs, the F genes start travelling across the pilus, dragging the rest of the genome behind them. Eventually, the pilus breaks, so most often the entire genome is not transferred. The bacterial genome can be measured, in minutes, from the origin of transfer with the amount of time it takes for a particular gene to be transferred from one bacterium to another indicating how far it is from the origin of replication.

- F′ – this is a large circular double-stranded DNA molecule that contains the fertility genes and a few other genes. These other genes are transferred very efficiently from one bacterium to the next because the length of the transferred DNA is short enough that it can move across the connection between the two bacterial cells before the pilus breaks.

The ability of the F′ episome to replicate and be maintained as a separate entity from the bacterial chromosomes would make it seem an ideal candidate to carry foreign DNA sequences. As we will see later, engineered versions of the F′ episome have indeed been modified to carry foreign DNA fragments. However, the size of the F′ episome precludes easy analysis and manipulation and its gene transfer properties can make it unstable.

In general, foreign DNA fragments need to be carried in a **vector** to ensure propagation and replication within a host cell. A vector is probably best described as autonomously replicating DNA sequences that can be used to carry foreign DNA fragments. All vectors are based on naturally occurring DNA sequences that can be replicated under particular circumstances. Most commonly used vectors are based either upon plasmids or bacteriophage lambda (λ). In general, vectors can be thought of as a series of discrete modules that provide requirements essential for efficient molecular cloning. A vector that is used predominantly for reproducing the DNA fragment is often referred to as a **cloning vector**, while if it is used for expressing a gene contained within the cloned DNA, it is called an **expression vector**.

A vector must possess the following characteristics to make it useful for molecular cloning:

Table 3.1. The general properties of commonly used vectors

Vector	Main uses	Maximum insert size (kb)	Example
Plasmid	General DNA manipulation Numerous specialized derivatives	10–20	pBR322, pUC18
λ (insertion)	Construction of cDNA libraries	~10	λgt11
λ (replacement)	Construction of genomic libraries	~23	λZAP, EMBL4
Cosmid	Construction of genomic libraries	~44	pJB8
M13	*In vitro* mutagenesis DNA sequencing	8–9	M13mp18
Phagemid	General DNA manipulation *In vitro* mutagenesis	10–20	pBluescript
YAC	Construction of genomic libraries	1000–2000	pYAC4
PAC	Construction of genomic libraries	75–90	pAd10SacBII
BAC	Construction of genomic libraries	130–150	pBAC108L

- the ability to self-replicate
- a selectable characteristic so that transformed cells may be recognized from untransformed cells.

Additionally, most vectors will contain at least one, and often multiple, restriction enzyme recognition sites so that DNA fragments can be cloned into the vector relatively easily. A huge array of different types of vector is available today, with many being highly specialized and designed to perform a specific function. In this chapter, I will discuss some of the general points of vector design, but will concentrate on vectors that are commonly used in cloning experiments (Table 3.1).

3.1 Plasmids

Plasmids are naturally occurring extra-chromosomal DNA fragments that are stably inherited from one generation to another in the extra-chromosomal state. Plasmids are widely distributed throughout prokaryotes and range in size from approximately 1500 bp to over 300 kbp. Most plasmids exist as closed-circular double-stranded DNA molecules that often confer a particular phenotype onto the bacterial cell in which they are replicated. That is, the plasmid will often carry a gene that encodes resistance to either antibiotics or heavy metals, or that produces DNA restriction and modification enzymes, that the bacterium would not normally possess. The replication of the plasmid is often coupled to that of the host cell in which it is maintained, with plasmid replication occurring at the same time as the host genome is replicated. Plasmids are often described as being either relaxed or stringent on the basis of the number of copies of the plasmid that are maintained within the cell. **Relaxed plasmids** (not to be confused with relaxed DNA, Chapter 1) are maintained at multiple copies per cell (10–200), while **stringent plasmids** are present at a single copy, or a low number of copies (1–2) per cell. At least part of the basis of this difference is the different mechanisms employed by plasmids in order to replicate themselves. In general, relaxed plasmids replicate using host derived proteins, while stringent plasmids encode protein factors that are necessary for their own replication.

Box 3.1. Naming genes and DNA

The names given to plasmids, genes and other DNA fragments may, at first glance, appear to be nothing more than a jumbled collection of letters and numbers. The process of naming genes and segments of DNA began by geneticists describing genes associated with visible phenotypes, e.g. in mouse coat colour *c* and *A* for the albino and agouti loci, respectively. Other gene names reflect biological function, for example, *Hbb* for the haemoglobin β-chain, and *Adh* for alcohol dehydrogenase enzymatic activity. *Drosophila* geneticists have brought the most whimsical approach gene naming – with names like *fushi tarazu* (*ftz*, from the Japanese words meaning 'segment deficient'), *spätzle* (*spz*, a type of German noodle), *dunce* (*dnc*), *forkhead* (*fkh*), *hedgehog* (*hh*) and *ether-a-go-go* (*eag*). The assignment of chromosomal locations for genes of unknown function developed soon after the establishment of successful metaphase spreads, chromosome banding methodologies, somatic cell hybrids, isozyme separation and the ability to associate genes and phenotypes with a particular site on the chromosome. It has also become common to use the same name for the gene as for the enzyme or other protein that it encodes. Often, the gene name is italicized whereas the gene product is not to distinguish between the two. This can, however, be a source of confusion, since there is not necessarily a one-to-one relationship between the two entities. Additionally, nomenclature used for one organism or species may be different in another. This non-systematic approach to gene naming can result not only in the same gene function in different organisms being designated differently, but also in the same gene from the same organism having several different names depending upon which research group is describing it.

Traditionally, recombinant plasmids tend to bear the initials of their creator(s) followed by a number that may indicate the numerical order in which the plasmids were produced, or perhaps has some deeper meaning. For example, the name of the plasmid pBR322 can be dissected into the following components:

- p – plasmid,

- BR – named by Paco Bolivar and Ray Rodrigues, who developed the plasmid (Bolivar *et al.*, 1977) and

- 322 – the number of the plasmid within their stock collection.

Plasmids constructed within my own laboratory have the nomenclature pRJRXXX, where XXX is a three-digit number indicating the order in which plasmids have been constructed so they can be easily identified in a laboratory plasmid list. Other plasmids are named for specific, but still comparatively obscure, reasons, e.g. the pUC plasmids are named after The University of California, while others have names that more readily indicate their predominant function, e.g. pYAC4. Although many attempts have been made to standardize nomenclature (White, Mattais and Nebert, 1998), historical names tend to be maintained. Indeed, the naming of DNA by its constructor does allow for wide variations that may not otherwise be seen.

Most plasmids in common use today are based upon the replication origin of the naturally occurring *E. coli* plasmid ColE1, or its very close relative pMB1 – see Box 3.1 for a description of plasmid names. ColE1 is a 6646 bp closed-circular DNA molecule that encodes a bacteriocin, colicin E1 (the product of the *cea* gene), and a resistance gene that allows the host bacterium to escape the effects of the bacteriocin (the product of the *imm* gene). Colicin E1 is a transmembrane protein that causes lethal membrane depolarization in bacteria (Konisky and Tokuda, 1979). The drug resistance gene codes for a protein that interferes with the action of colicin by inhibiting its ability to form a channel through the bacterial membrane (Zhang and Cramer, 1993). Bacteria harbouring the ColE1 plasmid can be distinguished from their counterparts that do not possess the plasmid by their ability to grow on plates containing colicin E1. However, screening for this type of growth is technically difficult and has long since been abandoned in favour of simpler antibiotic screening methods (see below).

The ColE1 plasmid is replicated in a relaxed fashion using the DNA polymerase provided from the host cell. Unlike the bacterial origin of replication (*OriC*), however, the replication of ColE1 proceeds in one direction only. The plasmid does not encode proteins that initiate the replication process, but it does code for several non-translated RNA molecules that are involved in the initiation of replication and one protein that plays a role in regulating the RNA molecules. The ColE1 DNA replication origin is in a region of DNA from which two RNA molecules (RNAI and RNAII) are constitutively transcribed from their own promoters (Figure 3.2). RNAII is complementary to the ColE1 origin of replication and binds to it to form a DNA–RNA hybrid (Cesareni *et al.*, 1991). The bound RNA II molecule is cleaved at the origin, by the host encoded enzyme RNase H, and serves as a primer from which DNA replication occurs. Through complementary base pairing, RNAII can bind on one DNA strand only, and hence the replication of the ColE1 plasmid is unidirectional.

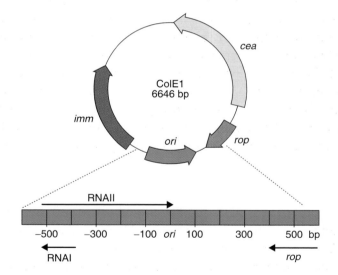

Figure 3.2. The ColE1 plasmid and origin of replication. The ColE1 plasmid is a double-stranded closed-circular DNA molecule, 6646 bp in length. It codes for, amongst other things, the bacteriocin colicin E1 (the product of the *cea* gene) and an immunity protein (the product of the *imm* gene) that prevents the toxic effects of the bacteriocin in cells harbouring the plasmid. DNA replication occurs in one direction only, as indicated by the direction of the *ori* arrow. Two non-translated RNA molecules, termed RNAI and RNAII, and the protein product of the *rop* gene, control the replication process – see the text for details. The relative positions of these replication control elements to the origin (*ori*) are shown. The arrows on the genes indicate the direction of transcription

Control of the replication process is achieved by another non-translated RNA molecule. RNAI is complementary to the 5′-end of RNAII, and the RNA duplex formed between RNAII and RNAI cannot serve as a replication primer. RNAI is a relatively short-lived species, and consequently the ColE1 plasmid is maintained at a level of about 15–20 copies per cell. The interaction between RNAI and RNAII is stabilized by the ROP protein (Helmer-Citterich *et al.*, 1988).

The mechanism of DNA replication employed by ColE1 has several important consequences. The arrangement of regulators results in plasmid incompatibility – that is, two plasmids with the same origin of replication cannot coexist in the same cell. RNAI of the resident plasmid prevents RNAII of the incoming plasmid from forming a primer, so it is not replicated. This is important for cloning experiments when a mixed population of plasmids, which is often the result of a ligation reaction, is transformed into bacteria. Individual transformants produced in this way will contain a single plasmid and not a mixed population. Plasmids with different replication origins (e.g. ColE1 and p15A) are able to co-exist within the same cell. An additional consequence of the ColE1 replication mechanism is that fresh protein synthesis is not required to initiate DNA replication. In the presence of antibiotics that block protein synthesis (e.g. chloramphenicol), chromosomal DNA replication is halted, but ColE1 based plasmids continue to be replicated and accumulate at high levels (1000–2000 copies per cell) (Clewell, 1972).

3.1.1   pBR322

ColE1, and its very close relative pMB1, have the potential to be useful cloning vectors, but they suffer from a number of disadvantages. Primarily, the difficulty in the identification of recombinant plasmids (those in which the plasmid DNA has been ligated to an insert) meant that they did not gain widespread usage. What was required was a plasmid that could be replicated in the same way as ColE1, but in which recombinants could be easily recognized.

The plasmid pBR322 (Figure 3.3) was the first widely used plasmid vector. It is a small plasmid (4363 bp) that was constructed using components from naturally occurring plasmids and other DNA fragments (Bolivar *et al.*, 1977). pBR322 contains the following components.

- *Origin of replication.* pBR322 carries the ColE1 replication origin and *rop* gene to ensure reasonably high plasmid copy number (15–20 copies per cell), which can be increased 200-fold by chloramphenicol amplification.

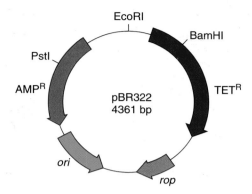

Figure 3.3. The plasmid pBR322. This plasmid contains the ColE1 origin of replication (*ori* and *rop*), together with two antibiotic resistance genes. The ampicillin resistance gene (AMP^R, or *bla* encoding β-lactamase) and the tetracycline resistance gene (TET^R) each contain a number of restriction enzyme recognition sites that occur only once within the plasmid sequence

- *Antibiotic resistance genes.* pBR322 carries two genes that can be used as selectable markers. The ampicillin resistance gene (termed *bla* or, more commonly, AMP^R) was cloned into the plasmid from the Tn3 transposon, and the tetracycline resistance gene (termed *tet* or TET^R) was cloned from the plasmid pSC101 (Bernardi and Bernardi, 1984).

- *Cloning sites.* The plasmid carries a number of unique restriction enzyme recognition sites. Some of these are located in one or other of the antibiotic resistance genes. For example, sites for PstI, PvuI and SacI are found within AMP^R, and sites for BamHI and HindIII are located within TET^R.

The antibiotic resistance genes in pBR322 allow for the direct selection of recombinants in a process called **insertional inactivation**. For example, if we want to clone a DNA fragment into the BamHI site of pBR322, then the insert DNA will interrupt the gene responsible for tetracycline resistance, but the gene for ampicillin resistance will not be altered. Transformed cells are first grown on bacterial plates containing ampicillin to kill all the cells that do not contain a plasmid. Those cells that grow on ampicillin are then replica plated onto medium containing both ampicillin and tetracycline. Those cells that grow in the presence of the ampicillin, but die under tetracycline selection, contain plasmids that have foreign DNA inserts (Figure 3.4). In other words, the insertion of a foreign DNA fragment into an antibiotic resistance gene inactivates the gene product and leads to antibiotic sensitivity.

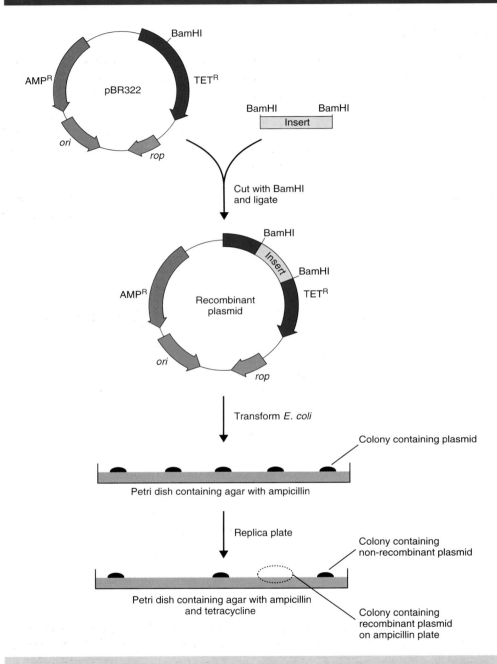

Figure 3.4. Insertional inactivation of antibiotic resistance genes in pBR322. If a DNA fragment is inserted into the BamHI restriction enzyme recognition site within the TET^R gene, the resulting recombinant plasmid, when transformed into bacteria, will still give rise to ampicillin resistance, but will be unable to promote bacterial growth on plates containing tetracycline. The TET^R gene will be functionally inactivated by the presence of the insert DNA

3.1.2 pUC Plasmids

pBR322 was a breakthrough for molecular biology as the first widely used plasmid for molecular cloning, but the double screening procedure required to identify recombinant clones was both time consuming and error prone. In 1982, a new series of plasmids were developed that permitted the identification of the foreign DNA containing cells in a single screening step. These are called the pUC plasmids (Vieira and Messing, 1982). They have three important additional features compared with pBR322 (Figure 3.5).

- *High copy number* – a mutation within the origin of replication produces 500–600 copies of the plasmid per cell without the need for chloroamphenicol amplification. The mutation, a G to A change one

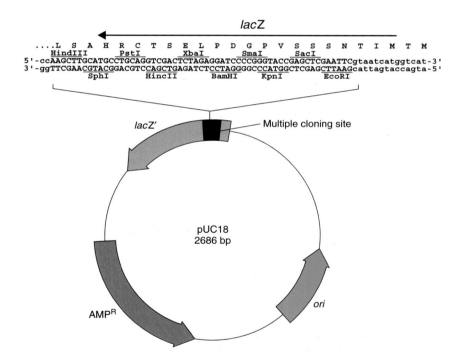

Figure 3.5. The pUC plasmids. pUC18 is a small plasmid that contains a mutated version of the ColE1 replication origin that promotes high-copy-number DNA replication. The plasmids also contain the ampicillin resistance gene (AMP^R) and the *lacZ'* gene, which encodes the first 63 amino acids of *lacZ* – the α-peptide. Embedded within the coding sequence of *lacZ'* are the recognition sites for a number of restriction enzymes. This multiple cloning site (or polylinker) is used to clone in DNA fragments. The presence of insert DNA will disrupt the function of the *lacZ* α-peptide and is used for screening. Different pUC plasmids differ in the composition of the multiple cloning site

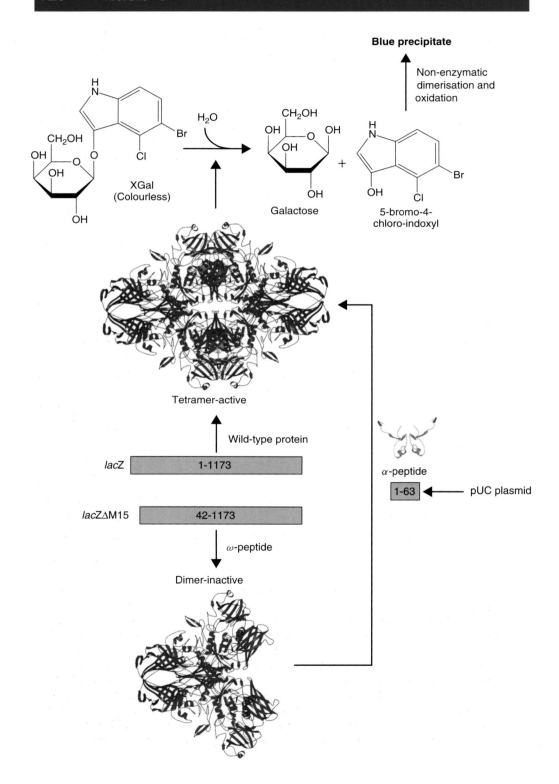

Blue precipitate

Non-enzymatic
dimerisation and
oxidation

H_2O

XGal
(Colourless)

Galactose

5-bromo-4-
chloro-indoxyl

Tetramer-active

Wild-type protein

*lac*Z   1-1173

*lac*ZΔM15   42-1173

ω-peptide

Dimer-inactive

α-peptide

1-63 ◄── pUC plasmid

nucleotide upstream of the initiation site of RNAI, reduces the level of the RNAI transcript and consequently results in an increase in DNA replication using RNAII as the primer.

- *Blue–white screening* – screening of this type is a special form of insertional activation that can be used during the primary selection of transformants, rather than requiring a second round of screening. It utilizes the amino-terminal portion of *E. coli* β-galactosidase (called the α-peptide) encoded by the vector in a form of intermolecular complementation that restores β-galactosidase activity to a defective enzyme (the ω-peptide) encoded by the host. The *E. coli* enzyme β-galactosidase is a large polypeptide (monomer = 1173 amino acids; 117 kDa) that is the product of the *lacZ* gene. The active form of the enzyme is a homo-tetramer (468 kDa). Certain mutations in the 5′ region of *lacZ* prevent subunit association of the resultant protein (ω-peptide) and the monomers lack enzyme activity (Ullmann, 1992). In some such mutants, subunit assembly (and enzyme activity) can be restored by the presence of a small (50- or so amino-acid) amino-terminal fragment of the *lacZ* product (the α-polypeptide) (Juers *et al.*, 2000). Such *lacZ* mutants are said to be subject to **α-complementation**. The product of the *lacZ*ΔM15 allele lacks amino acids 11–41 of wild-type β-galactosidase and is subject to α-complementation. Messing and co-workers took advantage of α-complementation in constructing the pUC plasmid series of cloning vectors (Vieira and Messing, 1982). These vectors carry a multiple cloning site (MCS) embedded in the sequence for the α-peptide gene fragment. The MCS does not alter the reading frame of *lacZ* or destroy the ability of the fragment to α-complement (Figure 3.6). The insertion of other DNA fragments into the MCS will, however, interrupt the coding sequence of the α-peptide, rendering it non-functional. If a chromogenic substrate (XGal, 5-bromo-4-chloro-3-indoyl-β-D-galactopyranoside) and a β-galactosidase

Figure 3.6. α-complementation and XGal staining. XGal (5-bromo-4-chloro-3-indolyl-β-D-galactoside) is cleaved by a functional β-galactosidase enzyme into galactose and a 5-bromo-4-chloro-indoxyl derivative. The indoxyl spontaneously dimerizes and oxidizes to form an insoluble blue dye (5,5′-dibromo-4,4′-dichloro-indigo; not shown). The β-galactosidase enzyme is the product of the *lacZ* gene and the active form of the enzyme is a tetramer of identical polypeptides. Certain mutants of *lacZ* (such as *lacZ*ΔM15) produce versions of the protein that do not include the extreme amino-terminal end of the 1173 amino acid polypeptide. Such derivatives, termed the ω-peptide, are unable to form the active tetramer and are not functional as β-galactosidase enzymes. The ω-peptide can be made functional by co-expressing the *lacZ* α-peptide (amino acids 1–63) in the same cell. The α-peptide promotes tetramer formation and restores enzymatic function

inducer (IPTG) are included in the plates on which the primary transformants are selected, non-recombinant molecules will catabolize the colourless substrate to give blue colonies, while recombinants will give white colonies.

- A multiple cloning site or polylinker. This is a synthetic piece of DNA that harbours the sequence of several unique restriction enzyme recognition sites. It was inserted within the portion of the vector encoding the β-galactosidase α-peptide in such a way that it does not affect its expression or function. However, inserting a foreign DNA fragment into any one of the polylinker restriction enzyme recognition sites invariably disrupts the activity of the α-peptide. Thus recombinant colonies remain white but non-recombinants turn blue.

The major cloning advantage of pUC plasmids over pBR322 is that foreign DNA fragments can be cloned into a variety of restriction enzyme sites and recombinants rapidly screened. Additionally, α-complementation requires only a small gene to be carried on the plasmid. The DNA encoding the α-peptide in the pUC-series of vectors is less than 400 bp in length. If the entire *lac*Z open reading frame were needed, over 3500 bp of DNA would need to be maintained on these vectors. In general, the stability of many plasmid vectors decreases as their size increases. This results in limiting the length of DNA that can be cloned into any particular plasmid. For example, the ability of bacterial cells to maintain recombinant pUC plasmids decreases significantly as their size approaches 15 kbp (Table 3.1). Since the α-polypeptide is small, the total size of the vector is minimized, allowing it to carry a correspondingly larger insert.

3.2 Selectable Markers

An essential feature of the plasmids we have discussed so far is the ability to accurately select for cells that have taken up the plasmid. Many such selection systems are currently available. The choice of selectable marker usually rests with the type of cell that is being transformed. Some of the markers will only function against prokaryotes, while others have a broader spectrum of action. Some of the commonly used selectable markers are listed below, together with their mechanism of action.

- *Ampicillin* – binds to and inhibits a number of enzymes in the bacterial membrane that are involved in the synthesis of the gram-negative cell wall. Therefore, proper cell replication cannot occur in the presence of ampicillin. The ampicillin resistance gene (AMP^R or *bla*) codes for the enzyme β-lactamase that is secreted into the periplasmic space of the bacterium, where it catalyzes hydrolysis of the β-lactam ring of the ampicillin. Thus, the gene product of the AMP^R gene destroys the antibiotic. Over time the ampicillin in a culture medium or petri dish may be substantially destroyed by β-lactamase. When this occurs, selective pressure to maintain the plasmid is lost and cell populations can arise that lack the plasmid.

- *Tetracycline* – binds to a protein of the 30S subunit of the ribosome and inhibits ribosomal translocation along the mRNA and thereby interferes with protein translation. The tetracycline resistance gene (TET^R) encodes a 399 amino acid outer membrane associated protein of gram-negative cells that prevents the antibiotic from entering the cell. Thus, the drug resistance gene does not destroy the antibiotic. Selective pressure will be maintained throughout the cell culture process to keep the plasmid containing the drug resistant gene.

- *Chloramphenicol* – binds to the ribosomal 50S subunit and inhibits protein synthesis. The chloramphenicol resistance gene (CM^R) codes for chloramphenicol acetyltransferase (CAT). The CAT protein is a tetrameric cytosolic protein that, in the presence of acetyl coenzyme A, catalyzes the formation of hydroxyl acetoxy derivatives of chloramphenicol that are unable to bind to the ribosome. As with ampicillin, the CM^R gene product destroys the antibiotic.

- *Kanamycin and neomycin* – bind to ribosomal components and inhibit protein synthesis. The KAN^R gene codes for a protein that is secreted into the periplasmic space and interferes with the transport of these antibiotics into the cell. Like tetracycline resistance, the KAN^R gene does not destroy the antibiotic.

- *Bleomycin and zeocin* – glycopeptide antibiotics that bind to DNA and inhibit DNA and RNA synthesis. They are active against most bacteria (including *E. coli*), eukaryotic microorganisms (e.g. yeast), plant cells and animal cells. The *Sh ble* gene from the bacterium *Streptoalloteichus hindustanus* encodes a small protein that confers resistance to zeocin by binding to the antibiotic (Gatignol, Durand and Tiraby, 1988).

- *Hygromycin B* – inhibits translation by interfering with ribosome translocation. The antibiotic is active against both prokaryotes and eukaryotes. The resistance gene (HYG^R, encoding hygromycin-B-phosphotransferase) inactivates the antibiotic by phosphorylation.

Plasmid based vectors are extremely widely used and have been adapted to serve a variety of functions. Many of these will be discussed in later chapters of this book, but here I will list a number of examples to give the reader a flavour of the diversity of plasmid use.

- *General cloning* – most DNA manipulation performed in the laboratory is carried out in plasmids. The ease of both use and storage of plasmid DNA molecules makes them a popular choice for most recombinant DNA experiments.

- *Shuttle vectors* – these plasmids contain not only the origin of replication and selectable marker for *E. coli*, but also functionally similar sequences for maintenance in other hosts. For example, plasmids for the cloning and expression of genes in the yeast *Saccharomyces cerevisiae* contain both replication origins and selectable markers for both *E. coli* and yeast. Most DNA manipulations will be performed using *E. coli* as a host, prior to transformation of the final DNA construct into yeast.

- *RNA production* – many plasmids have been designed so that the foreign DNA fragments cloned into then can be transcribed into RNA. Such plasmids contain the promoter sites for an RNA polymerase, e.g. those from the bacteriophages T3, T7 or SP6, such that RNA can be made *in vitro* using the purified RNA polymerase and the plasmid DNA. The RNA made by this method is often used as probes for hybridization in Northern blotting.

- *Protein production* – many plasmids contain promoter sequences to express the foreign genes that they contain. Often the expression is performed in *E. coli*, but, using the appropriate promoters, protein expression can be driven in almost any organism. High-level protein production could be driven from a strong promoter, while low-level production would be driven from weaker promoters. Levels of protein production may also be modulated by altering the copy number of the plasmid.

Given the wide range of plasmids that are available to researchers today, systems have been developed to move DNA fragments between a variety of plasmids with differing functions. For example, if you have cloned a gene, you might want to express the gene product at high levels in *E. coli*, while also expressing the protein in mammalian tissue culture cells and producing a tagged version of the protein to which monoclonal antibodies are available. Systems have therefore been devised for the shuttling of DNA fragments between

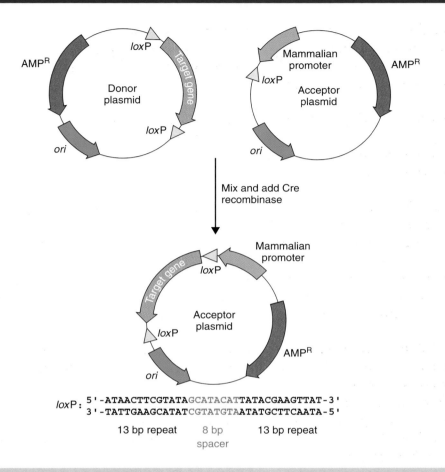

Figure 3.7. Gene shuffling between plasmids using recombination. Genes are transferred from the donor plasmid to the acceptor plasmid at the *lox*P sites using the Cre recombinase. See the text for details

vectors without having to use restriction enzymes. One such system is outlined in Figure 3.7. Here, the DNA fragment encoding the target gene is transferred from one plasmid to another by the action of the Cre recombinase. Cre is a 38 kDa recombinase protein from the bacteriophage P1 (Sternberg *et al.*, 1981). It mediates recombination between or within DNA sequences at specific locations called *lox*P sites (Abremski and Hoess, 1984). These sites consist of two 13 bp inverted repeats separated by an 8 bp spacer region. The 8 bp spacer region in the *lox*P site has a defined orientation that forces recombination to occur in a precise direction and orientation. Donor plasmids contain two *lox*P sites, which flank the target gene. Acceptor plasmid contain a single *lox*P site and elements to which the target gene will become fused. The target gene, once transferred, will become linked to the specific expression elements for which the acceptor vector has been designed. Furthermore, if the coding sequence for the gene of interest is in frame with the upstream *lox*P site in the donor vector, it will automatically be in frame with all peptides designed in the acceptor vector. An alternative donor and acceptor plasmid system is based upon site-specific recombination reactions mediated by phage λ (Karimi, Inze and Depicker, 2002). In this case, DNA fragments flanked by recombination sites (*att*) can be transferred into vectors containing compatible recombination sites (*att* × *att*P or *att*L × *att*R) in a reaction mediated by the λ recombination proteins.

The versatility of plasmids has lead to their widespread acceptance as the vectors of choice for many gene manipulation experiments. Plasmids do, however, suffer from a number of significant shortcomings. First, the efficiency at which the plasmid is transferred to a bacterial cell is very low. Plasmid DNA molecules must be transformed into competent bacterial cells (see Chapter 2), but this process is inefficient. At best, *E. coli* cells can be made competent to a level such that 1×10^9 transformed cells can be generated per microgram of plasmid DNA. For a typical plasmid, 1 µg represents about 1.5×10^{11} molecules. This means that the cells are taking up less than 1 per cent of the available DNA molecules. Second, the capacity of plasmids to carry large fragments of foreign DNA is limited (Table 3.1). Most plasmids become unstable if their overall size exceeds about 15 kbp. Plasmids larger than this tend to undergo recombination events, which can result in the reordering or elimination of DNA from them. Other types of vector have thus been developed to overcome these difficulties.

3.3 λ Vectors

In the early 1950s, André Lwoff described an astonishing property of *Escherichia coli*. When he irradiated certain strains of *E. coli* cells with a moderate dose of ultraviolet light, the bacteria stopped growing and, after about 90 min, the bacteria lysed and released many viral particles, called λ, into the culture media (Figure 3.8). The viruses, more commonly called bacteriophages or simply phages, are able to infect other *E. coli* cells that had not previously been infected by λ phages. Not all bacterial cells underwent this lytic phase when irradiated in this way. Most *E. coli* cells are relatively unaffected by such small ultraviolet doses. However, bacteria that had previously been exposed to phage λ, but had not undergone lysis, showed this remarkable property.

Upon infection by phage λ, *E. coli* cells will undergo one of two fates. Either cell lysis proceeds and newly synthesized phage particles will be released into the surrounding medium, or, alternatively, the phage can switch into a

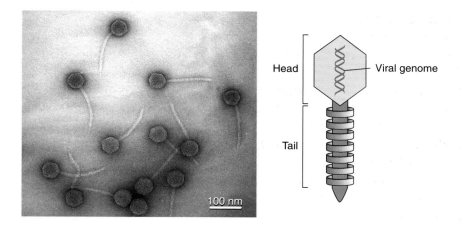

Figure 3.8. The structure of bacteriophage λ. An electron micrograph of λ phages that have been released upon bacterial lysis, and a diagrammatic representation of the overall structure of λ. The EM image is courtesy of Professor Ross Inman (University of Wisconsin) and is reproduced with permission. Other excellent EM images of λ can be found on Professor Inman's web site (http://www.biochem.wisc.edu/inman/empics/)

dormant lifestyle in which the phage DNA becomes integrated into the *E. coli* chromosome – in this case a lysogenic lifestyle is adopted. The life cycle of λ phage is shown diagrammatically in Figure 3.9. It is the lysogenic bacteria that show rapid lysis upon ultraviolet radiation. In the laboratory, λ phage growth and replication is monitored on petri dishes (Figure 3.10). The phage is mixed with *E. coli* cells is a soft agar solution called top agar. The mixture is poured onto the surface of a nutrient agar plate and incubated to allow bacterial growth. λ growth will cause the death (lysis) of the *E. coli* cells surrounding the site of initial infection. Such sites are observed on the plates as somewhat turbid plaques in the bacterial lawn. λ DNA can be purified from the phage particles contained with these plaques.

The genetics and molecular biology of bacteriophage λ have been extensively studied – for further information on λ, readers are directed to the excellent text by Mark Ptashne (Ptashne, 1992). The DNA contained within the phage is a linear double-stranded molecule 48 502 bp in length. The extreme 5′- and 3′-ends of the λ genome have 12 bases that are single-stranded – called the cohesive or *cos* ends. These sequences are complementary and can anneal with each other to form a circular double-stranded DNA molecule (Figure 3.11). Functionally related genes of λ are generally clustered together on the λ genome, except for the two positive regulatory genes *N* and *Q*. Genes on the left-hand side of the conventional genome map code for head and tail proteins of the phage particle. These are followed by genes whose protein products

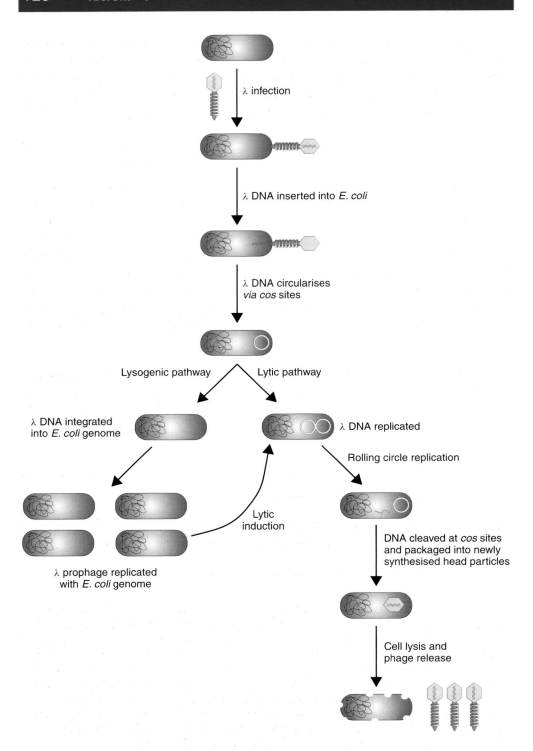

λ infection

λ DNA inserted into *E. coli*

λ DNA circularises *via cos* sites

Lysogenic pathway   Lytic pathway

λ DNA integrated into *E. coli* genome   λ DNA replicated

Rolling circle replication

Lytic induction

λ prophage replicated with *E. coli* genome

DNA cleaved at *cos* sites and packaged into newly synthesised head particles

Cell lysis and phage release

are concerned with recombination and the processes of lysogeny in which the circularized phage chromosome is inserted into the host chromosome and is stably replicated as a prophage. To the right of the map are genes concerned with transcriptional regulation and prophage immunity to superinfection (N, cro, cI), followed by the genes for DNA synthesis, late function regulation (Q) and host cell lysis.

Our extensive knowledge of λ phage and the ways in which the lytic and lysogenic life cycles are regulated have made λ an ideal vector to carry foreign DNA fragments. The major advantage of λ based vectors over plasmids is the efficiency at which the phage can infect *E. coli* cells. As we have already discussed, the transformation of plasmid DNA into bacteria is not an efficient process, whereas λ infection is a very efficient way to introduce DNA into a bacterial cell. To understand how λ can be exploited as a vector, it is important to have a basic knowledge of the phage itself. λ phage infection and lysis occurs in number of defined steps. Infection occurs as a result of the adsorption of the λ phage particle to the bacterial cell by binding to the maltose receptor. The λ genomic DNA is injected into the cell and almost immediately circularizes. At this point it can enter one of two pathways.

- *Lysogenic pathway.* The phage DNA becomes integrated into the bacterial genome (*via* homologous recombination between *attP* and the bacterial genomic *attB* site) and is replicated along with the bacterial DNA. The prophage DNA remains integrated until it is induced to enter the lytic pathway.

- *Lytic pathway.* Large-scale production of bacteriophage particles (proteins and DNA) occurs that eventually leads to the lysis of the cell.

The decision as to whether lysis or lysogeny occurs is the result of the activity of the cII protein. Active cII is required for the transcription of the cI repressor

Figure 3.9. λ life cycle. Upon infection, bacteriophage λ attaches to the surface of a bacterial cell, and its DNA enters the bacterium. Almost immediately, the λ DNA circularizes. The DNA can then enter either the lysogenic or the lytic pathway. During lysogeny, the λ DNA integrates into the *E. coli* chromosome and is replicated, along with the host DNA, such that the prophage is passed onto subsequent generations. In the lytic phase, the λ DNA does not integrate, but is immediately replicated and transcribed to produce new phage particles. Eventually, bacterial cell lysis occurs and the newly formed phage are released into the surrounding medium. The lysogenic prophage may be induced into the lytic cycle by, for example, treatment with UV light. In this case the λ DNA loops out of the *E. coli* genome and the lytic pathway is initiated

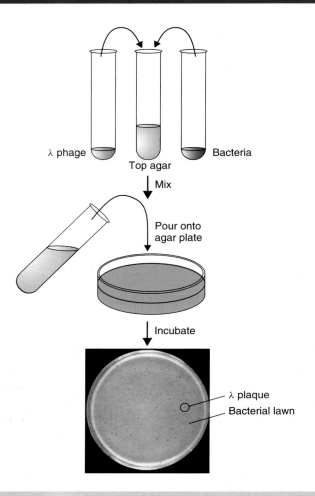

Figure 3.10. λ plaques. λ phage is grown in the laboratory on a lawn of bacterial cells. The bacteria and the λ phage particles are mixed with liquid, but cool, top agar. The mixture is then poured onto an already set agar plate where the top agar is allowed to solidify. The plate is then incubated for 12–16 h at 37 °C. λ plaques form as turbid circles in the bacterial lawn

and for some of the genes required for phage DNA integration into the *E. coli* chromosome. Active cII results in the adoption of the lysogenic pathway, while inactive cII results in the lytic pathway being followed. The cII protein is relatively unstable and is susceptible to cleavage and destruction by bacterial proteases. Environmental conditions influence the activities of these proteases. When grown in rich medium, for example, the proteases are generally active, such that cII is degraded and λ lysis occurs. Under conditions of *E. coli* starvation, the proteases are less functional and, consequently, λ will more

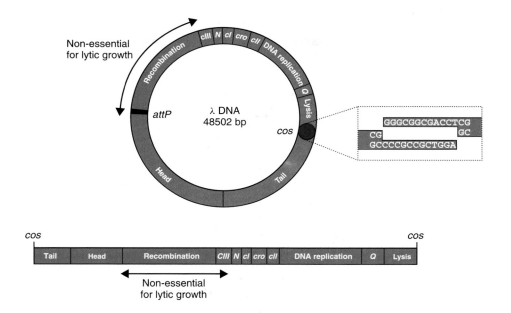

Figure 3.11. The circular and linear forms of the λ genome. λ DNA exists in a linear form in the bacteriophage and in a circular form upon entering the bacterium. The switch from the linear to the circular form occurs through complementation of the overhanging DNA ends at the *cos* sites. Many of the genes required for the integration of λ into the host chromosome, or for new phage replication and assembly, are grouped together on the λ chromosome. Some of these genes, or sets of genes, are shown. A region of the genome that is not required for lytic growth is indicated

frequently lysogenize. This behaviour makes sense, for in starved cells there will be less of the components necessary to make new phage particles.

The lytic pathway is characterized by a series of transcriptional events that produce different sets of proteins that are required for replication of the phage DNA and the production of new phage particles.

- *Early transcription.* Transcription of the *N* and *cro* genes occurs. This transcription is subject to repression by the product of the *cI* gene and in a lysogen this repression is the basis of immunity to superinfection.

- *Delayed early transcription.* The N protein product binds to the bacterial RNA polymerase and promotes transcription of the phage genes involved in DNA replication.

- *Replication.* Early replication proceeds from a single origin of replication site. Later replication proceeds *via* a rolling circle mechanism to produce long concatamers of the phage DNA that are connected to each other at the *cos* sites.

- *Late transcription*. The protein product of the *cro* gene builds up to a critical level and then stops early transcription. The product of the *Q* gene activates transcription, resulting in the production of the proteins required for the head and tail of the mature phage particle, and those required for bacterial cell lysis.

Finally, phage assembly occurs when a unit length of DNA is placed into the assembled head by cleavage of the concatameric DNA at the *cos* sites. The tail is added and the mature phage particle is completed. Upon cell lysis, approximately 100 newly synthesized phage particles are released from a single infected bacterial cell.

Wild-type λ DNA contains few unique restriction enzyme recognition sites into which foreign DNA fragments could be cloned, and is consequently not wholly suitable as a vector to carry such sequences. Additionally, the packaging of DNA into the λ phage is size limited. Efficient packaging will only occur with DNA fragments representing between 78 and 105 per cent of the wild-type genome size (37–51 kbp). These limits pose severe restrictions upon the amount of DNA that can be cloned into the phage genome. Two important developments, however, suggested that λ might be suitable as a cloning vector. Firstly it was determined that the gene products required for recombination could be removed from the λ genome and the lytic life cycle could still be completed and plaques would form. The remaining DNA, often referred to as the left-hand and right-hand arms of the λ genome, is capable of providing all necessary functions for the lytic pathway to occur. Secondly, naturally occurring restriction enzyme recognition sites could be eliminated without loss of gene function, which permitted the development of vectors with unique sites for the insertion of foreign DNA.

λ vectors could thus be constructed that lacked the genes required for recombination, and therefore could only enter the lytic cycle, but were capable of carrying much larger foreign DNA inserts. Two basic types of λ vector have been developed:

- *insertional vector* – DNA is inserted into a specific restriction enzyme recognition site;

- *replacement vector* – foreign DNA replaces a piece of DNA (stuffer fragment) of the vector.

The advantage of replacement vectors is that they are capable of carrying larger DNA inserts. For example, λEMBL4 is a 42 kbp vector that contains 14 of kbp stuffer DNA between the left-hand and right-hand arms of λ. The ligation of just the λ arms would generate a 28 kbp λ genome. This is too small to be packaged into a λ particle. The insertion of foreign DNA between the two λ arms will, however, enable the genome to attain a suitable packaging size. The packaging size limit means that λEMBL4 is capable of holding foreign DNA fragments up to approximately 23 kbp in size (Figure 3.12).

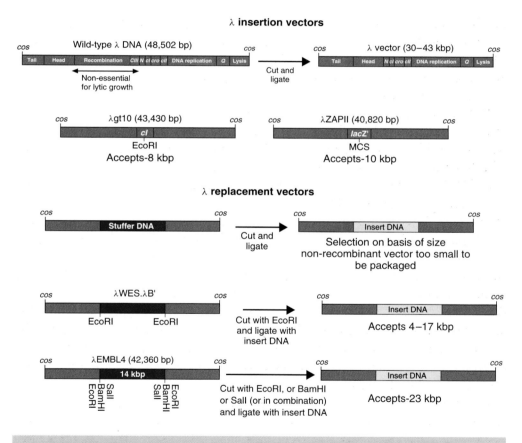

Figure 3.12. λ insertion and replacement vectors. All λ vectors have regions non-essential for lytic growth removed to increase the amount of DNA that will be packaged into the mature λ phage. Two λ insertion vectors are shown. λgt10 contains a unique EcoRI restriction enzyme recognition site in the cl gene. Recombinants will form clear rather than turbid plaques. λZAPII contains a multiple cloning site (MCS) in the *lacZ'* gene and recombinants are identified using blue–white screening. Recombinants in λ replacement vectors are the only phages that will grow; if the two λ ends are ligated in the absence of insert DNA, the DNA is too small to be packaged

The size limitations of λ packaging thus provide a mechanism to ensure that foreign DNA has been inserted in between the λ arms to form a recombinant. Several other basic strategies have been devised to identify λ phage recombinants.

- *Inactivation of the cI gene.* Several λ phage vectors (e.g. λgt11) have unique restriction enzyme recognition sites contained within the *cI* gene. Phages in which the *cI* gene has been disrupted by foreign DNA insertion have an altered morphology, in which the plaques produced appear 'clear' as opposed to turbid. Screening of this type is technically difficult and requires a deal of skill on the part of the observer.

- *Blue–white screening.* λ phage vectors (e.g. λZAP) have been constructed to contain the *lacZ′* gene expressing the α-fragment of β-galactosidase. Screening for recombinant phages can then be preformed in *E. coli* cells

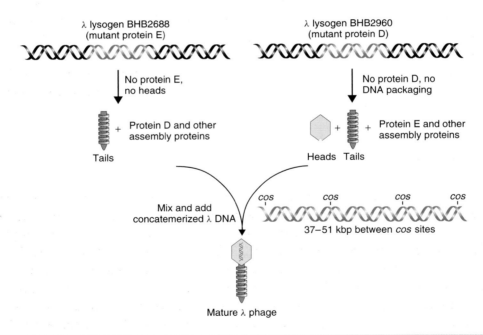

Figure 3.13. *In vitro* packaging of λ phage particles. Two different λ lysogens are used to produce the various components required for the packaging of λ particles. One of these lysogens (BHB2688) has a defective *E* gene, which results in no heads being produced. The other (BHB2960) has a defective *D* gene, resulting in a defect in DNA packaging. Mixing cell lysates of the two will result in an extract that is able to package concatamerized λ DNA. The multimerized DNA (37–51 kbp) will be cleaved at the *cos* sites and packaged into a mature λ phage particle

expressing the ω-fragment of β-galactosidase in a similar way to screening for recombinants in pUC based plasmids.

Once a recombinant λ genome has been constructed, the problem arises of how to get the DNA into a viral particle so that it can be replicated in *E. coli* cells (Figure 3.13). Normal *in vivo* packaging of λ DNA involves first making pre-heads, structures composed of the major capsid protein encoded by gene *E*. A unit length of λ DNA is then inserted into the pre-head, with the unit length being prepared by cleavage of concatamerized λ genomes at neighbouring *cos* sites. A minor capsid protein D is then inserted in the pre-heads to complete head maturation, and the products of other genes serve as assembly proteins, ensuring joining of the completed tails to the completed heads. λ packaging of recombinant genomes can occur *in vitro* by utilizing two *E. coli* strains that bear λ lysogens containing different defects in the packaging pathway. A defect in producing protein E, resulting from a mutation introduced into gene *E*, prevents pre-heads being formed in strain BHB2688. A mutation in gene *D* prevents maturation of the pre-heads, with enclosed DNA, into complete heads in strain BHB2690. The components of the BHB2688/BHB2690 mixed lysate, however, complement each other's deficiencies and provide all the products for correct packaging (Figure 3.13). Consequently, recombinant λ genomes can be constructed *in vitro* and packaged into mature λ phage particles before being propagated and replicated in *E. coli* cells.

3.4 Cosmid Vectors

The only DNA requirements for *in vitro* packaging into λ phage are the presence of two *cos* sites that are separated by 37–51 kbp of intervening sequence. Cosmids were developed in light of this observation, and are simply plasmids that contain a λ phage *cos* site (Collins and Brüning, 1978). Figure 3.14 shows the overall architecture of a cosmid vector and a cloning scheme for the insertion of foreign DNA. As plasmids, cosmids contain an origin of replication and a selectable marker. Cosmids also possess a unique restriction enzyme recognition site into which DNA fragments can be ligated. After the packaging reaction has occurred, the newly formed λ particles are used to infect *E. coli* cells. The DNA is injected into the bacterium like normal λ DNA and circularizes through complementation of the *cos* ends. The lack of other λ sequences means, however, that λ infection will not proceed beyond this stage. The circularized DNA will, however, be maintained in the *E. coli* cell as a plasmid. Therefore selection of transformants is made on the basis of antibiotic resistance and bacterial colonies (rather than plaques) will form that contain

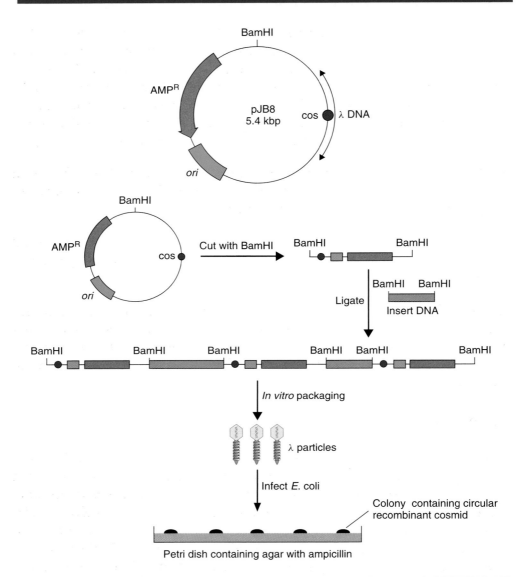

Figure 3.14. Cloning using a cosmid vector. The overall architecture of a cosmid vector, pJB8, is shown, together with a scheme for the insertion of foreign DNA into a cosmid. Since the cosmid lacks other λ genes, when the DNA is inserted into the *E. coli* cell it is maintained as a plasmid and selected for on the basis of antibiotic resistance

the recombinant cosmid. Since λ phage particles can accept between 37 and 51 kbp of DNA, and most cosmids are about 5 kbp in size, between 32 and 47 kbp of DNA can cloned into these vectors. This represents considerably more than could be cloned into a λ vector itself.

Cosmids, like plasmids, are very stable, but the insertion of large DNA fragments can mean that recombinant cosmids are difficult to maintain in a bacterial cell. Repeat DNA sequences are common in eukaryotic DNA, and DNA rearrangements can occur *via* recombination of the repeats present on the DNA inserted into the cosmid. The major difficulty in working with cosmids is, however, the production of linear, ligated DNA fragments in which the cosmid and insert are concatamerized together. Two basic problems exist.

- Ligation reactions of cosmid and insert DNA, like those shown in Figure 3.14, will generate circular DNA molecules that are unable to participate in the *in vitro* packaging reaction.

- More than one insert DNA molecule can be ligated between each cosmid DNA fragment. This could give a false impression of the DNA organization of the insert.

These difficulties can be overcome by cutting the cosmid with two different restriction enzymes to generate left-hand and right-hand ends that cannot religate to each other (Ish-Horowicz and Burke, 1981). Suitable phosphatase treatment of the insert DNA ensures that multiple inserts cannot be ligated to the cosmid DNA (see Chapter 2).

3.5 M13 Vectors

M13, and its very close relatives f1 and fd, are filamentous *E. coli* bacteriophages. M13 is a male-specific lysogenic phage with a circular single-stranded DNA genome 6407 bp in length (Figure 3.15). M13 phage particles have dimensions of about 900 nm × 9 nm and contain a single-stranded circular DNA molecule (designated as the + strand). M13 infects bacteria that harbour the F pilus. The phage particle absorbs *via* one end to the F pilus, and the single-stranded phage DNA enters the bacterium (Figure 3.16). Very rapidly, the single-stranded DNA is converted into double-stranded (replicative form, RF) DNA by the synthesis of a complementary DNA strand (the – strand) using bacterial DNA polymerase. The RF form of the phage genome is rapidly multiplied until about 100 RF molecules are present within the bacterium. Transcription of the viral genes occurs to produce proteins required for the assembly of new viral particles. The production of a virally encoded single-stranded binding protein (the protein product of gene 2) eventually forces asymmetric replication of the RF DNA. This results in only one viral DNA strand being synthesized (the + strand). These single-stranded DNA molecules are assembled into new viral particles, and are released from the cell without

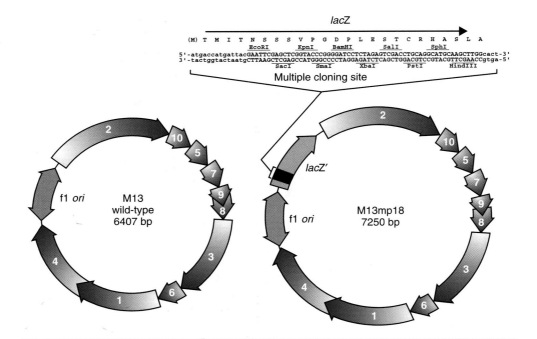

Figure 3.15. The genomes of wild-type M13 and an engineered derivative, M13mp18. The wild-type M13 genome encodes 10 open reading frames that are all transcribed in the clockwise direction. Replication of the genome initiates bi-directionally from a specific sequence between genes 2 and 4. M13mp18 additionally bears the *lacZ'* gene for blue–white screening of recombinants. Embedded within this gene is a multiple cloning site providing a number of unique restriction enzyme recognition site to aid the cloning of foreign DNA fragments into the vector. The maps shown here represent the double-stranded from, or replicative form (RF), of the vector that exists within the *E. coli* host. Viral particles contain only a single strand of the DNA

cell lysis occurring. Up to 1000 phage particles can be released into the medium per cell per generation. M13 phage infection does not result in bacterial cell death and, consequently, M13 infections appear as turbid plaques. The *E. coli* cells around the site of infection have not been killed, but they grow more slowly due to the burden placed upon them by producing phage particles. The M13 origin of replication (called the f1 *ori*) contains two overlapping, but distinct, DNA sequences that act to control the synthesis of DNA. These sites – the f1 initiator and the f1 terminator – signal the beginning and end of DNA replication. The initiator is recognized by the protein product of gene 2, which nicks the + strand in the RF DNA. The nick indicates the position at which unidirectional rolling-circle DNA replication will commence. The newly formed + strand is cleaved at the terminator sequence, again by the protein

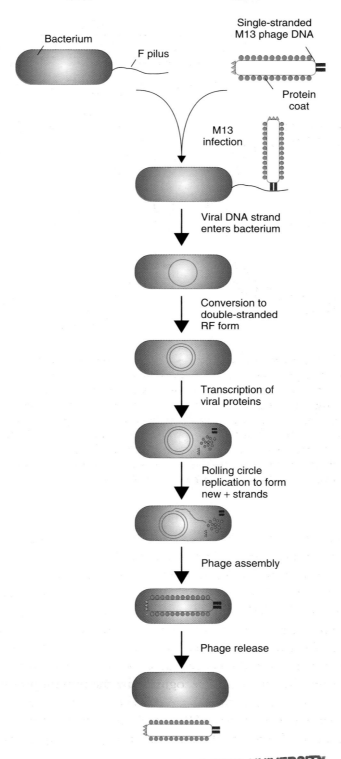

product of gene 2. Following cleavage, the two ends of the + strand are ligated to form the single-stranded genome.

The switch between the double-stranded RF form and the single-stranded + form of the M13 viral genome made it an ideal candidate for exploitation as a vector. As we will see in later chapters, the single-stranded DNA produced in the phage particle have led to great advances in mutagenesis *in vitro* (Chapter 7) and DNA sequencing (Chapter 8). Unlike λ, M13 does not have a non-essential region that can be deleted prior to the insertion of foreign DNA. However, there is an intergenic region between the origin of replication and gene 2 (Figure 3.15) into which foreign DNA fragments may be inserted. M13 vectors were developed in the late 1970s when the *lacZ'* gene (encoding the α-peptide of β-galactosidase) was inserted into the M13 genome (Messing *et al.*, 1977). Subsequently, the same polylinker and α-peptide fragments as the pUC plasmid series were engineered into M13 and naturally occurring restriction enzyme recognition sites were eliminated (Yanisch-Perron, Vieira and Messing, 1985; Norrander, Kempe and Messing, 1983). The RF form of M13 vectors can be isolated by standard plasmid DNA preparation procedures and foreign DNA can be inserted into them as if they were conventional plasmids.

The specific use of M13 vectors is as an aid to the formation of single-stranded DNA. Once a foreign DNA fragment has been cloned into M13, large amounts of the single-stranded form can be easily isolated from the mature phage that are extruded from infected *E. coli* cells. The main difficulty with vectors of this type is that they tend to be unstable when DNA fragments larger than a few kilobases are inserted into them (Zinder and Boeke, 1982).

3.6   Phagemids

Phagemids are plasmids that contain the f1 phage origin of replication for the production of single-stranded DNA. Phagemids are generally small plasmids so that they have the ability to accept larger DNA inserts than M13-based vectors. Phagemids were originally developed in the early 1980s, when it was found

Figure 3.16. The M13 life cycle. The single-stranded M13 genome is encased by coat proteins. Bacterial infection occurs when the phage particle attaches to the *E. coli* pilus and the single DNA strand is injected into the host. The DNA is immediately converted to a double-stranded form and is replicated and transcribed to produce viral proteins. The build-up of viral protein 2 eventually forces asymmetric DNA replication to produce single DNA strands. These are packaged into new viral particles, which are secreted from the bacteria without cell lysis occurring

that the insertion of the f1 origin of replication could be cloned into pBR322 to drive the production of single-stranded DNA (Dotto and Horiuchi, 1981; Dotto, Enea and Zinder, 1981). The f1 replication origin was not sufficient to direct single-stranded DNA production, but if a bacterium carrying a phagemid was superinfected with a functional wild-type M13 or f1 **helper phage**, then the production of single-stranded phagemid DNA would occur. The phagemid single-stranded DNA would be packaged into viral particles and secreted into the surrounding medium in the same way that M13 phage particles are produced. Additionally, it was found that cloning the f1 origin in the reverse orientation would lead to the production of the opposite strand of DNA (Dente, Cesaveni and Cortese, 1983). Thus, single-stranded DNA representing either strand of a cloned fragment could be produced after cloning into a suitable phagemid vector. Phagemids have the advantage that, in the absence of a helper phage, double-stranded DNA can be isolated as a normal plasmid. Moreover, the lack of additional phage genes the vectors need to carry means that their small size has an increased capacity for carrying larger foreign DNA fragments.

Other phagemids have been developed that take advantage of various aspects of plasmids, λ phage and M13 phage. We have already seen that the f1 replication origin is composed of an initiator and a terminator. In the wild-type M13 phage genome these sequences overlap with each other, such that replication initiates and then terminates after the full circular genome has been replicated. The initiator and terminator elements may be separated from each other to provide starting and ending points for DNA replication on a linear DNA molecule. A λ insertional vector, λZAP, was constructed such that the left-hand and right-hand λ arms were connected *via* the DNA sequence of a phagemid beginning with the f1 initiator and ending with the f1 terminator (Short *et al.*, 1988). This vector (shown in Figure 3.17) has the ability to function as a λ phage for, for example, the construction of a cDNA library. However, the foreign DNA can be excised from the λ phage in the form of a plasmid after superinfection with a wild-type M13 based phage. λZAP contains all the λ DNA sequences required for lytic growth, and in between these is the DNA sequence needed for plasmid replication and selection (the ColE1 *ori* and AMP^R). Additionally, λZAP contains the *lacZ'* gene and multiple cloning site sequence in a similar fashion to the pUC plasmids. The plasmid sequences in the vector begin with the f1 initiator and end with the f1 terminator. Vectors bearing foreign DNA can be selected by blue–white screening of λ plaques, and the insert DNA can be isolated in the form of a plasmid when bacteria harbouring the λ phage are super-infected with an f1 helper phage. Proteins produced by the helper phage will result in DNA replication between the f1 initiator and terminator. The

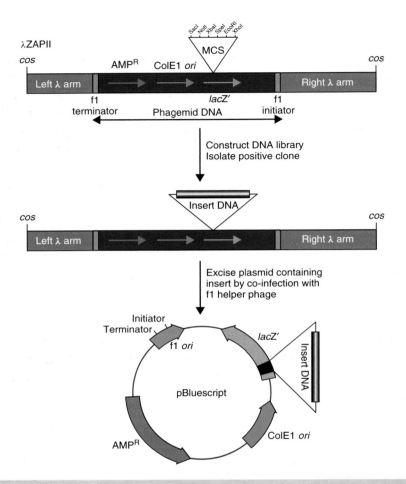

Figure 3.17. The *in vivo* excision of phagemid DNA from a λ phage vector. λZAPII is a sophisticated λ phage vector containing the elements of the λ and M13 phages as well as the sequences required for stable phagemid production. The DNA sequence for the entire pBluescript phagemid is contained within the λ vector between the f1 initiator and terminator. Foreign DNA inserted into the multiple cloning site (MCS) of λZAPII can be recovered in the form of a phagemid. Bacteria harbouring the λ phage are superinfected with an M13 based phage to drive DNA replication of sequences between the f1 initiator and terminator. The M13 phages produced using this DNA can be used to infect F' *E. coli* cells and double-stranded plasmid DNA isolated

single-stranded DNA will circularize and will be packaged as an M13-like phage and secreted from the cell. The introduction of the M13 phage particles into an F' *E. coli* strain and selection on ampicillin will result in the formation of colonies containing the recombinant plasmid, which can then be isolated as double-stranded plasmids.

3.7 Artificial Chromosomes

The major limitation of most of the vectors that we have discussed so far is the size limit of the DNA that can be cloned into them. Natural eukaryotic chromosomes consist of hundreds or thousands of genes, together with DNA elements required for chromosomal stability and function such as **telomeres** and **centromeres**. Telomeres, which consist of DNA and protein, are located at the ends of chromosomes and protect them from damage. Centromeres are segments of highly repetitive DNA that are essential for the proper control of chromosome distribution during cell division. A logical extension of vector design to clone very large DNA fragments is, therefore, to reconstruct an autonomously replicating chromosome into which DNA fragments may be cloned. Cloning in this way is conceptually similar to cloning in λ phage – with the reconstruction of a replication competent DNA molecule – except that the scale of the foreign DNA that can be cloned is much greater.

3.7.1 YACs

Yeast artificial chromosome (YAC) vectors allow the cloning, within yeast cells, of fragments of foreign genomic DNA that can approach 500 kbp in size. These vectors contain several elements of typical yeast chromosomes, including the following.

- *A yeast centromere (CEN4).* The yeast centromere is specified by a 125 bp DNA segment. The consensus sequence consists of three elements: a 78–86 bp region with more than 90 per cent AT residues, flanked by a conserved sequence on one side and a short consensus sequence on the other (reviewed by Clarke (1990)).

- *Yeast autonomously replicating sequence (ARS1).* Yeast *ARS* elements are essentially origins of replication that function in yeast cells autonomously from the replication of yeast chromosomal replication origins.

- *Yeast telomeres (TEL).* Telomeres are the specific sequences (5'-TGTGGGTGTGGTG-3') that are present at the ends of chromosomes in multiple copies and are necessary for replication and chromosome maintenance.

- *Genes for YAC selection in yeast.* The vector has a functional copy of *URA3*, a gene involved in uracil biosynthesis, and *TRP1*, a gene involved in tryptophan biosynthesis, that allow selection of yeast cells that have taken up the vector. The YAC is transformed into a host yeast cell that is defective in these biosynthetic pathways, and transformants are identified by their ability to complement the nutritional defect.

- *Bacterial replication origin and a bacterial selectable marker.* In order to propagate the YAC vector in bacterial cells, prior to insertion of genomic DNA, YAC vectors usually contain the ColE1 *ori* and the ampicillin resistance gene for growth and analysis in *E. coli*.

The cloning of DNA fragments into a YAC is shown diagrammatically in Figure 3.18. The YAC is cleaved using restriction enzymes to generate two 'arms' that each have a telomere sequence at the end. One of the arms contains an autonomous replication sequence (*ARS1*), a centromere (*CEN4*) and a

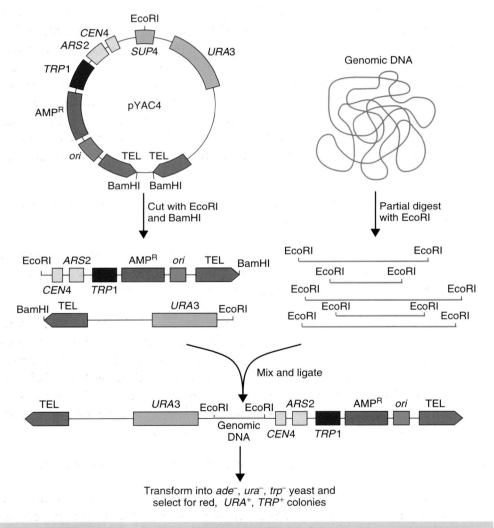

Figure 3.18. Cloning of very large DNA fragments into a YAC vector. See the text for details

selectable marker (*TRP1*). The other arm contains a second selectable marker (*URA3*). Large DNA fragments (>100 kbp) are then ligated between the two arms (Anand, Villasante and Tyler-Smitu, 1989). The insertion of foreign DNA into the cloning site inactivates the suppressor tRNA gene *SUP4*, expressing tRNA^{Tyr}, in the vector DNA. In an *ade2–ochre* host yeast cell, the expression of *SUP4* results in the formation of white colonies, while in those in which it has been insertionally inactivated will give rise to red yeast colonies (Burke, Carle and Olson, 1987). Yeast cells that are mutated in the *ADE2* gene product (coding for the enzyme phosphoribosylamino-imidazole-carboxylase) have a block in the adenine biosynthetic pathway, causing an intermediate to accumulate in the vacuole. This intermediate gives the cell a red colour. The recombinant YACs are therefore transformed into a yeast strain that has defects in its chromosomal copies of the *ura3, trp1* and *ade2* genes. Transformants are identified as those red colonies that grow on media lacking both uracil and tryptophan. This ensures that the cell has received an artificial chromosome with both telomeres (because of complementation of the two nutritional mutations) and the artificial chromosome contains insert DNA (because the cell is red).

There are difficulties associated with working with YACs. Some of these are listed below.

- Very large DNA molecules are very fragile and prone to breakage, leading to problems of rearrangement.

- It is estimated that between 10 and 60 per cent of clones in YAC genomic libraries are chimaeric, i.e. regions from different parts of the genome become joined in a single YAC clone (Green *et al.*, 1991).

- Clones tend to be unstable, with their foreign DNA inserts often being deleted. Naturally occurring repetitive DNA sequences are rare in the yeast genome, and the insertion of such sequences from, say, human DNA inserts appears to increase the recombination frequency within the YAC. This may make the YAC unstable. Interestingly, however, larger YAC vectors are more stable in yeast than shorter ones, which consequently favours cloning of large stretches of DNA (Smith, Smyth and Moir, 1990).

- There is a high rate of loss of the entire YAC during mitotic growth.

- It is difficult to separate the YAC from the other host chromosomes because of their similar size. Separation requires sophisticated pulsed-field gel electrophoresis (PFGE).

- The yield of DNA is not high when the YAC is isolated from yeast cells.

3.7.2 PACs

To overcome some of the problems associated with using cosmid or YAC systems, a method for cloning and packaging DNA fragments using a bacteriophage P1 system has been developed that offers the ability to clone large genomic DNA fragments of between 70 and 95 kbp in size. P1 bacteriophage has a much larger genome than λ phage (in the range of 110–115 kbp), and vectors have been designed with the essential replication components of P1 incorporated into a plasmid (Ioannou *et al.*, 1994). Upon infecting *E. coli*, bacteriophage P1 may either express its lytic functions, producing 100–200 new bacteriophage particles and lysing the infected bacterium, or the infecting bacteriophage may repress its lytic functions, and the bacteriophage genome is maintained as a large, stable, low-copy plasmid. P1 phage has two replication origins – one to control lytic DNA replication, and the other to maintain the plasmid during non-lytic growth. During the lytic cycle, new phage DNA is produced and cleaved at a *pac* site prior to insertion into phage particles.

The cloning of foreign DNA fragments into a P1 vector, or P1 artificial chromosome (PAC), is shown in Figure 3.19. The PAC vector is digested with the restriction enzymes *Sca*I and BamHI to generate two vector arms: a short and a long arm. Genomic DNA is partially digested with MboI (recognition sequence 5′-GTAC-3′, yielding BamHI-compatible sticky ends) and size selected on a sucrose gradient. Fragments between 70 and 95 kb in length are isolated and ligated in between the vector arms to generate a series of linear molecules. If ligation occurs between two short arms, the resulting molecule will contain neither the P1 replication origins nor the KAN^R gene, and will be non-viable. If both arms are long there will be no *pac* site, and no packaging into the phage heads will occur. The only viable recombinant will consist of the insert sequence flanked by both a short and long arm. Phage P1 uses a 'head-full' packaging strategy and can accommodate a total DNA length of approximately 110–115 kbp. This means that any inserts longer than 95–100 kbp will result in the truncation of the packaged DNA before both *lox*P sites are inserted into the phage, and the molecule will be unable to circularize upon transfection into the host. Once injected into the *cre*$^+$ *E. coli* host cell, the Cre protein circularizes the DNA at the *lox*P sites, and DNA then replicates using the plasmid origin of replication. The original vector BamHI restriction enzyme site, into which the foreign DNA was inserted, is located within the bacterial *sacB* gene (encoding levansucrase). The expression of this gene is toxic to *E. coli* cells growing on sucrose. Thus sucrose growth provides a mechanism of positive selection for those PACs containing inserts. Propagation of *E. coli* cells harbouring the recombinant PAC on media containing sucrose permits growth of colonies with DNA inserts.

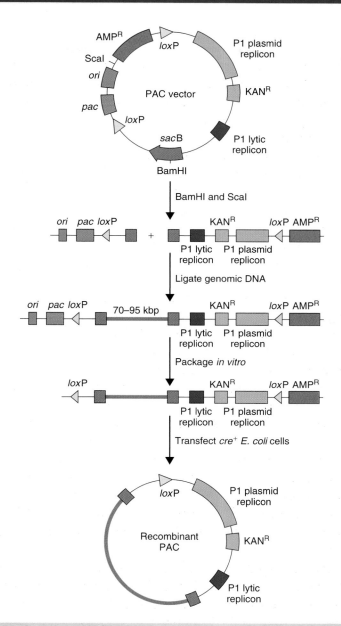

Figure 3.19. Cloning into a PAC vector. The PAC vector contains the P1 bacteriophage plasmid and lytic replicons, together with a *pac* cleavage site to allow DNA assembly into phage particles. Additionally, the vector contains the pUC origin of replication (*ori*) and ampicillin resistance gene for the propagation and selection of the vector itself. These sequences are lost in the recombinant PAC. Large DNA inserts are cloned into the *sac*B gene (whose function can be selected against) and packaged *in vitro* into P1 phage particles. Transfection of the P1 phage particles into an *E. coli* cell harbouring a copy of the Cre recombinase will result in circularization of the recombinant PAC at the *lox*P sites. The circular form is then maintained at low copy number using the P1 plasmid replicon

The recombinant PACs are maintained as plasmids within the *E. coli* using kanamycin resistance as a selection marker. Plasmid copy number can be increased more than 25-fold by isopropyl β-D-thiogalactopyranoside (IPTG) induction of a *lac* promoter controlled high-copy P1 lytic replicon that is present within the recombinant PAC (Pierce *et al.*, 1992). The recombinant DNA molecules are then isolated as plasmids using traditional methods.

Using the P1 DNA packaging system, genomic DNA from 70 to 95 kb can be readily cloned and manipulated. Improvements in vector design have allowed the production of PACs that can accommodate 130–150 kbp inserts. The major advantages of the P1 DNA packaging method over other genomic cloning methods are

- the large size of the DNA fragments that may be inserted into the vectors,

- no rearrangement or deletion of methylated DNA occurs because of the use of restriction-minus host strains and

- recombinant DNA is easily recovered as plasmids for further screening and manipulation.

3.7.3 BACs

Bacterial artificial chromosomes (BACs) are engineered versions of F′ plasmids (Shizuya *et al.*, 1992). BACs are capable of carrying approximately 200 kbp of inserted DNA sequence, and the F-factor origin of replication (*oriS*) maintains their level at approximately one copy per cell. In addition to *oriS*, BACs contain four F-factor genes required for replication and maintenance of copy number, *rep*E, *par*A, *par*B and *par*C. The overall architecture of a typical BAC is shown in Figure 3.20.

In addition to the F-factor genes, pBeloBac11 also contains a selectable antibiotic resistance maker (CAM^R) and the *lacZ′* gene harbouring a multiple cloning site for the blue–white screening of BACs containing inserts (Kim *et al.*, 1996b). Additionally, the BAC contains a λ *cos* site (*cos*N) and a *lox*P site. These sites are used for specific cleavage of the insert containing BAC during restriction mapping. The *cos*N site can be cleaved using λ terminase (Rackwitz *et al.*, 1985), while the *lox*P site can be cleaved by the Cre protein in the presence of an oligonucleotide to the *lox*P sequence (Abremski, Hoess and Stanbers, 1983). Additional BACs have been constructed that contain the recognition sites for extremely rare-cutting restriction enzymes. For example, I-*Sce*I is an intron encoded restriction enzyme from the mitochondria of the

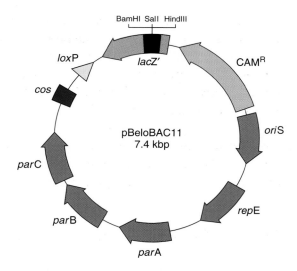

BamHI SalI HindIII

loxP

lacZ'

CAM^R

cos

pBeloBAC11
7.4 kbp

oriS

parC

repE

parB

parA

Figure 3.20. The structure of a BAC vector. *See the text for details*

yeast *Saccharomyces cerevisiae* (Monteilhet *et al.*, 1990). Its large recognition sequence (5'-TAGGGATAACAGGGTAAT-3') does not occur once in the human genome, and is consequently very useful for linearizing the vector without cleaving the insert DNA fragments.

The DNA inserted into a BAC appears to be very stable. It can survive intact for many hundreds of generations in *E. coli* cells, and appears to be less prone to rearrangements and deletions when maintained in a recombination defective *E. coli* host cell. The main drawback of using BAC vectors is that they are present in only one or two copies per cell. This can complicate isolation and screening.

3.7.4 HACs

Human artificial chromosomes (HACs) have been constructed that can survive for extended periods in tissue culture cells (Harrington *et al.*, 1997) (Figure 3.21). As we have already seen, three elements are required for the stability of linear chromosomes – centromeres, telomeres and an origin of replication. The human telomere repeat sequence (5'-TTAGGG-3') is well known, but it is distinct from its yeast equivalent and the two are not interchangeable. A YAC will not function as a chromosome in human cells. To aid in the isolation of human centromere and replication origin sequences, HACs have been constructed that can be transfected into and maintained within

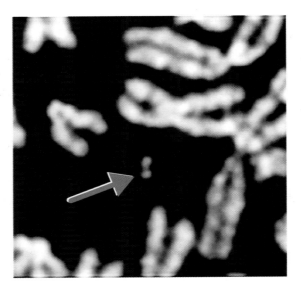

Figure 3.21. A human artificial chromosome in tissue culture cells. A synthetic α satellite containing microchromosome formed by transfection of α satellite DNA and telomeric sequences into tissue culture cells. A clonal line was isolated and shown to contain a HAC (denoted by the arrow) derived from transfected DNA and not from truncation of endogenous chromosomes. Reprinted, with permission, from Willard (2000). Copyright (2000) American Association for the Advancement of Science

human cells (Henning *et al.*, 1999). This approach identified multiple repeats of a 171 bp DNA sequence (called an **α satellite repeat**) contained within a 3 million base pair DNA fragment of the human X chromosome that functions as a centromere (Schueler *et al.*, 2001). The repetitive nature of these sequences makes them difficult to study and makes the identification of the centromere itself extremely hard.

The human genome contains multiple origins of replication. The average human chromosome contains approximately 150×10^6 bp of DNA, while DNA polymerase functions maximally at about 3000 replicated bases per minute. Were replication to begin at a single site, each chromosome would take over a month to be replicated, rather than the hour it actually takes. Multiple replication origins mean that there are many places on the eukaryotic chromosome where replication can begin and that the process of complete replication proceeds at a more rapid rate. The sequence of the human replication origin is very degenerate (Vashee *et al.*, 2001), but the development of HACs should allow a more precise mapping of these regions.

HACs have great potential as tools for both basic research and medical therapy. Artificial chromosomes may ultimately lead to gene therapy vectors (see Chapter 13) with some advantages over existing viral based vectors.

- They exist as extrachromsomal elements and so would not result in insertional mutagenesis.

- They should have no size constraint on the amount on DNA that they could carry.

- By virtue of differences between centromere behaviour in mitosis and meiosis, they might be designed not to function in the germ line.

These possibilities remain for the future and will depend on having a much greater understanding of chromosome function than is currently available.

4 Polymerase chain reaction

Key concepts

♦ PCR is the amplification of specific DNA sequences *in vitro*

♦ PCR requires two primers – one that is complementary to each strand of DNA – and a DNA polymerase

♦ Repetitive heating and cooling cycles amplify the DNA between the two primer binding sites to yield large quantities of replicated DNA

Since its inception in the mid-1980s (Saiki *et al.*, 1985; Mullis and Faloona, 1987), it is not unreasonable to say that the impact of the polymerase chain reaction (PCR) on molecular biology, forensic science and the diagnosis of human genetic diseases has been immense. The process itself is an extremely straightforward extension of the properties of DNA replication, but it has been utilized in a variety of different ways to make many cloning and DNA manipulation experiments both easier and, in many cases, possible for the first time. In its simplest terms, PCR is the repetitive copying of a section of double-stranded DNA. The process is shown diagrammatically in Figure 4.1. The power of PCR is derived from its ability to replicate large amounts of a specific section of DNA in a short period of time. At its limits, PCR is able to rapidly amplify a specific region of a single DNA molecule *in vitro* to yield sufficient quantities that can be cloned, sequenced, analysed by restriction mapping etc. The impact of PCR was underscored in 1993 with the award of the Nobel Prize in Chemistry to Kary Mullis for his contributions toward the development of the procedure. Mullis has written about his exciting discovery (Mullis, 1990), and an interesting autobiography of Mullis can be found on the Nobel prize web site (http://www.nobel.se/chemistry/laureates/1993/mullis-autobio.html).

PCR involves two oligonucleotide **primers**, usually between 17 and 30 nucleotides in length, which flank the DNA target sequence that is to be copied. One of the primers is the same sequence as one strand of the DNA

Analysis of Genes and Genomes   Richard J. Reece
© 2004 John Wiley & Sons, Ltd   ISBNs: 0-470-84379-9 (HB); 0-470-84380-2 (PB)

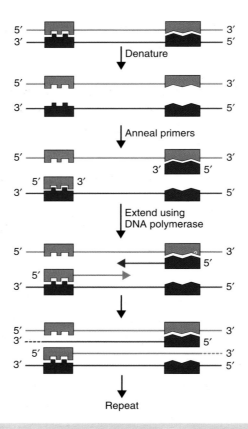

Figure 4.1. The steps of a PCR experiment. The two DNA strands of the target DNA molecule, shown in red and blue to differentiate them, are denatured, or separated, by heating. The boxed regions depict unique sequences within the target DNA to which the oligonucleotide primers will bind. Once the strands are separated, they are then cooled in the presence of oligonucleotides that are complementary to each strand. This results in the annealing of the oligonucleotides to their complementary DNA sequence. The oligonucleotides are designed such that their 3'-ends face each other. The oligonucleotides are then extended using DNA polymerase in the presence of the four deoxynucleotide triphosphates. This cycle of denaturing, annealing and extension is repeated 20–30 times to result in a massive amplification of the DNA in between the two oligonucleotide binding sites

(say, the sense strand), while the other primer is the same sequence as the other DNA strand (the antisense strand). The sense strand primer will bind, through complementary base pairing interactions, to the antisense strand and will initiate DNA synthesis of a new sense strand. Similarly, the antisense primer will bind to the sense strand of the DNA and will initiate the synthesis of a new antisense strand. The PCR reaction is split into three separate stages, each of which is performed at a different temperature (Figure 4.1). The cycle of denaturation–annealing–extension is repeated 20–30 times in order to achieve satisfactory amplification of a specific DNA sequence. The three steps in each PCR cycle are the following.

- *Denaturation*. The two strands of the target DNA molecule are separated into its component strands by heating. As we have seen in Chapter 1, DNA can be reversibly denatured by a cycle of heating and cooling. This step is most often performed at 94 °C.

- *Annealing*. The two target strands are then allowed to cool in the presence of the oligonucleotide primers. One of the primers recognizes and binds to one of the target DNA strands, and the other primer recognizes and binds to the other strand. The primers are designed such that the free 3′-end of each primer faces the other one, and so DNA synthesis proceeds on both strands through the region between the two primers. The temperature at which annealing of the primers to the template DNA occurs depends upon the length and sequence of the primer, and the level of specificity required in a particular PCR reaction – in much the same way as we saw for Southern blotting in Chapter 2. Typically a temperature in the range of 45–60 °C would be chosen.

- *Extension*. A DNA polymerase binds to the free 3′-end of each of the bound oligonucleotides and uses dNTPs to synthesize a new DNA strand in a 5′ to 3′ direction. The first PCR experiments utilized the Klenow fragment of DNA polymerase I as the replication enzyme but, because of the heat denaturation step, fresh enzyme had to be added to during each cycle. The breakthrough came with the introduction of Taq DNA polymerase from the thermophilic bacterium *Thermus aquaticus* (Lawyer *et al.*, 1989). Taq DNA polymerase is resistant to high temperatures – it can withstand the 94 °C denaturing step and still retain full activity. This means that the enzyme does not need to be replenished during the PCR reaction, and that the heating and cooling cycles required for PCR to proceed can be fully automated in specially designed heating blocks. Furthermore, Taq DNA polymerase has a temperature optimum for DNA replication of 72 °C. The high temperature at which the extension reaction can be performed means that the specificity of primer annealing is not compromised.

After one round of PCR replication, a newly synthesized copy of each DNA strand is produced, as illustrated in Figure 4.1. The site of initiation of DNA synthesis is dictated by the binding of the oligonucleotide primers to the template DNA. Specific hybridization, through base pair hydrogen bonding, will precisely position each oligonucleotide to sequences that are complementary on the template strand. What is not so obvious, however, is how DNA synthesis is terminated. If we inspect the results of a typical specific PCR reaction on an agarose gel (Figure 4.2), we see that a single, discrete band is formed. This suggests that the DNA fragments produced are homogenous and that they begin and end at the same point. To understand how this can occur, we need to look at the DNA fragments produced during several cycles of PCR (Figure 4.3).

During the first PCR cycle, the two target DNA strands are separated and DNA replication initiates at the point of primer binding. The two newly synthesized DNA strands at the end of cycle 1 (shown in green in Figure 4.3) will each have defined 5'-ends, as dictated by the site of primer binding, but ill defined 3'-ends. DNA synthesis will not terminate at a specific point, but will only stop during the heat denaturation step of cycle 2. Each of the DNA strands present at the end of cycle 1 will proceed into the next cycle of PCR,

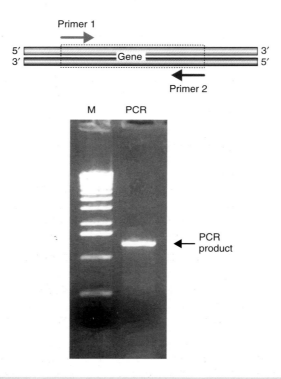

Figure 4.2. PCR amplification of a gene from genomic DNA. Two oligonucleotide primers were designed to flank the gene in the human genome. A PCR reaction was performed for 25 cycles, and one-10th of the total reaction was run on an agarose gel adjacent to a series of DNA size standards (M). The gel was stained with ethidium bromide and photographed under UV light

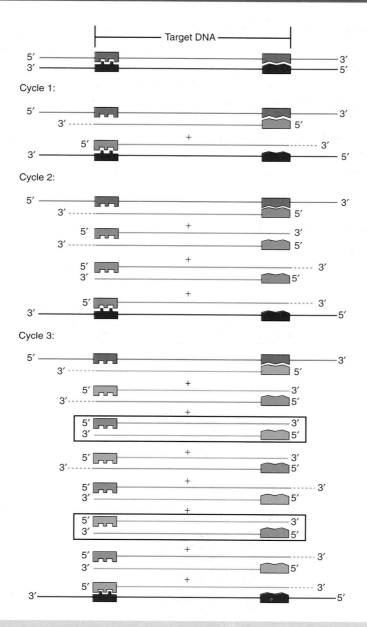

Figure 4.3. The products formed during the first three cycles of a PCR experiment. The target DNA molecule (depicted in red and blue) contains sequences that are complementary to the two oligonucleotides. The oligonucleotides act as initiation points for DNA replication. The products from each cycle of PCR are all templates for DNA replication in subsequent cycles. Shown here are the products of the first PCR cycle (green), the second cycle (purple) and the third cycle (brown). Completion of the third cycle results in the formation of two double-stranded DNA molecules – boxed – whose 5'- and 3'-ends match exactly to the ends of the oligonucleotide primers. Subsequent cycles will result in the exponential increase of this type of DNA molecule

and each will act as a template for the binding of new primers. In cycle 2, primer binding to both the original template strands and the strands synthesized during cycle 1 will occur. Primer binding to the original template strands will result in the formation of the same products that were made during cycle 1.

Table 4.1. The theoretical yield of correctly formed double-stranded target DNA molecules during a PCR experiment beginning with a single target DNA molecule

Cycle number	Double-stranded target molecules	Larger double-stranded molecules	Total DNA strands replicated
1	0	2	2
2	0	4	4
3	2	6	8
4	8	8	16
5	22	10	32
6	52	12	64
7	114	14	128
8	240	16	256
9	494	18	512
10	1 004	20	1 024
11	2 026	22	2 048
12	4 072	24	4 096
13	8 166	26	8 192
14	16 356	28	16 384
15	32 738	30	32 768
16	65 504	32	65 536
17	131 038	34	131 072
18	262 108	36	262 144
19	524 250	38	524 288
20	1 048 536	40	1 048 576
21	2 097 110	42	2 097 152
22	4 194 260	44	4 194 304
23	8 388 562	46	8 388 608
24	16 777 168	48	16 777 216
25	33 554 382	50	33 554 432
26	67 108 812	52	67 108 864
27	134 217 674	54	134 217 728
28	268 435 400	56	268 435 456
29	536 870 854	58	536 870 912
30	1073 741 764	60	1073 741 824

However, primer binding to the DNA strands produced in cycle 1, followed by replication, will result in the formation of a DNA strand with both a defined 5'-end and a defined 3'-end. This occurs because DNA replication will terminate when there is no more DNA sequence to copy. Thus, at the end of cycle 2, two DNA strands are formed (shown in purple) that have a defined 5'- and a defined 3'-end. These are, however, base-paired to DNA fragments that have ill defined 3'-ends. Again, the products from cycle 2 of the PCR process will go forward into cycle 3 and, again, each DNA strand can be used as a template for primer binding. At the end of cycle 3, two double-stranded DNA molecules are formed that have 5'- and 3'-ends beginning and ending at the positions of primer binding within the original target DNA sequence. These are the boxed sequences shown in Figure 4.3.

Beyond cycle 3 of the PCR, the repeating cycles of heat denaturation, primer annealing and extension will result in the exponential accumulation of specific target fragments of DNA (Table 4.1). After 25 cycles, typical for many PCR experiments, an amplification of about 30 million-fold is expected, and amplifications of this order are actually attained in practice. All PCR reactions are 'contaminated' with small quantities of DNA fragments that have incorrectly formed ends, but the massive amplification of specific DNA fragments means that these are almost insignificant.

There are several factors that need to be considered when attempting to amplify a specific DNA sequence using PCR. These include the choice of DNA polymerase used, the template DNA, the overall reaction conditions and the sequence of the oligonucleotide primers. We will briefly discuss each of these issues below with particular reference as to how genetic engineering experiments may be influenced, but interested readers are directed to a variety of texts that deal with the practical aspects of PCR in greater depth (McPherson and Møller, 2000; Innis, Gelfand and Sninsky, 1999).

4.1 PCR Reaction Conditions

A typical PCR experiment will contain the following components:

- DNA (0.01–0.1 μg)
- primer 1 (20 pmol)
- primer 2 (20 pmol)
- Tris-HCl (20 mM, pH 8.0)
- $MgCl_2$ (2 mM)

- KCl (25 mM) or KCl (10 mM) and $(NH_4)_2SO_4$ (10 mM)
- deoxynucleotide triphosphates (50 μM each of dATP, dCTP, dGTP, dTTP)
- thermostable DNA polymerase (2 units)
- a total reaction volume of 50–100 μL.

We will discuss the polymerase, the input DNA and the primers in greater detail below. Of the reagent components of the reaction (Tris, KCl and $MgCl_2$), the concentration of magnesium ions in the reaction plays a significant role in the success of a PCR reaction (Figure 4.4). Magnesium is required for the DNA polymerase to function, but the specificity of any particular PCR reaction is dependent upon the concentration of magnesium used. At low concentrations of magnesium, the reaction fails because the polymerase is insufficiently active. At high concentrations of magnesium, the reaction loses specificity and multiple products are produced. The optimum magnesium concentration needs to be determined empirically for each separate PCR primer set, but will usually be in the range of 1–5 mM. The buffer and salt components of the reaction (Tris and KCl) are usually held constant, although some protocols reduce the level of KCl to encourage DNA polymerase to remain on the template for longer and achieve a greater length of amplified product (Foord and Rose, 1994).

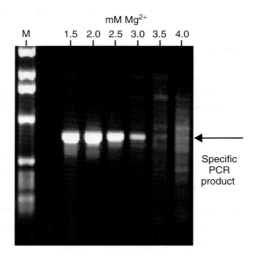

Figure 4.4. The effect of magnesium concentration on the efficiency and specificity of a PCR experiment. A PCR experiment was set up containing different concentrations of magnesium chloride. After PCR, the products were separated on an agarose gel and stained with ethidium bromide. The size of the PCR product obtained was compared with a set of DNA size standards (M). Reproduced from *Critical Factors for Successful PCR* by permission, of Qiagen GmbH

Once the PCR reaction has been set up, it is often covered with a layer of mineral oil to prevent evaporation of the sample during heating – alternatively, a PCR machine with a heated lid will prevent evaporation – before being subjected to the varying temperatures that will promote the denaturation, annealing, and extension components of a PCR cycle. Typical cycling conditions for a PCR experiment might be

- 94 °C, 30 s – denaturation
- 60 °C, 30 s – annealing
- 72 °C, 1 min – extension.

The denaturing and annealing steps are relatively short, but are sufficient to allow breaking and reforming of the hydrogen bonds between DNA strands. Early protocols included an initial denaturing step (94 °C, 2 min) to ensure that the initial template DNA was fully single stranded. This is usually not included now since long exposures to high temperature will induce nicks in the template DNA. The length of the amplified PCR product dictates the time allowed for the extension step of the reaction. Most of the polymerases used in PCR will replicate DNA *in vitro* at a rate of approximately $500-1000$ bp min^{-1} (Table 4.2). The time that the PCR experiment is incubated at the extension temperature is adjusted depending upon the length of the expected product.

Table 4.2. The properties of different types of thermostable DNA polymerase

	Taq DNA polymerase	Tfl DNA polymerase	Pfu DNA polymerase	Tli DNA polymerase	Tgo DNA polymerase
Organism	*Thermus aquaticus*	*Thermus flavus*	*Pyrococcus furiosis*	*Thermococcus litoralis*	*Thermococcus gorgonarius*
PCR optima:					
Extension time/kb (min)	1	1	2	2	2
Extension temperature (°C)	70–75	70–74	70–75	70–75	70–75
[Magnesium] (mM)	1–4	1–4	2–4	2–4	2–4
pH @ 25 °C	7.0–7.5	7.0–7.5	8.0–9.0	7.0–7.5	7.0–7.5
[dNTP] (mM)	40–200	40–200	40–200	40–200	200
[Primers] (mM)	0.1–1.0	0.1–1.0	0.1–1.0	0.1–1.0	0.1–1.0
5′–3′ exonuclease activity	Yes	Yes	No	No	No
3′–5′ exonuclease activity	No	No	Yes	Yes	Yes
Approximate error rate (errors per base replicated)	5.0×10^{-4}	5×10^{-5}	1.3×10^{-6}	2.8×10^{-6}	2.0×10^{-6}
PCR product ends	3′-A	3′-A	Blunt	Blunt	Blunt

The number of PCR cycles that are performed during an individual experiment depends upon both the amount of initial DNA template in the reaction and the amount of DNA required after the amplification process. In general, to avoid replication errors (see below), as few cycles as possible will be performed. This will usually be in the range of 17–25 cycles. After the cycles are complete, many PCR protocols include a final extension step (72 °C for 5 min) to ensure that all of the DNA in the reaction has been replicated into a double-stranded form. This final extension may also increase cloning efficiency of the PCR products (Li and Guy, 1996).

4.2 Thermostable DNA Polymerases

The bacterium *Thermus aquaticus* was first discovered in several hot springs in the Yellowstone National Park (Brock and Freeze, 1969). It has since been found in thermal habitats throughout the world. The organism has a temperature tolerance range between about 50 and 80 °C, and its optimum growth temperature is around 70 °C. Taq DNA polymerase is a monomeric enzyme with a molecular weight of 94 kDa that is isolated from the organism (Chien, Edgar and Trela, 1976; Lawyer *et al.*, 1989). The enzyme itself is thermostable; it replicates DNA at 74 °C and remains functional even after incubation at 95 °C. The enzyme includes a 5′ to 3′ polymerase activity and a 5′ to 3′ exonuclease activity, but it lacks a 3′ to 5′ exonuclease (proofreading) activity. The lack of proofreading activity means that if an incorrect base is inserted into the extending polynucleotide chain, it cannot be removed and consequently Taq DNA polymerase is error prone and will introduce mutations into amplified PCR products. In *in vitro* assays, Taq DNA polymerase misincorporates a base every 10^4–10^5 bases replicated (Barnes, 1992; Cline, Braman and Hogrefe, 1996). The wide variation in the estimated error rate is due, in part, to the methods used to assess the introduction of mutations. At its worst estimated level, with an error rate of 1×10^{-4} errors per base pair replicated, if a 1 kb sequence is amplified for 25 cycles with Taq then approximately 10 per cent of the amplified products will contain mutations. However, since mutations occurring in one cycle will be amplified in later cycles, the actual mutational frequency may vary from experiment to experiment. This level of error introduction does not, however, affect the affect the outcome of a PCR experiment. If the PCR is being performed merely to identify the presence or absence of a gene within a particular target DNA molecule, then the success of the reaction will not be affected by the introduction of errors into the amplified sequence. However if the amplified gene is to be studied functionally, then PCR errors may significantly affect the experiment. The problem of error introduction does mean, however, that PCR products should be subjected to DNA sequence analysis before they are used in cloning experiments. Additionally, several independent PCR clones should be chosen to ensure that the sequence obtained is representative.

Another functional aspect of Taq DNA polymerase that impinges upon the sequence of the final PCR product is the tendency of the enzyme to incorporate a deoxynucleotide (often an adenosine) in a template-independent manner on the 3'-end of the newly synthesized DNA strand. A consequence of this activity is that PCR products produced by Taq do not have blunts ends, but have a single 3' A residue overhang. This property has been exploited to aid the cloning of PCR products (see below).

Since the discovery of Taq DNA polymerase, a number of other thermostable DNA polymerases have been described and have been used in PCR experiments. While Taq remains the most commonly used of the thermostable enzymes for PCR, polymerases from other sources have different properties that make them useful for certain applications (Table 4.2).

Some of the other thermostable DNA polymerases, e.g. Pfu polymerase isolated from the organism *Pyrococcus furiosis*, do possess a 3' to 5' exonuclease proofreading activity, and so their mutation rate is reduced. Using the example above, if a 1 kb segment of DNA is amplified over 25 cycles with Pfu polymerase, then only 0.1 per cent of the amplified products will contain mutations. Additionally, some of the other thermostable DNA polymerases produce blunt-ended PCR products (Table 4.2).

The 5' to 3' exonuclease activity of Taq DNA polymerase means that the enzyme is able to degrade the oligonucleotide primers within the PCR reaction. This is particularly relevant during the first denaturing step of cycle 1, when the oligonucleotides are not bound to the DNA template, and the polymerase is free in solution. During the first heating cycle, the temperature of the PCR mix rises from room temperature (or 4 °C if the reaction was set up on ice) to 94 °C. This means that, at some point, the temperature within the tube will be 72 °C – the optimum for the polymerase – but the enzyme will be unable to replicate DNA since none of the oligonucleotides are bound to the template DNA. Passing through the temperature of the enzyme without replication occurring will tend to result in primer degradation, and subsequent inefficient PCR. To overcome this problem, and to prevent non-specific PCR products being synthesized prior to cycling, Taq DNA polymerase can be added to the reaction mix already at 94 °C. This 'hot start' increases both the yield and specificity of the PCR reaction. Alternatively, Taq DNA polymerase can be mixed with a specific antibody that binds to the enzyme and inhibits its activity (Kellogg *et al.*, 1994).

The antibody – enzyme complex inhibits replication at low temperatures, but the complex irreversibly dissociates at high temperature, after which the enzyme is unhindered in its function.

The existence of a number of thermostable DNA polymerases with varying properties has led to the 'blending' of polymerases for specific functions. For instance, Taq DNA polymerase produces high yields of PCR product, but is error prone and has a maximum PCR product size of about 5–7 kbp. Pfu DNA polymerase, on the other hand, is much less error prone, but still has difficulty efficiently producing PCR products over 7 kbp. A mixture of the two polymerases (15 parts Taq and 1 part Pfu) has been found to efficiently amplify DNA fragments up to 35 kbp in length with high fidelity (Barnes, 1994).

4.3   Template DNA

Almost any DNA sample can be used as a template for a PCR reaction, including linear, closed-circular and supercoiled plasmid DNA, genomic DNA, cDNA etc. The source of the DNA is immaterial, since PCR is merely a sequence directed event. The only requirement is that the primer binding sites, and the sequence between them, are intact. DNA samples over 7000 years old have successfully been used in PCR experiments (Lawlor *et al.*, 1991).

In the cases we have looked at so far, and for the sake of clarity, we have considered the amplification of a single target DNA molecule. While this is certainly achievable (Li *et al.*, 1988), in practice larger amounts of DNA are commonly used. When very small amounts of DNA are used, contamination of the PCR reaction can become a major problem. The massive amplification properties of PCR mean that even the slightest DNA contamination can ruin an experiment. Contamination may come from a variety of sources, including the researcher who is performing the experiment, the tubes and tips that are used to set up the reaction and even the enzymes and buffers used in the reaction itself. In a typical PCR experiment between 0.1 and 1 µg of genomic DNA would be added to the reaction so that a relatively low number of PCR cycles can be performed and still sufficient material produced for further experiments. This reduces the likelihood of contaminating sequences interfering with the desired amplification. How many copies of the target sequence does this amount of DNA correspond to? If you add 1 µg of human genomic DNA to a PCR reaction, this is equivalent to $1 \times 10^{-6}/(6.4 \times 10^9 \times 650) = 2.4 \times 10^{-19}$ mol, since human DNA contains approximately 6.4×10^9 bp of DNA and the average molecular weight of a base pair is 650 Da. Therefore 1 μg of human DNA corresponds to 2.4×10^{-19} mol $\times 6 \times 10^{23}$ (Avogadro's number) = approximately 144 000 molecules. That is, a single gene will be represented 288 000 times in 1 μg of genomic diploid DNA. An eight million-fold amplification of a 1000 bp segment of this genomic DNA, that should be achieved after 25 PCR cycles, will generate almost 10 μg of that 1000 bp DNA fragment. This amount is sufficient that a small fraction of the PCR reaction (typically 10 μL of a 50 μL total reaction) can be readily visualized by ethidium bromide staining after agarose gel electrophoresis. This amount is also more than sufficient for most cloning procedures and DNA sequencing protocols.

4.4 Oligonucleotide Primers

The success, or otherwise, of a PCR experiment is almost wholly dependent upon the oligonucleotide primers. The primers need to be designed such that one recognizes the sense strand of the DNA to be replicated (i.e. is the same sequence as the antisense strand) while the other recognizes the antisense strand of the target DNA (i.e. is the same sequence as the sense strand). This is shown in Figure 4.5. Typically, primers will have the following characteristics.

- They will be between 17 and 30 nucleotides in length – sufficient to allow unique annealing to a single sequence within a genome.

- They will have a GC content of approximately 50 per cent.

- The annealing temperatures of the pair of primers – calculated from the equation $2(AT) + 4(GC)$ – used in a single experiment should be approximately equal.

- Sequences with long runs of a single nucleotide should be avoided to prevent binding of the primer to repetitive sequences in the target DNA.

- Individual primers should not contain sequences that are complementary. For example, a primer of the sequence 5′-GAGATCGATGCATCGATCTC-3′ may appear a good choice for a PCR primer (20 nucleotides long, 50 per cent GC content and not containing repetitive sequences), but it is palindromic and will form a hair-pin structure if the 5′-end binds to the 3′-end. This secondary structure is undesirable, and will effectively remove the primer from the PCR reaction so amplification of the target will not occur.

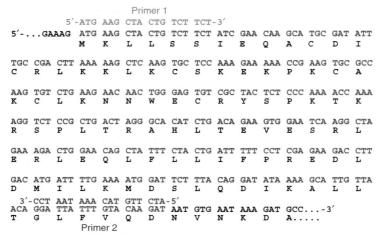

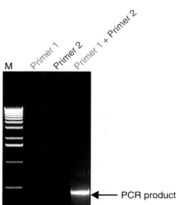

Figure 4.5. The location of primers to amplify part of the *GAL4* gene from the *Saccharomyces cerevisiae* genome. (a) The DNA sequence, and corresponding amino acid sequence, of part of the yeast *GAL4* gene. The DNA sequence encoding the first 100 amino acids of the Gal4p protein is shown in brown. The locations of primers to amplify this sequence are shown. Note that primer 1 is the same sequence as the sense strand and will therefore bind to the antisense strand and the action of DNA polymerase will make a new sense strand. Similarly, primer 2 is the same sequence as the antisense strand, so that it recognizes the sense strand and will lead to the formation of a new antisense strand. (b) The result of a PCR experiment performed with either primer on its own, or with a combination of both primers. The PCR product was visualized after ethidium bromide staining an agarose gel

- There should be no complementarity between the two primers or the 3′ ends of a single primer. For example, the following two primers again appear to be good choices: 5′-GATCGATCGATACGTGATCC-3′ and 5′-CGTAGCTAGCTAGGATCACG-3′. However, the 3′-ends of the primers are complementary to each other and primer dimers can form, which will be replicated during the first cycles of the PCR reaction:

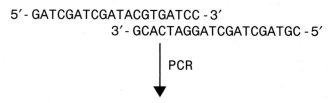

Most primers that conform to the above criteria can be made to work in PCR experiments, but a number of freely available software packages, including several of which that are web based, have been written to aid the primer design process (Rozen and Skaletsky, 1998).

4.4.1 Synthesis of Oligonucleotide Primers

Most oligonucleotides are made on commercial nucleic acid synthesizers using phosphoramidite chemistry. Oligonucleotide phosphoramidite synthetic chemistry was introduced nearly 20 years ago (McBride and Caruthers, 1983). The building blocks used for synthesis are DNA phosphoramidite nucleosides (sometimes called monomers). These are modified to prevent branching or other undesirable side reactions from occurring during synthesis. They are modified at the 5'-end (with a dimethoxytrityl group) and at the 3'-end (with a β-cyanoethyl protected 3'-phosphite group), and may also include additional modifiers to protect reactive primary amines in the nucleoside ring structure.

The phosphoramidite approach to oligonucleotide synthesis proceeds in four steps that are schematically outlined in Figure 4.6. Automated synthesis is performed on solid supports, usually polystyrene. The polystyrene is loaded into a small column that serves as the reaction chamber. A loaded column is attached to reagent delivery lines on a DNA synthesizer and the chemical reactions proceed under computer control. Bases are added to the growing chain in a 3' to 5' direction (opposite to enzymatic synthesis by DNA polymerases). Synthesis is begun using polystyrene that is already derivatized with the first base, which is attached *via* an ester linkage at the 3'-hydroxyl (Figure 4.6(a)).

Primer synthesis initiates with cleavage of the 5'-trityl group (Figure 4.6(b)) by brief treatment with acid. Monomer activated by tetrazole is coupled to the available 5'-hydroxyl (Figure 4.6(c)) and the resulting phosphite linkage is oxidized to phosphate by treatment with iodine (Figure 4.6(d)). This completes one 'cycle' of oligonucleotide synthesis.

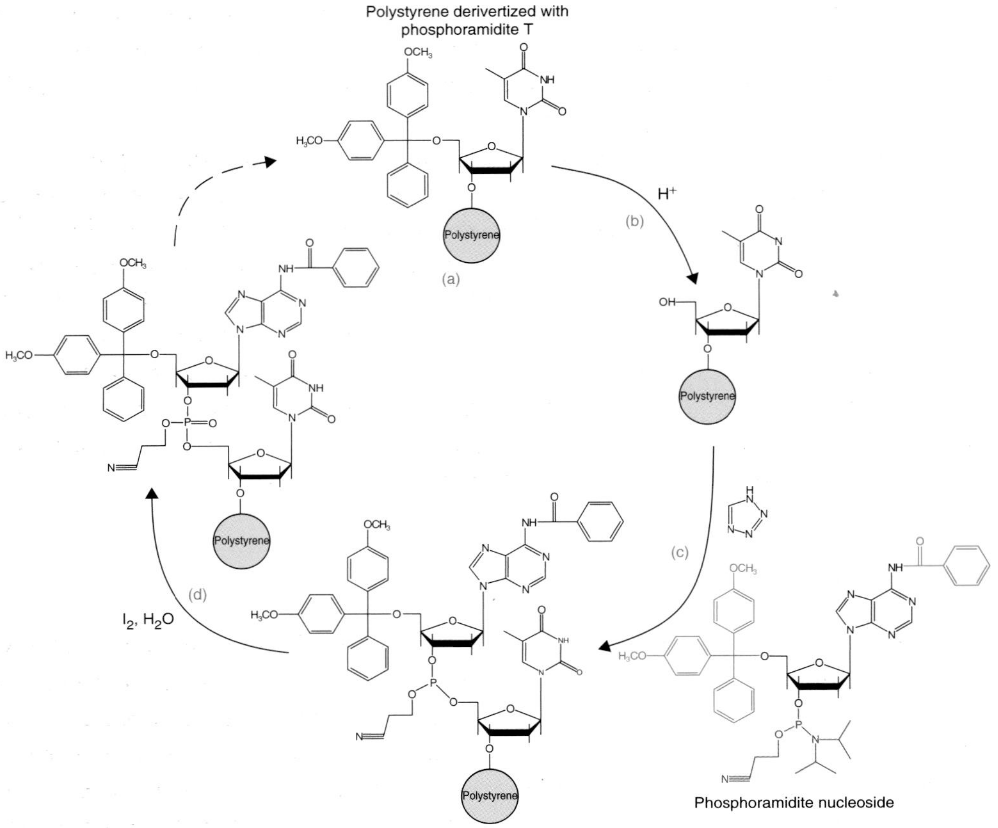

Figure 4.6. The synthetic production of oligonucleotide primers. The production of 5′-AT-3′. Phophoramidite nucleosides are modified with a dimethoxytrityl protecting group on the 5′-end (red) and a β-cyanoethyl protected 3′-phosphite group (blue). Additionally, other modifiers (green) protect primary amines occurring elsewhere in the molecule. See the text for details of the reaction cycle

The nucleoside condensation reaction is highly efficient, with less than 1 per cent of the 5′-hydroxyl groups not reacting with the incoming nucleoside. To prevent these unreacted molecules participating in subsequent reactions, and resulting in unwanted truncation deletions, the unreacted 5′-OH groups are blocked by acetylation (capping) with acetic anhydride before the oxidation step. The efficiency of coupling ensures that primers up to about 60 nucleotides in length can be manufactured routinely.

After synthesis is complete, the oligonucleotide is cleaved from the solid support with concentrated ammonium hydroxide at room temperature. Continued incubation in ammonia at elevated temperature will deprotect the phosphorus *via* β-elimination of the cyanoethyl group and also removes the protecting groups from the heterocyclic bases. The finished oligonucleotide can be purified from contaminating chemicals by precipitation, and the full-length sequence is usually isolated by HPLC purification.

The great advantage to chemical primer synthesis is that any sequence can be manufactured rapidly and relatively cheaply (<$0.5 per base). The sequence of a primer may also be mixed. For example, with the primer outlined in Figure 4.6, if a 1:1 mixture of dT and dC monomers were added to the reaction at stage C, then the resulting primer could have the sequence of either 5′-TA-3′ or 5′-CA-3′. The final product would be an equal mixture of the two species. By controlling the amount of each monomer added at the condensation step, the primer can be **doped** to any specified concentration at any position except the 3′-end. Additionally, modified bases (e.g. those containing phosphothioates, or labelled with biotin) can be added to the primer sequence.

4.5 Primer Mismatches

The oligonucleotide primers that are used in a PCR experiments need not match the target sequence exactly. This is particularly relevant when trying to make mutations, or deliberate changes, in the amplified DNA sequence or when attempting to search for gene sequences that are homologous to one already known. We will discuss some of the more straightforward uses of PCR-based mutagenesis here, but will save some of the more elegant mutagenic strategies for Chapter 7. The only place within the primer sequence that must match the target sequence exactly is the extreme 3′-end of the primer. If the 3′-end of the primer does not precisely match the target sequence, then the polymerase will not efficiently extend the primer. A consequence of this is that the PCR will be inefficient, or will fail completely. This property has, however, been exploited in the diagnosis of point mutations within genes (see below).

To think about how PCR might be used to introduce mutations into amplified products, we need to think about the primers themselves. The primers initiate the DNA replication process, but are themselves incorporated into each strand of the final amplified product. Consequently, any base changes between the primer and the template DNA will be carried forward into the amplified product. Since we cannot introduce mutations at the 3′-end of the primer, a favourite location to introduce changes is the 5′-end of the primer (Figure 4.7).

In this case, we are amplifying the same sequence as shown in Figure 4.5. However, this time the oligonucleotide primers contain additional sequences at their 5′-ends. In the case of primer 1, it contains the recognition sequence for the EcoRI restriction enzyme at the 5′-end of the sequence used to recognize

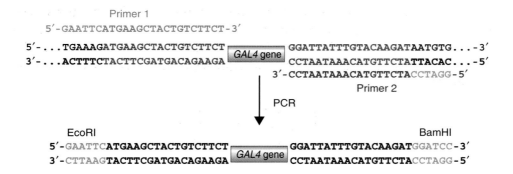

Figure 4.7. Primers to amplify part of the *GAL4* gene from the *Saccharomyces cerevisiae* genome and to include restriction enzyme recognition sites. The final PCR product contains the sequences present in the primers

the *GAL4* gene. This primer and the primer shown in Figure 4.5 will bind to the same template DNA sequence with approximately the same affinity. The EcoRI recognition sequence does not match the template sequence exactly, but the mismatches are not sufficient to prevent specific primer binding. The EcoRI recognition sequence will, however, be incorporated into the final PCR product as shown in Figure 4.7. Primer 2 in this figure contains the recognition site for the BamHI restriction enzyme, which will also be incorporated into the final product. Cloning of the final product now becomes straightforward after cutting with the two restriction enzymes – having first ensured that there are no EcoRI or BamHI restriction enzyme sites in the PCR product itself! Often restriction enzymes require more DNA than just their recognition site in order to cleave efficiently. Therefore three to six additional residues are usually added to the 5′-end of the primer before the restriction enzyme recognition site. These are often G or C (termed a **GC clamp**) to provide the maximum level of annealing between the two DNA stands and efficient cleavage by restriction enzymes.

Any mismatches between the primer and the template DNA will be carried forward into the final PCR product. Therefore, deliberate mutations may be introduced into the final PCR product by altering the sequence of the primer. This is particularly important if you want to alter the coding sequence of a gene to change the amino acid sequence of a protein, or if you want to alter the codon usage of a gene to, for example, introduce a restriction enzyme recognition site without altering the amino acid sequence of the resulting protein, or so that preferred codons are used if the resultant protein is to be expressed at high levels.

The second major use of mismatched primers is in the search for genes encoding a particular protein, and in the search for homologous genes. The isolation and characterization of a protein is common in biochemistry. For example, you may isolate a protein and sequence its amino terminal end to find the following amino acids: Met–Ile–Trp–Pro–Phe. The degeneracy of the genetic code means that this amino acid sequence could be encoded for by the following DNA sequences:

```
         Met-Ile-Trp-Pro-Phe
      5'-ATG-ATA-TGG-CCA-TTC-3'
              C       C   T
              T       T
                      G
```

With just the protein sequence at hand, it is impossible to tell which codon will be used in a particular gene to encode an individual amino acid. Therefore to PCR amplify the gene encoding this protein sequence, we must design a degenerate primer. The primer must be able to bind to all possible combinations that could encode the protein sequence. The primer below could be synthesized to perform this function:

```
      5'-ATG-ATA-TGG-CCN-TTC-3'
              C           T
              T
```

where N is an equimolar mixture of A, T, C and G. The primer above would be produced as a mixture of 24 different primers. Only one of these 24 combinations will be a perfect match to the protein coding sequence, but others will differ by only a single nucleotide from the target sequence and may still promote efficient PCR. For the case shown here, an additional primer recognizing sequences 3' to those shown here is required for PCR to proceed.

An alternative to mixing all four nucleotides at one position in a primer is to use **inosine**. Inosine is a purine, which occurs naturally in tRNAs, that can form base pairs with C, T and A. The inosine–A pairing will not fit correctly in double-stranded DNA as a purine-purine pairing, so there will be an energetic penalty to pay when the helix bulges out at this pairing. Inosine can be used in primers at positions where any of the four bases might be required. Each use of inosine thus reduces the degeneracy of the primer pool fourfold. Inosine–G mismatches may occur, but precise base pairing at other positions in the primer may overcome such a problem. Using inosine in the primers requires that the DNA polymerase used in the PCR experiment is capable of synthesizing DNA over an inosine-containing template. Taq polymerase can do this, but some other polymerases (e.g. Vent and Pfu) are unable (Knittel and Picard, 1993).

Many genes occur in families with similar amino acid sequences (and consequently similar gene sequences) encoding a similar function within a different

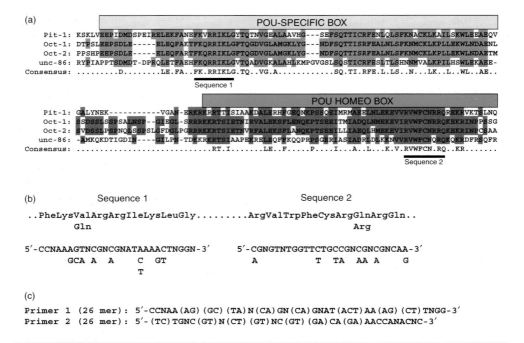

Figure 4.8. The use of degenerate PCR primers to identify novel POU genes. (a) Sequence alignment of the Pit-1, Oct-1, Oct-2 and unc-86 proteins. Sequences present in at least two of the proteins are coloured, while sequences present in all four proteins are written below as the consensus. (b) The nine amino acid sequences that are most highly conserved within the POU specific box and the POU homeo-box, and the DNA sequences that may encode them. (c) The sequence of primers used to identify additional POU-domain proteins

protein. PCR can be used to identify additional family members once some have already been characterized. To illustrate this, it is easiest to look at a specific example. The POU (pronounced 'pow', to rhyme with 'how') domain containing proteins are a family of transcription factors that share a highly conserved DNA binding domain (Figure 4.8(a)). The family derives its name from the first four members to be characterized – three mammalian proteins Pit-1, Oct-1 and Oct-2, and the *unc-86* gene from the nematode *Caenorhabditis elegans* (Herr *et al.*, 1988). There are two highly conserved portions of the POU domain – termed the POU specific domain and the POU homeodomain. The latter has a high degree of homology with *Drosophila* homeobox containing developmental control proteins. Mutations in these genes have dramatic effects on the development of an organism. For example, mutations in the mouse Pit-1 gene result in a failure in pituitary gland development and the production of dwarf mice (Li *et al.*, 1990).

To identify other members of the POU-domain family, primers were designed to the most conserved regions of the POU specific region and of the POU homeobox (Figure 4.8(b)). These primers were highly degenerate, based on the degeneracy of the genetic code, but have been successfully used to identify additional POU domains from a variety of sources (Lillycrop *et al.*, 1992). The two primers shown here (Figure 4.8(c)) were found to amplify an approximately 400 bp DNA fragment that was somewhat heterologous. The heterogeneity is due to differences in length of individual genes encoding POU-domain proteins that are amplified from DNA samples. This approach led to the identification of many new genes, including Brn-3 family members that are highly expressed in neuronal cells and plays a role in the differentiation and survival of sensory and motor neurons (Erkman *et al.*, 1996).

4.6 PCR in the Diagnosis of Genetic Disease

The power of PCR to amplify specific DNA sequences has made it a valuable tool in the diagnosis of genetic defects and mutations. Since PCR requires prior knowledge of the DNA sequence that is to be amplified, the site of the mutation must be known before PCR analysis can be attempted. The great advantage of PCR based methods is the small amount of DNA required to make a diagnosis. Samples of blood, or cells from the inside of the cheek, provide sufficient material to make adult diagnosis, while small samples of chorionic villi can be used to made a diagnosis *in utero*.

PCR is used to detect both insertion/deletion mutations and point mutations. A wide range of techniques have been developed to detect mutations by PCR. Here we will concentrate on a couple of examples, but the reader should be aware that many alternatives also exist.

- *Insertion/deletion mutations.* Waardenburg syndrome, an inherited autosomal dominant disease that is characterized by a combination of deafness and abnormal pigmentation. The disease is responsible for over 2 per cent of the cases of adult deafness, and is often associated with a frontal white blaze of hair and white eyelashes. Certain types of Waardenburg syndrome are caused by mutations in the PAX-3 gene, a transcription factor involved in regulating embryonic development. One of the first mutations found within PAX-3 from a Waardenburg syndrome patient was an 18 bp deletion in the DNA encoding the DNA binding domain of the transcription factor (Tassabehji *et al.*, 1992). This deletion can be detected in other Waardenburg syndrome patients by PCR. Primers can be designed to flank the site of the deletion (Figure 4.9). Amplification of the wild-type sequence will yield a large DNA fragment (156 bp in Figure 4.9), while amplification of the mutant sequence will yield a smaller DNA fragment (138 bp in Figure 4.9). Bands of these sizes can be easily separated on polyacrylamide or agarose gels. Thus, the presence of the mutation can be determined by the size of the PCR product obtained. In this case, because the disease is dominant, most sufferers will be heterozygotes, having one copy of the wild-type gene and one copy of the mutant gene. PCR amplification of a heterozygote will yield two different sized DNA fragments (156 and 138 bp in Figure 4.9).

(a)
```
                     Primer 1
5'-CAG GGC CGC GTC AAC CAG CTC GGC-3'
5'-GGC CAG GGC CGC GTC AAC CAG CTC GGC GGC GTT TTT ATC AAC GGC AGG CCG CTG
    G   Q   G   R   V   N   Q   L   G   G   V   F   I   N   G   R   P   L

CCC AAC CAC ATC CGC CAC AAG ATC GTG GAG ATG GCC CAC CAC GGC ATC CGG CCC TGC
 P   N   H   I   R   H   K   I   V   E   M   A   H   H   G   I   R   P   C

GTC ATC TCG CGC CAG CTG CGC GTG TCC CAC GGC TGC GTC TCC AAG ATC CTG-3'
 V   I   S   R   Q   L   R   V   S   H   G   C   V   S   K   I   L
                            3'-AGG GTG CCG ACG CAG AGG TTC TAG-5'
                                              Primer 2
```

(b)

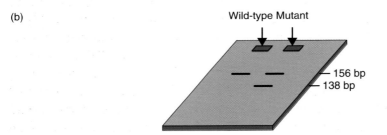

Figure 4.9. PCR to detect a pathological deletion mutation. A. The wild-type DNA sequence of the sense strand of part of exon 2 from the human Pax-3 gene. The sequence shown represents the first part of the paired box, a highly conserved element that functions as a transcription factor. Certain patients suffering from Waardenburg syndrome contain a deletion of the sequence shown in bold. This results in the deletion of seven amino acids (MAHHGIR) and the formation of a new codon (AGG, which encodes arginine). The net result is the deletion of six amino acids (MAHHGI) from the protein. Primers are designed that flank the site of the deletion. (b) PCR amplification of a normal individual with primers 1 and 2 will yield a 156 bp DNA fragment. Amplification of the DNA from an individual suffering from this Waardenburg syndrome mutation will yield two DNA fragments – the 156 bp band and a smaller 138 bp band. The smaller band is derived from the chromosome that contains the 18 bp deletion. Since Waardenburg syndrome is a dominant trait, most sufferers are heterozygotes

- *Point mutations.* The method described above cannot be used to detect point mutations (base changes) within a gene. Flanking primers would simply produce the same sized DNA fragment for both mutant and wild-type individuals. The PCR products could be subjected to DNA sequence analysis (Chapter 8), but this would be time consuming. What is required is a PCR based system to identify single-base mutations. Allele-specific PCR exploits the property that, in order to be efficiently extended by DNA polymerase, a primer needs to have a correctly base paired 3'-end (Newton *et al.*, 1989; Wu *et al.*, 1989). The allele-specific PCR amplification of a mutation within the β-globin gene causing sickle cell anaemia is outlined in Figure 4.10. To detect point mutations within a gene, three primers must be designed to take part in two PCR reactions. One primer (primer 3 in Figure 4.10) is common to both reactions, while the other primers (primers 1 and 2) detect the presence of either the wild-type or the mutated DNA sequence. The amplification of the wild-type DNA sequence with primers 1 and 3 will proceed. However, if primers 2 and 3 are used, the reaction will fail because of the mismatch between the wild-type sequence and the extreme 3'-end of primer 2. Similarly, amplification of the mutant DNA sequence with primers 1 and 3 will fail, whereas PCR will proceed normally with primers 2 and 3 using the mutant sequence as the template. The results of such an analysis are shown diagrammatically in Figure 4.10(b). Here, we are comparing a wild-type DNA sequence with that of an individual who is homozygous for the mutation. If an individual were heterozygous, then both PCR reactions would proceed normally since both alleles would be present in the template DNA. To ensure that all PCR reactions are working correctly, a set of control primers – which amplify a region of an unrelated gene – are usually included in the PCR experiment so that one band should always be present, the control PCR, and the presence or absence of an additional band is searched for.

4.7 Cloning PCR Products

As we have already seen, it is possible to clone PCR products by the insertion of extra sequences on the 5'-ends of primers such that restriction enzyme recognition sites are incorporated. It is also possible to clone PCR products directly, taking advantage of the terminal tranferase activity of Taq DNA polymerase to add a template-independent A residue to the 3'-end of PCR products. A consequence of this activity is that most of the DNA molecules amplified using Taq polymerase possess single 3'-A overhangs. In a process termed **TA cloning**, these can be ligated to a linearized 'T-vector', which has

(a)

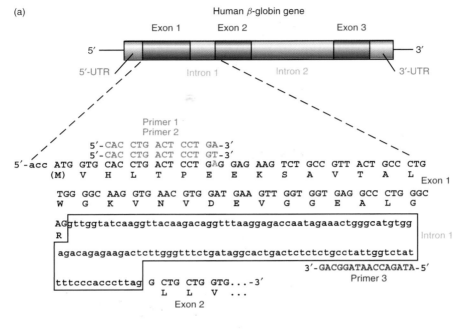

Human β-globin gene

Exon 1   Exon 2   Exon 3

5′   3′

5′-UTR   Intron 1   Intron 2   3′-UTR

Primer 1
Primer 2

5′-CAC CTG ACT CCT GA-3′
5′-CAC CTG ACT CCT GT-3′
5′-acc ATG GTG CAC CTG ACT CCT GAG GAG AAG TCT GCC GTT ACT GCC CTG
(M)   V   H   L   T   P   E   E   K   S   A   V   T   A   L
                                                                    Exon 1

TGG GGC AAG GTG AAC GTG GAT GAA GTT GGT GGT GAG GCC CTG GGC
W   G   K   V   N   V   D   E   V   G   G   E   A   L   G

AGgttggtatcaaggttacaagacaggtttaaggagaccaatagaaactgggcatgtgg
R   Intron 1

agacagagaagactcttgggtttctgataggcactgactctctctgcctattggtctat

3′-GACGGATAACCAGATA-5′
Primer 3

tttcccacccttag G CTG CTG GTG...-3′
L   L   V ...
Exon 2

(b)

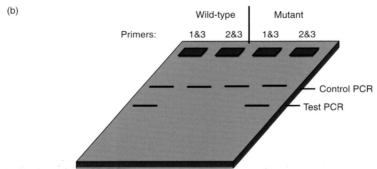

Wild-type   Mutant

Primers: 1&3   2&3   1&3   2&3

Control PCR

Test PCR

Figure 4.10. PCR to detect a point mutation. (a) Sickle cell anemia can be caused by a single base pair mutation, an A to T transversion converting Glu6 to Val, in the human β-globin gene. In individuals that are homozygous for this mutation, the substitution in the β-globin subunit of haemoglobin results in reduced solubility of deoxyheamoglobin and erythrocytes that assume irregular shapes. (b) The identification of a single base change in DNA using allele-specific PCR

single 3′-T overhangs on both ends to allow direct, high-efficiency cloning of PCR products (Zhou, Clark and Gomez–Sanchez, 1995; Marchuk *et al.*, 1991). The complementarity between the PCR product 3′-A overhangs and vector 3′-T overhangs aids efficient ligation that does not occur with blunt-ended DNA molecules. The TA cloning strategy is shown diagrammatically in Figure 4.11.

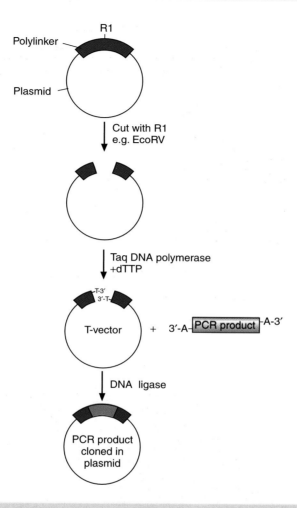

Figure 4.11. TA cloning of PCR products. A plasmid vector is cut with a restriction enzyme that results in the formation of blunt ends, e.g. EcoRV. The cut vector is then treated with Taq DNA polymerase in the presence of deoxythymine triphosphate (dTTP). The terminal tranferase activity of the polymerase catalyses the addition of a single dT residue onto the 3′-hydroxyl ends of the DNA. The resulting T-vector is mixed with a Taq-generated PCR product and the two are joined using DNA ligase. As a result the PCR product is cloned into the T-vector

4.8 RT–PCR

As we have discussed above, PCR involves the amplification of double-stranded DNA. DNA, however, does not necessarily have to be the starting material for PCR. Reverse transcription – polymerase chain reaction (RT–PCR) has been

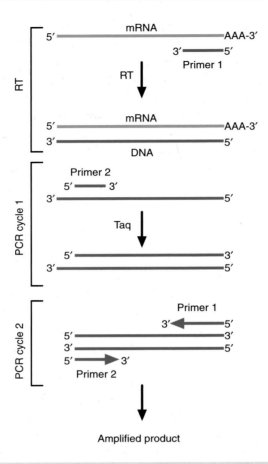

Figure 4.12. RT–PCR. The enzyme reverse transcriptase uses a complementary oligonu-cleotide to prime DNA synthesis from an RNA molecule. The single strand of DNA produced is then used as a template for the synthesis of a second DNA strand, and then for amplification by PCR

devised as a method of RNA amplification and quantitation after its conversion to DNA. RT–PCR can be used for cloning, cDNA library construction and probe synthesis. The technique consists of two parts (Figure 4.12) – the synthesis of DNA from RNA by reverse transcription (RT) and the subsequent amplification of a specific DNA molecule by polymerase chain reaction (PCR). The RT reaction uses an RNA template (typically either total RNA or polyA$^+$ RNA), a primer (random or oligo dT primers), dNTPs, buffer and a reverse transcriptase enzyme (which we will discuss more in Chapter 5) to generate a single-stranded DNA molecule complementary to the RNA (cDNA). The cDNA then serves as a template in the PCR reaction. During the first cycle of

PCR, the single DNA strand is made double stranded through the binding of another, complementary, primer and the action of Taq DNA polymerase.

Like other methods of mRNA analysis, such as northern blots and nuclease protection assays, RT–PCR can be used to quantify the amount of mRNA that was contained in the original sample. This type of analysis is particularly important for monitoring changes in gene expression. However, because PCR amplification is exponential, small sample-to-sample concentration and loading differences are amplified as well. Even large differences in target concentration (100-fold or more) may produce the same intensity of band after 25 or 30 PCR cycles. Therefore, RT–PCR requires careful optimization when used for quantitative mRNA analysis. Quantitation usually takes one of two forms – relative or absolute.

- Relative quantitation compares transcript abundance across multiple samples, using a co-amplified internal control for sample normalization. Results are expressed as ratios of the gene specific signal to the internal control signal. This yields a corrected relative value for the gene specific product in each sample. These values may be compared between samples for an estimate of the relative expression of target RNA in the samples.

- Absolute quantitation, using competitive RT–PCR, measures the absolute amount (e.g. 5.3×10^5 copies) of a specific mRNA sequence within a sample. Dilutions of a synthetic RNA (containing identical primer binding sites, but slightly shorter than the target RNA) are added to the sample and are co-amplified with the target. The PCR product from the endogenous transcript is then compared with the concentration curve created by the synthetic competitor RNA.

4.9   Real-time PCR

Quantitative real-time RT–PCR combines the best attributes of both relative and competitive RT–PCR in that it is accurate, precise, high throughput and relatively easy to perform. Real-time PCR automates the otherwise laborious process of relative RT–PCR by quantitating reaction products for each sample in every cycle. Real-time PCR systems rely upon the detection and quantitation of a fluorescent reporter, whose signal increases in direct proportion to the amount of PCR product in a reaction. In the simplest form, the reporter is the double-strand DNA-specific dye SYBR® Green (Wittwer *et al.*, 1997). SYBR Green binds double-stranded DNA, probably in the minor groove, and, upon excitation, emits light. Thus, if the dye is included in a PCR reaction, as a

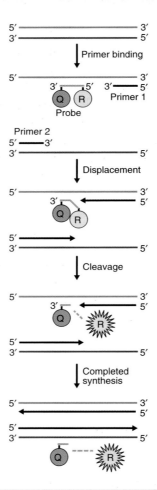

Figure 4.13. TaqMan® real-time PCR quantification. Three primers are used during the PCR process – two of these (primers 1 and 2) dictate the beginning of DNA replication on each DNA strand, and the third (the probe) binds to one strand in between. The probe contains two modified bases – a fluorescent reporter (R) at its 5'-end and a fluorescence quencher (Q) at its 3'-end. As DNA replication proceeds, the extended product from primer 1 displaces the 5'-end of the probe and the exonuclease activity of the polymerase cleaves the fluorescent reporter from the probe. The separation of the reporter from the quencher allows it to fluoresce. The amount of fluorescence is proportional to the amount of PCR product being made and is measured during each PCR cycle

PCR product accumulates the fluorescence increases. The advantages of SYBR Green are that it is inexpensive, easy to use, and sensitive. The disadvantage is that SYBR Green will bind to any double-stranded DNA in the reaction, including primer dimers and other non-specific reaction products, which can result in an over-estimation of the target concentration. For single PCR product reactions with well designed primers, SYBR Green can work extremely well, with spurious non-specific background only showing up in very late cycles.

The alternative method for quantifying PCR products is TaqMan®, which relies on fluorescence resonance energy transfer (FRET) of hybridization probes for quantitation (Figure 4.13). TaqMan probes are oligonucleotides that contain a fluorescent reporter dye, typically attached to the 5′ base, and a quenching dye, typically attached to the 3′ base. The probe is designed to hybridize to an internal region of a PCR product. When irradiated, the excited reporter dye transfers energy to the nearby quenching dye molecule rather than fluorescing, resulting in a non-fluorescent substrate. During PCR, when the polymerase replicates a template on which a probe is bound, the 5′-3′ exonuclease activity of the polymerase cleaves the probe (Holland *et al.*, 1991). This separates the fluorescent and quenching dyes and FRET no longer occurs. Fluorescence increases in each PCR cycle, proportional to the rate of probe cleavage, and is measured in a modified thermocycler. Real-time PCR is a powerful quantitative tool, but the cost of reagents and equipment is much higher than that of standard PCR reactions.

4.10 Applications of PCR

The polymerase chain reaction has revolutionized molecular biology by allowing the amplification and characterization of minute amounts of nucleic acids. As well as being of use to basic scientists, this technique is of immense importance in medicine for the identification of mutations within small amounts of human DNA, and to pathologists, who routinely need to detect and characterize small amounts of infectious micro-organisms. A detailed description of the many and varied uses to which PCR has been applied is beyond the scope of this text, but a few of the major applications of PCR are listed below:

- molecular cloning
- DNA sequencing
- archaeology
- forensics
- amplification of unknown sequences
- clinical pathology
- genetic diagnosis
- characterizing unknown mutations

- fingerprinting/population analysis
- genome analysis
- quantitative PCR of RNA or DNA.

We will touch on some of these topics in later chapters but, again, interested readers are directed toward more dedicated literature (McPherson and Møller, 2000; Innis, Gelfand and Sninsky, 1999).

5 Cloning a gene

Key concepts

- DNA libraries are pools of recombinant DNA molecules
- Genomic libraries contain fragments of all DNA sequences present in the genome
- cDNA libraries contain DNA copies of mRNA and are tissue and developmental stage specific. Their formation is dependent on an RNA-dependent DNA polymerase enzyme, reverse transcriptase
- PCR based libraries negate the requirement for cloned DNA fragments and can undergo subtraction to isolate genes that are differentially expressed

Genomes contain an enormous amount of DNA (Table 5.1). Consequently, each gene contained within a genome represents only a tiny fraction of the genome size itself. All traditional DNA cloning strategies are composed of four parts: the generation of foreign DNA fragments, the insertion of foreign DNA into a vector, the transformation of the recombinant DNA molecule into a host cell in which it can replicate and a method of selecting or screening clones to identify those that contain the particular recombinant we are interested in. In this chapter we will address some of the particular problems and issues with the first two of these steps in the formation of DNA libraries. A DNA library is simply a collection of DNA fragments.

There are several different types of library that we will consider here. DNA fragment libraries are designated as being either a **genomic DNA library** or a **cDNA library**. Most traditional methods of library construction involve the physical cloning of various DNA fragments into a suitable vector. However, as we will see later, DNA fragments that are not cloned (e.g. those derived

Analysis of Genes and Genomes   Richard J. Reece
© 2004 John Wiley & Sons, Ltd   ISBNs: 0-470-84379-9 (HB); 0-470-84380-2 (PB)

Table 5.1. Some fully sequenced genomes

Organism	Genome size (bp)	Number of chromosomes (haploid where appropriate)	Number of protein-coding genes	Reference
Methanococcus jannschii (archaebacteria)	1.7×10^6	1	1750	(Bult et al., 1996)
Haemophilus influenzae (virus)	1.8×10^6	1	1850	(Fleischmann et al., 1995)
Bacillus subtilis (bacterium)	4.2×10^6	1	4100	(Kunst et al., 1997)
Escherichia coli (bacterium)	4.6×10^6	1	4288	(Blattner et al., 1997)
Streptomyces coelicolor (actinomycete)	8.7×10^6	1	7825	(Bentley et al., 2002)
Schizosaccharomyces pombe (yeast)	1.4×10^7	3	4824	(Wood et al., 2002)
Saccharomyces cerevisiae (yeast)	1.4×10^7	16	6184	(Goffeau et al., 1997)
Caenorhabditis elegans (worm)	0.97×10^7	6	19000	(The C. elegans Sequencing Consortium, 1998)
Drosophila melanogaster (fruit fly)	1.2×10^8	4	13500	(Adams et al., 2000)
Anopheles gambiae (malaria mosquito)	2.7×10^8	5	13700	(Holt et al., 2002)
Arabidopsis thaliana (plant)	1.2×10^8	5	25498	(The Arabidopsis Genome Initiative, 2000)
Oryza sativa (rice)	4.3×10^8	12	~50000	(Yu et al., 2002; Goff et al., 2002)
Mus musculus (mouse)	2.5×10^9	21	~22000	(Mouse Genome Sequencing Consortium, 2002)
Homo sapiens (human)	2.9×10^9	23	~32000	(Lander et al., 2001; Venter et al., 2001)

from PCR products) are becoming increasingly important in genetic engineering experiments.

A genomic DNA library should contain representative copies of all the genetic material of an individual organism. Libraries such as this are organism specific. That is, a library constructed from any tissue from within a single organism should contain the same DNA fragments as those derived from any other tissue. However, libraries generated from different organisms, e.g. those derived from mouse and rat, will be different. Genomic DNA libraries should contain all of the genetic material, whether that material is expressed in a particular tissue type or developmental stage or not. Genomic libraries will therefore contain all DNA sequences: expressed genes, non-expressed genes, exons and introns, promoter and terminator regions and intervening DNA sequences.

cDNA libraries are constructed by the conversion of mRNA from a particular tissue sample into DNA fragments that can be cloned into an appropriate vector. cDNA libraries thus contain only the coding sequence of genes expressed in a tissue sample together with small regions of the 5′ and 3′ untranslated portions of the gene. Consequently, cDNA libraries isolated from different tissues of the same organism may be radically different in their composition. The genes expressed in one tissue type or developmental stage may well be different from those expressed in another tissue type or developmental stage. Additionally, the composition of a cDNA library reflects the relative abundance of mRNA in the original tissue sample. Highly expressed genes will be represented in the library multiple times, whereas genes expressed at a low level will be represented in the library less frequently.

5.1 Genomic Libraries

The smallest unit of DNA within a genome is the chromosome. Even in the simplest organisms, however, chromosomes contain an enormous quantity of DNA. For example, the *E. coli* chromosome contains some 4.6 Mbp (4 600 000 bp) of DNA (Table 5.1). This amount of DNA is far too large to be cloned into any of the vectors currently available (Chapter 3). Therefore it is necessary, and indeed desirable, to fragment the DNA before it is cloned into an appropriate vector. A 'divide and conquer' strategy comes into play here, whereby relatively small fragments of the genome can be assigned a specific function whereas the whole genome is somewhat impenetrable. The method of fragmentation plays an important role in the quality of the final library. Ideally, the genomic DNA should be broken up into random and overlapping fragments prior to cloning. Such cleavage would ensure that the library contains representative copies of all DNA fragments present within the genome, and that fragment bias is not encountered by the cleavage of DNA at specific sites only. There are two basic mechanisms for cleaving DNA that are used in the construction of genomic libraries.

(a) *Mechanical shearing.* Purified genomic DNA is either passed several times through an narrow-gauge syringe needle or subjected to sonication to break up the DNA into suitable size fragments that can be cloned. Typically, an average DNA fragment size of about 20 kbp is desirable for cloning into λ based vectors. Mechanical methods such as these have the advantage that DNA fragmentation is random, but suffer from the fact that large quantities of DNA are required, and that the average DNA fragmentation size may be quite variable.

(b) *Restriction enzyme digestion.* Restriction enzymes, such as EcoRI, often recognize 6 bp DNA sequences and cleave the DNA within the recognition sequence. On average, a 6 bp DNA sequence will occur approximately every 4000 bp within DNA. Complete digestion of genomic DNA with EcoRI will generate DNA fragments that are generally too small to be useful in genomic library construction. Other restriction enzymes, e.g. NotI, recognize and cleave 8 bp recognition sequences. Such sequences will occur much less commonly within DNA (approximately once every 65 kbp). However, restriction enzyme cleavage to produce DNA fragments suffers as a consequence of the recognition sites themselves. If, by chance, a gene that we would like to clone contains multiple recognition sites for a particular restriction enzyme, then the fragments generated after enzyme digestion may be too small to clone, and consequently the gene may not be represented within a library. To overcome this problem genomic DNA libraries are usually constructed by digesting the genomic DNA with restriction enzymes in such a way that the digestion does not go to completion (Figure 5.1). Partial restriction digests will ensure that not all DNA recognition sequences are cut and, consequently, that the library produced should contain copies of genes that may possess multiple restriction enzyme recognition sequences. In practice, restriction digestion is normally performed using a restriction enzyme, or often two, that recognize and cleave very commonly occurring sequences. For example, as shown in Figure 5.2, high-molecular-weight genomic DNA is partially cleaved with a mixture of the restriction enzymes HaeIII and AluI. Each of these restriction enzymes recognizes a 4 bp DNA sequence. Their recognition sequences should therefore occur, on average, approximately every 256 bp within genomic DNA. The partial digestion, however, limits the number of restriction enzymes sites that are actually cut and leads to

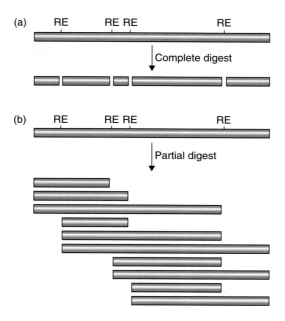

Figure 5.1. The complete and partial digestion of a DNA fragment using a restriction enzyme. (a) Complete digestion ensures that all restriction enzyme recognition sites (RE) are cut. (b) Partial digestion results in the cleavage of a random subset of the recognition sites. Partial digestion will generate a variety of products as indicated

the formation of genomic DNA fragments of a suitable size for cloning. DNA fragments produced in this manner have blunt ends since both HaeIII and AluI cut DNA in a blunt-ended fashion:

```
HaeIII: 5'-GG|CC-3'     AluI: 5'-AG|CT-3'
        3'-CC|GG-5'           3'-TC|GA-5'
```

The blunt ended DNA fragments can prove problematical when attempts are made to clone them. As we have already seen (Chapter 2), the ligation of sticky-ended DNA is considerably more efficient than that in which the DNA ends are blunt. Consequently, it is desirable to generate genomic fragments that contain sticky ends in the cloning process. This can be achieved in one of two ways.

- *Linkers or adaptors.* As shown in Figure 5.2, the blunt ended DNA fragments can be ligated to a series of oligonucleotides that either contain the recognition sequence for a restriction enzyme (linkers) or possess one blunt end for ligation to the genomic DNA and an overhanging sticky end for cloning into particular restriction sites (adaptors). In the case shown here,

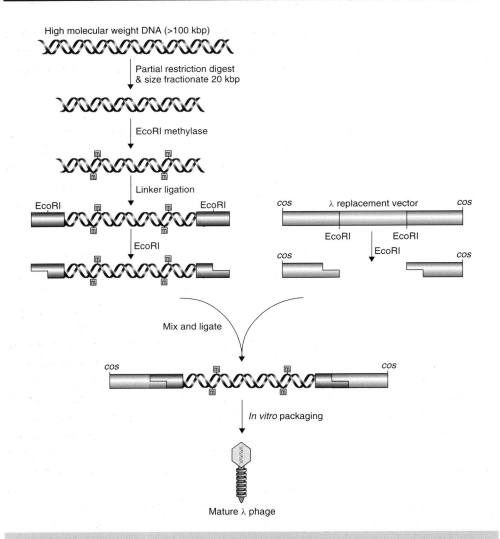

Figure 5.2. The construction of a genomic DNA library. See the text for details

the DNA fragments are first protected from restriction enzyme cleavage by treatment with a specific DNA methylase (Maniatis *et al.*, 1978). Treatment of the DNA fragments with the EcoRI methylase, in the presence of S-adenosylmethoinine, will result in the methylation of the internal-most A residue within the EcoRI recognition sequence (5′-GAATTC-3′). DNA modified in this fashion is unable to be cleaved by the restriction enzyme (see Figure 2.1). The oligonucleotide linkers are then added to the methylated DNA in large excess in the presence of high concentrations of DNA ligase. Subsequent treatment with the EcoRI restriction enzyme will result in DNA cleavage only within the linker molecules which are the only ones that contain non-methylated EcoRI restriction enzyme recognition sequences. The resulting DNA fragments can then be cloned into the EcoRI restriction site of a suitable vector.

- *Restriction enzymes that generate sticky ends.* The genomic DNA may be initially digested with a commonly occurring restriction enzyme that generates sticky ends. For example, digestion on genomic DNA with the restriction enzyme Sau3AI (recognition sequence 5′-GATC-3′) generates DNA fragments that are compatible with the sticky end produced by BamHI (recognition sequence 5′-GGATCC-3′) cleavage of a vector. The ease of this second approach makes its use far more prevalent.

Once the DNA fragments are produced, there are cloned into a suitable vector. Often this will be a λ based vector but, as we have seen in Chapter 3, a variety of vectors are available for cloning large DNA fragments. The recombinant vector and insert combinations are then grown in *E. coli* such that a single bacterial colony or viral plaque arises from the ligation of a single genomic DNA fragment into the vector. *E. coli* cells infected with either a λ phage or transformed with a plasmid DNA are unable to support the replication of additional DNA molecules of the same type. Consequently, each λ plaque or bacterial colony contains multiple copies of the same recombinant DNA molecule. A library of these molecules is produced by pooling colonies or plaques such that sufficient are present to ensure that each genomic DNA fragment is represented at least once within the library. The main advantage of cloning large DNA fragments is that fewer individual clones must be pooled together to form a representative library. A pertinent question to ask here is how many individual colonies or plaques must be pooled to ensure that a library is truly representative of the genomic DNA from which it was made. The answer to this depends upon both the size of the genome from which the library is made and upon the average size of the cloned DNA fragments within the library. For example, if a library of the *E. coli* genome (4.6 Mbp) were constructed containing 5 kbp fragments, then the fraction of the genome size compared to the average individual cloned fragment size (f) would give the lowest possible number of clones that the library must contain:

$$f = \frac{\text{genome size}}{\text{fragment size}} = \frac{4\,600\,000\text{ bp}}{5000\text{ bp}} = 920$$

Therefore, an *E. coli* genomic library of this size would require at least 920 independent clones. Using the same calculation, a human genomic library containing similar sized inserts would require at least 580 000 independent

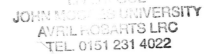

recombinants to construct a representative library. If the fragment size is increased to 20 kbp, as is common for λ vectors, then the human library must contain at least 145 000 independent recombinant clones to be representative. The ratio of genome size to fragment size is, however, an under-estimate of the complexity required for the construction of a library. Libraries must contain a much larger number of recombinant clones than this since some sequences are invariably under-represented either by chance sampling error, or as a consequence of the DNA sequence itself – perhaps the cloned DNA is relatively toxic to the host cell in which the recombinant vector is replicated, or they contain sequences that are difficult to clone, e.g. highly repetitive DNA. In the mid-1970s, Clarke and Carbon derived a formula relating the probability (P) of including any DNA sequence in a random library of N independent recombinants (Clarke and Carbon, 1976):

$$N = \frac{\ln (1 - P)}{\ln \left(1 - \dfrac{1}{f} \right)}$$

where f is the ratio of the genome size to the fragment size described above, and ln is the natural log. Therefore, to achieve a 99 per cent probability ($P = 0.99$) of including any particular sequence of random human genomic DNA in a library of 20 kbp fragments, $N = 6.7 \times 10^5$. In practice, most human genomic libraries will contain over one million independent recombinant clones.

   The pooling together of either recombinant plaques or bacterial colonies generates a **primary library**. The recombinant clones are simply washed off the growth plates and combined into a suitable test-tube. The library should contain a representative copy of each DNA molecule from which it was produced. Of course, it is possible that some DNA molecules cannot be incorporated within the library. Certain DNA sequences may be toxic to E. coli. The foreign DNA may

- be fortuitously expressed in E. coli and the protein or protein fragment may be harmful to bacterial growth,

- act as a binding site for E. coli proteins and sequester them in such a way that they are unable to perform their natural function or

- be highly repetitive and eliminated from bacterial cell through recombination.

The primary library is usually of a low titre and is often quite unstable. To increase both its stability and its titre, the library is often subjected to an amplification step. That is, the collection of phages or bacterial colonies is plated out once more, and the resulting progeny collected to form an **amplified library**. The amplified library usually has a much larger volume than the primary library, and consequently may be screened many, or even hundreds, of times. Pooled collections of λ phages can be stored almost indefinitely. Bacterial cells harbouring plasmids are more difficult to store and there is often a high degree of recombinant clone loss upon resurrection of frozen bacterial cells. Amplification of the library is essential if the library is to be screened multiple times. However, it is possible that the amplification process will result in the composition of the amplified library not truly reflecting the primary one. As we have already discussed, certain DNA sequences may be relatively toxic to *E. coli* cells; as a consequence bacteria harbouring such clones will grow more slowly than other bacteria harbouring DNA sequences that do not affect bacterial growth. Such problematic DNA sequences may be present in the primary library, but will be lost, or under-represented, after the growth phase required to produce the amplified library.

5.2 cDNA Libraries

Not only are the genomes of higher-eukaryotic organisms big, but also only a small fraction of the DNA contained within them codes for genes. The Human Genome Sequencing Project (Chapter 9) has estimated that genes constitute only about 1.5 per cent of the DNA contained within the genome. The knowledge of the entire genome sequence is important to understand the potential of a cell, i.e. the proteins that it could potentially produce, but perhaps more important is knowledge of the protein content that individual cells actually produce. All cells within an individual organism are derived from the same genome sequence, but the way in which the genome is transcribed and translated is unique to individual cell types, and to the individual developmental stages of each cell. Although many of the genes expressed by different cell types will be the same, e.g. the genes encoding the enzymes of the TCA cycle, some will also be different, e.g. some of the genes expressed within a skin cell will be different to those of a muscle cell. These differentially expressed genes, and the proteins that they produce, define each individual cell type. Thus, the mRNA that is contained within a cell gives us a snapshot of the genes being expressed within that cell at any particular time. mRNA actually represents only a small fraction of the total RNA contained within a cell (Table 5.2).

Most eukaryotic protein coding genes are transcribed by RNA polymerase II and the resulting mRNA is usually subjected to a number of post-transcriptional modifications, including the additions of a 7-methylguanosine cap at the 5′-end, and the addition of 100–200 adenine residues (a poly(A) tail) at the 3′-end

Table 5.2. The distribution of RNA molecules within cells. In eukaryotes, RNA polymerase II is responsible for the production of approximately 60 per cent of newly synthesized transcripts. Due to its instability, however, mRNA accumulates at a level of 10 per cent or less (Brandhorst and McConkey, 1974)

RNA type	Relative abundance (%)	
	E. coli	Mammalian
mRNA	5	10
tRNA	15	15
rRNA	80	75

of the transcript (see Figure 1.27). Additionally, the mRNA undergoes splicing to remove the introns so that the translation of a single contiguous message can occur.

The problem with mRNA is, of course, that it cannot be maintained in stable vectors and is difficult to manipulate. Consequently, a DNA copy (called complementary DNA, or **cDNA**) of the mRNA is required before a library can be constructed. The conversion of RNA to DNA is dependent upon the action of **reverse transcriptase**, an enzyme found in retroviruses that is responsible for the conversion of their RNA genome into a DNA copy prior to integration into host cells (Figure 5.3). David Baltimore and Howard Temin first discovered the enzyme independently in 1970 (Temin and Mizutani, 1970; Baltimore, 1970). Reverse transcriptase is an RNA-dependent DNA polymerase that, like

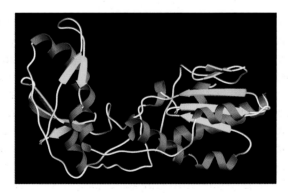

Figure 5.3. Reverse transcriptase. The X-ray crystal structure at 1.8 Å resolution of a catalytically active fragment of reverse transcriptase from Moloney murine leukemia virus (MMLV-RT) (Georgiadis *et al.*, 1995). The enzyme is an RNA-dependent DNA polymerase that is used in the conversion of mRNA into cDNA

all other DNA polymerases, catalyses the addition of new nucleotides to a growing chain in a 5′ to 3′ direction. Reverse transcriptases generally have two types of enzymatic activity.

- *DNA polymerase activity.* In the retroviral life cycle, reverse transcriptase produces a DNA copy from RNA only but, as used in the laboratory, it will transcribe both single-stranded RNA and single-stranded DNA templates with essentially the same efficiency. In both cases, an RNA or DNA primer is required to initiate synthesis.

- *RNaseH activity.* RNaseH is a ribonuclease that degrades the RNA from RNA–DNA hybrids, such as those formed during reverse transcription of an RNA template. RNaseH functions as both an endonuclease and exonuclease to hydrolyse its target molecules.

All retroviruses encode their own reverse transcriptase (RT), but the commercially available enzymes used in cDNA library construction are derived either from Moloney murine leukemia virus (MMLV-RT) or from Avian myeloblastosis virus (AMV-RT), after purification of the enzyme from virally infected cells or following expression in and purification from *E. coli*. Both enzymes have the same fundamental activities, but differ in a number of characteristics, including temperature and pH optima. MMLV-RT is a single polypeptide of 71 kDa in size, while AMV-RT is composed of two polypeptide chains 64 kDa and 96 kDa in size. Most importantly, MMLV-RT has a very weak RNaseH activity compared to AMV-RT, which gives it an obvious advantage when being used to synthesize DNA from long RNA molecules.

The process of producing a double-stranded cDNA copy of an mRNA molecule is shown in Figure 5.4. The presence of a polyA tail is unique to mRNA, and provides a mechanism of distinguishing and isolating mRNA from the more abundant rRNA and tRNA molecules. mRNA can be physically isolated from its more abundant relatives by passing total RNA over a column to which polymers of deoxythymidine (oligo-dT) are bound. RNA molecules that do not contain multiple adenine residues will be unable to adhere to such a column and will flow straight through the column. mRNA molecules, on the other hand, will bind through complementary base pairing to the column and will be eluted only when the concentration of salt flowing through the column is lowered.

The cloning of cDNA is initiated by mixing short (12–18 base) oligonucleotides of dT with purified mRNA such that the oligonucleotide will anneal to the polyA tail of the RNA molecule. Reverse transcriptase is then added and uses the oligo-dT as a primer to synthesize a single strand of cDNA in the presence of

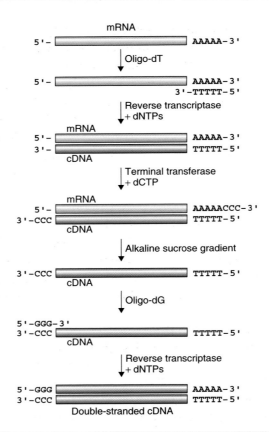

Figure 5.4. The construction of a cDNA library. See the text for details

the four deoxynucleotide triphosphates (dNTPs). The resulting molecules will be double-stranded hybrids of one cDNA and one mRNA molecule. An oligo-dT primer used to make a cDNA strand will have heterologous ends. The primer can pair at numerous positions throughout the polyA tail and consequently will yield cDNA fragments of different lengths which may have been derived from the same mRNA molecule. To overcome this problem, anchored oligo-dT primers are often employed. In addition to the 12–18 base dT sequence, anchored primers are constructed such that the extreme 3'-end contains either a G, A, or C residue (Liang and Pardee, 1992). Such primers ($5'$-$T_{12-18}V$-$3'$, where V = G, A, or C) will only efficiently initiate DNA replication if they are paired at the extreme 5'-end of the polyA tail, when the G, A, or C residue can base pair with the nucleotide immediately preceding the polyA sequence.

The production of the second DNA strand, like all DNA replication, requires a primer to initiate DNA synthesis. However, beyond the polyA tail, mRNA molecules produced from different genes will be different. Therefore, a mechanism is required to initiate DNA synthesis at sequences corresponding to the 5'-end of the mRNA. Early cDNA cloning strategies involved the formation of a hair-pin in the newly synthesized cDNA strand, which would serve as a self-priming structure for the formation of the second strand. The hair-pin would be subsequently removed from the double-stranded cDNA by treatment with S1 nuclease (Efstratiadis *et al.*, 1976). However, such methods invariably resulted in the loss of sequences at the 5'-end of genes, and so the second DNA strand is usually synthesized following either nick translation or homopolymer tailing.

- *Nick translation.* RNAse H is used to partially digest the RNA component of the RNA–DNA hybrids (Gubler and Hoffman, 1983). The remaining RNA is then used as a primer for fresh DNA synthesis using DNA polymerase I in the presence of the four dNTPs and finally DNA ligase is used to seal any remaining nicks in the DNA backbone. The resulting double-stranded cDNA molecule can subsequently be cloned into a suitable vector.

- *Homopolymer tailing.* The RNA–DNA hybrids formed after the first cDNA strand synthesis are treated with the enzyme **terminal transferase** in the presence of a single deoxynucleotide triphosphate. Terminal deoxynucleotidal transferase (TdT) is a template independent polymerase that catalyses the addition of deoxynucleotides to the 3'-ends of DNA molecules (Chang and Bollum, 1986). TdT activity was initially identified by the analysis of immunoglobin (VDJ) recombination in which extra nucleotides were found to be inserted into the joined segments that were not present in either segment before joining (Alt and Baltimore, 1982). TdT is found at high concentration in the thymus and bone marrow where such recombination events occur, but is commercially available as a recombinant protein over-produced in and purified from *E. coli*. DNA (and RNA) molecules incubated with TdT in the presence of dCTP will have multiple C residues added to their 3'-ends (Figure 5.4). Prior to the synthesis of the second DNA strand, the RNA of the RNA–DNA hybrids must be removed to provide a single-stranded template for new DNA synthesis. This can be achieved easily by treating the hybrids with alkali. RNA is hydrolysed into ribonucleotides around pH 11, while DNA is resistant to hydrolysis up to about pH 13 (Watson and Yamazaki, 1973). Increasing the pH to about 12 therefore results in the hydrolysis of the RNA, but not the DNA. Full-length cDNA strands are separated from the ribonucleotides on the basis of their size using sucrose gradient centrifugation. The resulting cDNA strands will have multiple C residues at their 3'-ends and multiple T residues at their 5'-ends (Figure 5.4). Second-strand cDNA synthesis is then initiated using

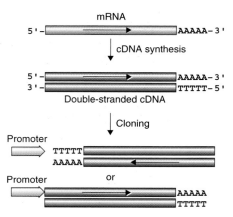

Figure 5.5. cDNA that is to be expressed must be cloned in a defined orientation so that the promoter element to which it is attached will initiate the transcription of the sense strand of the DNA, rather than the antisense strand

an oligo-dG primer that will bind, through complementary base pairing, to the newly formed polyC sequence. Reverse transcriptase, performing the role of a DNA-dependent DNA polymerase, in the presence of the four dNTPs will produce the second cDNA strand.

Homopolymer tailing has an additional advantage in that both the 5'- and 3'-ends of the original mRNA are tagged with specific and known sequences in the resulting double-stranded cDNA. This can be immensely helpful when cloning cDNA fragments in a specific orientation is required, e.g. during the expression of the cDNA. mRNA molecules are directional. The 5'-end represents the beginning of the gene sequence, and the 3' polyA tail occurs at the end of the gene sequence. Therefore, if we want to express the cDNA in, for instance, bacterial cells, it is important to ensure that only the sense strand of the cDNA is transcribed. If the antisense strand is cloned downstream of a bacterial promoter, then the resulting transcript (if produced at all) will not encode the intended protein (Figure 5.5).

5.3   Directional cDNA Cloning

The synthesis of cDNA using modified oligonucleotides to initiate each strand of DNA synthesis allows the insertion of unique restriction enzyme recognition sites at either end of the cDNA so that cloning of the cDNA fragments can only occur in one direction (Figure 5.6). In the example shown here, the oligo-dT primer also contains additional sequences at the 5'-end that encode a XhoI

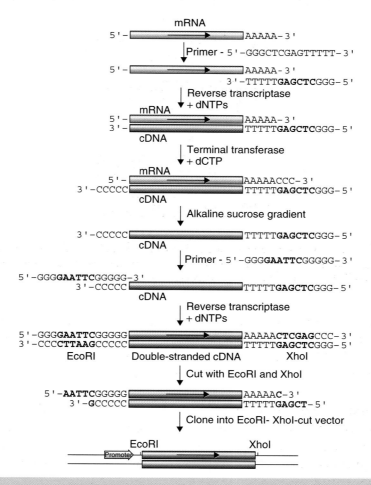

Figure 5.6. Directional cDNA cloning. Modified primers initiate DNA synthesis and result in the insertion of restriction enzyme recognition sequences at the 5'- and 3'-ends of the cDNA

restriction enzyme recognition site (5'-CTCGAG-3'). As we discussed for PCR in Chapter 4, the primer initiates DNA synthesis and is itself incorporated into the extended product. Thus a XhoI restriction enzyme recognition site will be incorporated into the 3'-end of the cDNA. Similarly, the primer used to initiate the second cDNA strand contains, in addition to the oligo-dG sequence, an EcoRI restriction enzyme recognition site (5'-GAATTC-3') at its 5'-end. Consequently, the produced cDNA will contain an EcoRI site its 5'-end and a XhoI site at its 3'-end. The placement of these sites means that the cDNA can be cloned directionally. A plasmid bearing a suitable promoter followed by, in order, an EcoRI and a XhoI restriction enzyme recognition site will accept the cut cDNA fragments in one orientation only. Thus the promoter will drive the expression of the gene encoded by the cDNA and not the reverse orientation of the opposite strand.

An obvious problem of cutting cDNA with restriction enzymes is that the cDNA itself may contain restriction enzyme recognition sites. Strategies to overcome this problem similar to those we have already encountered during the construction of genomic libraries can also be employed here. Additionally, the inclusion of methylated forms of various deoxynucleotides during the synthesis of the cDNA will protect the newly synthesized DNA strands from cleavage by certain restriction enzymes. For example, cDNA produced in the presence of methylated dCTP will be resistant to cleavage by XhoI by virtue of the presence of the methylated C residues (Endow and Roberts, 1977). Alternatively, rare-cutting restriction enzyme recognition sites, e.g. the recognition sequence for NotI (5'-GCGGCCGC-3'), may be added to the ends of the cDNA fragments to reduce the likelihood of enzyme cleavage within the cDNA itself.

The initiation of cDNA synthesis using oligo-dT primers has been immensely successful in the construction of a variety of cDNA libraries. The approach does, however, have limitations. An oligo-dT primer is suitable only for reverse transcription of mRNA molecules with poly(A) tails. Prokaryotic RNA and some eukaryotic mRNAs do not have polyA tails (Adesnik *et al.*, 1972). Therefore prokaryotic cDNA libraries cannot be produced by this method, and in eukaryotic libraries some sequences may never be present. Additionally, initial priming at the 3'-end of transcripts will tend to result in the formation of libraries that are enriched with DNA fragments representing the 3'-ends of genes – long transcripts may therefore not be fully represented within the library. These problems can be addressed by using random primers to initiate the first strand of cDNA synthesis. The random primers are usually six to nine nucleotides in length and are synthesized to be a mixture of all possible bases at each position (5'-NNNNNN-3'). Random primers will hybridize at random positions along the mRNA and will serve as starting points for DNA synthesis. cDNA cloned by this method, following the synthesis of the second strand, is unlikely to be full length, but will generate DNA fragments that are more representative of the starting mRNA. Methods have been devised to clone full-length cDNAs starting from a fragment that may have been isolated from a random-primed library (Frohman, Duch and Martin, 1988).

5.4 PCR Based Libraries

The construction of high-quality cDNA libraries is both time consuming and technically difficult. The stability and permanency of a library in which the cDNA fragments have been physically cloned into a vector, coupled with the ability to screen it multiple times, makes these libraries popular choices for isolating cDNA clones. In many cases, however, the need to construct a cloned cDNA library can be bypassed by the analysis of PCR products formed from mRNA. This type of approach is only possible if screening of the DNA fragments (Chapter 6) is performed using nucleic acid hybridization and is not applicable when functional analysis of the encoded protein is required. Nevertheless, PCR-based libraries are easy and rapid to both construct and screen. PCR-based libraries are constructed using a combination of reverse transcriptase and PCR (**RT–PCR**) (Mocharla, Mocharla and Hodes, 1990). RT–PCR is both sensitive and versatile. The technique can be used to determine the presence or absence of a transcript, to estimate expression levels and to clone cDNA products without the necessity of constructing and screening a cDNA library.

A generalized overall scheme for the production of an RT–PCR library from a mixed population of unknown mRNA molecules is shown in Figure 5.7. Most RT–PCR protocols employ reverse transcriptase to produce the first cDNA strand (Murakawa *et al.*, 1988). The production of a single strand of cDNA is

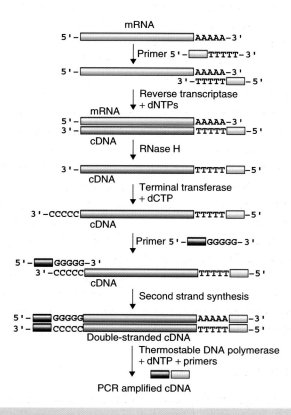

Figure 5.7. RT–PCR to produce an amplified cDNA library

sufficient prior to the progression of the PCR stage, where second-strand cDNA synthesis and subsequent PCR amplification is performed using a thermostable DNA polymerase – e.g. Taq DNA polymerase (see Chapter 4). In addition to their DNA-dependent DNA polymerase activity, some thermostable DNA polymerases (e.g. *Thermus thermophilus* (*Tth*) DNA polymerase) possess a reverse transcriptase activity in the presence of manganese ions. This has led to the development of protocols for single-enzyme reverse transcription and PCR amplification (Myers and Gelfand, 1991). Systems have also been developed in which the reverse transcriptase reaction and PCR are performed in the same buffer to eliminate secondary additions to the reaction mix to decrease both hands-on time and the likelihood of introducing contaminants into the reaction (Wang, Cao and Johnson, 1992). Such systems are ideal for the amplification of mRNA molecules whose sequence is already known using highly specific primers, but the construction of an amplified representative library requires additional steps to ensure that each mRNA molecule within the population is represented within the library. Several methods have been devised to amplify all potential mRNA species within a sample. The method outlined in Figure 5.7 utilizes many of the same elements as we have already seen in cDNA library construction. The first cDNA strand is synthesized using reverse transcriptase from an oligo-dT primer to which additional, unique sequences have been added at the 5′-end. The mRNA strand of the RNA–DNA hybrid is removed by treatment with RNaseH, prior to the addition of multiple C residues at the 3′-end of the DNA molecule using terminal transferase. The second cDNA strand is synthesized using an oligo-dG primer that, again, has unique sequences at its 5′-end. The thermostable DNA polymerase that will be used for the subsequent PCR reaction may also be used to perform the synthesis of the second cDNA strand. The unique sequences at the 5′- and 3′-ends of the resulting double-stranded cDNA are then used as primer binding sites for a PCR reaction using primers containing these sequences. The resulting PCR products will contain a huge number of copies of each cDNA molecule produced in the RT reaction.

5.5   Subtraction Libraries

As we have discussed earlier, many of the mRNA molecules produced by different cells will be the same. For example, almost all cells need to produce the enzymes required for glucose metabolism, and many of the intracellular protein components of all cells, are identical. Therefore, we might want to just concentrate on the differences between cell types to identify genes that are distinctive to a cell type, developmental stage or particular environmental stress. The advantage of PCR-based cDNA libraries is that they are amenable to

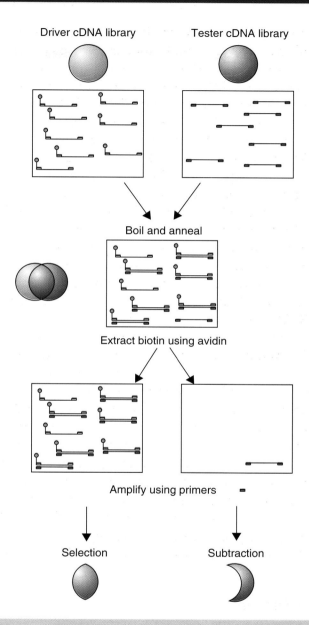

Figure 5.8. Subtractive hybridization of PCR-based libraries. Two PCR-based cDNA libraries are constructed. The driver library is labelled with biotin, while the tester library is not. Hybridization of the two libraries ensures that common sequences will be biotin labelled, while sequences that are unique to the tester library will not. The biotin labelled sequences can be removed from the mixture *via* binding to avidin, thereby enriching the sequences unique to the tester library. Enriched sequences can be amplified, by PCR, using primers unique to the tester library

the removal (subtraction) of sequences that are common between two separate libraries. This gives an enrichment of the unique sequences and allows these to be studied more readily. A mechanism by which this type of subtraction can occur is shown in Figure 5.8.

Two PCR based cDNA libraries are constructed from different mRNA samples. One of the libraries (the driver) is produced using an oligonucleotide that has a biotin moiety chemically added to it. Biotin is a cofactor required for enzymes that are involved in ATP-dependent carboxylation reactions, e.g. acetyl-CoA carboxylase and pyruvate carboxylase (Figure 5.9). Biotin deficiencies in animals are rare, but can be observed following excessive consumption of raw eggs (Baugh, Malone and Butterworth, 1968). The binding of an egg-white protein, called avidin, to biotin prevents its intestinal absorption (Figure 5.9). The complex formed between biotin and avidin is extraordinarily stable (binding constant, $K_d \sim 1 \times 10^{-15}$ M) and avidin can effectively sequester

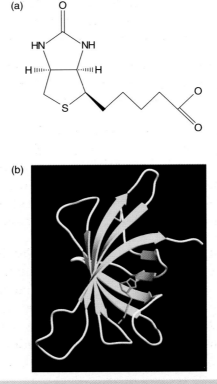

Figure 5.9. The structure of biotin and the avidin–biotin complex. (a) The molecular structure of biotin. (b) The binding of the biotin molecule to avidin. Shown here is an avidin monomer with a biotin molecule (blue) bound (Pugliese *et al.*, 1993). The functional avidin molecule is a homotetramer

biotin from all but the most dilute of solutions (Green, 1963). Biotin can be chemically coupled to phosphoramidites (see Figure 4.6) through its carboxyl group and thus can be incorporated into chemically synthesized oligonucleotide primers at either the 5′- or 3′-ends or throughout the primer in association with one of the bases. Additionally, biotin may be added to DNA fragments after synthesis using photoactivateable biotin analogues (Barr and Emanuel, 1990). Oligonucleotides and DNA fragments produced that are labelled with biotin have a high affinity for avidin.

The driver cDNA library is denatured, by heating, to separate its individual strands, and then hybridized in large excess to the denatured tester cDNA library (Luqmani and Lymboura, 1994). The majority of the double-stranded cDNA molecules produced by this will be reformed copies of the driver cDNA library since it is present in excess. However, several hybridization possibilities can occur between the driver cDNA and the tester cDNA:

- cDNA molecules only present in the driver cDNA will be unable to bind to a complementary partner from the tester cDNA and will reform as double-stranded DNA with each strand being derived from the driver library.

- cDNA molecules present in both libraries may hybridize to each other to produce hybrids in which one strand is derived from the driver cDNA and one strand is derived from the tester cDNA.

- cDNA molecules only present in the tester cDNA will be unable to bind to a complementary partner from the driver cDNA and will reform as double-stranded DNA, with each strand being derived from the tester library. Importantly, because these molecules contain no driver cDNA, they will not be labelled with biotin. All other DNA species described above will contain at least one DNA strand that possesses a biotin molecule.

Physical separation of the hybrid cDNA molecules by passing the cDNA molecules through a column to which avidin is attached will result in the adherence of all DNA species that possess one or more strands of the driver cDNA. Therefore, only DNA species unique to the tester cDNA library will not adhere to the avidin column. PCR amplification of the column flow-through and the material retained by the column (after elution with free biotin) will result in the formation of a subtraction and a selection library, respectively. The subtraction library contains sequences unique to the tester library that are not present in the driver library, and the selection library contains shared or common sequences that are present in both libraries (Figure 5.10).

Library subtraction is an extremely powerful technique for the enrichment of sequences present only in the tester cDNA library. For example, Brady *et al.*

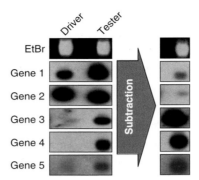

Figure 5.10. Enrichment of DNA molecules during subtractive hybridization. The subtraction of the driver library from the target library results in the elimination of common sequences (genes 1 and 2) and the enrichment of sequences that are more abundant in the target library (genes 3–5). Shown here are agarose gels of the libraries stained with ethidium bromide (EtBr), and then blotted and probed with the DNA sequences for five separate genes. The intensity of the band in the probed gels indicates the abundance of the gene in the library. Images courtesy of Gerard Brady and Abdulla Bashein, Epistem Ltd

analysed the mRNA content of single hemopoietic precursor cells during the differentiation pathway of blood cells from progenitors in the bone marrow (Brady *et al.*, 1995). They identified 29 differentially expressed genes that were activated during different stages of the differentiation pathway.

5.6   Library Construction in the Post-genome Era

The main purpose of constructing a library is to identify individual genes. One might think therefore that when the full DNA sequence of a genome is known (Table 5.1) the need for libraries might wane. This view is, however, only partially correct. Knowledge of the sequence of, say, the human genome does not tell us the function of the majority of genes. Libraries, such as those described in this chapter, will continue to play an important role in gene identification and functional assignment. Several commercial and non-profit organizations provide access to all, or most, of the genes present within some fully sequenced genomes (Table 5.1). For example, the Human Genome Mapping Project (http://www.hgmp.mrc.ac.uk) and the Mammalian Gene Collection (http://mgc.nci.nih.gov) will provide, at nominal cost, a plasmid containing a cDNA copy of one of a large number of human and mouse genes. These services negate the need to clone already identified genes, or full-length versions of cDNA molecules when partial molecules have been isolated from a library, but do not impact on the process of gene identification. Libraries continue to form the lynch-pin of gene identification.

6 Gene identification

Key concepts

♦ Screening relies on a unique property of a clone in a library

♦ If the DNA within the clone is not expressed, then gene identification on the basis of DNA sequence alone must be performed

♦ Expression libraries can be screened for protein sequence (using antibodies) or by protein function

♦ Protein interaction screening is used to identify pairs of proteins, and protein complexes, that physically associate

Attempting to find the single specific recombinant clone within a library of, say, one million different clones seems like a daunting task. In Chapter 3 we looked at ways in which recombinant clones themselves could be selected for, e.g. the use of antibiotic resistance markers or physical size as a method for selecting recombinant verses non-recombinant vectors. However, identifying the individual sequence or function of the recombinant portion of the clone is more difficult altogether. What is required is some sort of selection process by which one molecule can be distinguished from another. The selection of an individual recombinant can be based upon either its sequence or some function of the polypeptide that is encoded by that DNA sequence. The selection process can be based on one of four criteria:

- DNA sequence of the clone

- protein sequence of the encoded polypeptide

- a biochemical function of the polypeptide or

- the ability of the polypeptide encoded by the recombinant clone to interact with other polypeptides.

Analysis of Genes and Genomes   Richard J. Reece
© 2004 John Wiley & Sons, Ltd   ISBNs: 0-470-84379-9 (HB); 0-470-84380-2 (PB)

Here, we will look at each of these methods in turn. Advances in DNA sequencing technology (see Chapter 9) have led to a decrease in the use of nucleic acid hybridization techniques to identify genes, but many of the techniques we will talk about here to identify proteins and protein function have become increasingly popular.

6.1   Screening by Nucleic Acid Hybridization

The complementary base pairing of one nucleic acid strand with another can be used to identify recombinant clones that contain DNA sequences that are identical, or similar, to that of a probe sequence. A difficulty that readers will recognize with this type of screening is that you need to know something about the DNA sequence you want to find before a probe can be designed to search for that sequence. We have encountered this problem before when looking at the design of PCR primers (see Chapter 4). Nevertheless, hybridization screening provided the backbone of gene identification for many years. The main advantages of this type of screening are that it does not rely on the expression of the cloned DNA fragments within the library and it can be applied to almost any vector system into which a library has been cloned.

A generalized scheme for the identification of recombinant clones based solely on their DNA sequence is outlined in Figure 6.1. This scheme is based

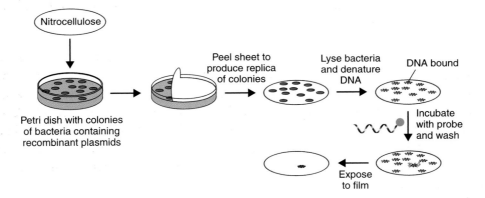

Figure 6.1. Screening for DNA sequences by nucleic acid hybridization. A sheet of nitrocellulose (or nylon) membrane is placed on top of an agar plate to generate a replica of the bacterial colonies. DNA from the bacteria, which includes the recombinant library, is attached to the membrane after the bacteria have been lysed. The denatured (single-stranded) DNA can then be used as a template for the binding of complementary, radiolabelled, DNA sequences. The binding of these sequences to the membrane can be analysed by exposure of the washed membrane to X-ray film

on that originally described by Grunstein and Hogness, but includes several later modifications, which allow the screening of a large number of bacterial colonies from a single plate, and also utilizes a filter lift procedure (Grunstein and Hogness, 1975; Grunstein and Wallis, 1979; Hanahan and Meselson, 1980). The basic idea of this type of screening procedure is to capture the DNA contained within each clone on to a nitrocellulose (or nowadays nylon) filter so that it can be used as a fixed point for the binding of other, complementary, DNA molecules. The colonies to be screened are grown on agar plates that contain the appropriate antibiotics etc. to allow the growth of cells containing the recombinant molecules. When the bacterial colonies have grown, a sheet of nylon is placed on top of them and then lifted off to produce a replica version of the plate. A portion of each bacterial colony will adhere to the filter sheet and will be removed from the agar plate along with the nitrocellulose. The nylon replica is then treated in various ways to lyse the bacteria and firmly attach the DNA to the sheet. Typically, the following steps are carried out:

- the nylon sheet is treated with alkali (e.g. 0.5 M NaOH) to initiate both bacterial cell lysis and DNA denaturation;

- upon neutralization, the sheet is treated with proteases (e.g. proteinase K) in order to remove the protein and leave the denatured DNA bound to the membrane and

- the sheet is then baked at 80 °C, or treated with UV light, to firmly adhere the DNA to the membrane.

This results in a nitrocellulose sheet containing a denatured (single-stranded) DNA copy of the bacterial colonies originally present on the agar plate. The single-stranded DNA replica can then be used as a template to bind other DNA molecules. The hybridization of a labelled single-stranded probe to the DNA on the nitrocellulose membrane will reveal the location of colonies on the original dish that contain identical, or at least similar, DNA sequences. The probe can be any single-stranded nucleic acid sequence and does not need to match the target sequence precisely.

An example of a recombinant clone isolated using this approach is the cDNA encoding a protein involved in plasmin production. Plasmin is a potent serine proteinase that has important functions in diverse physiological processes in mammals, such as degradation of extracellular matrix proteins, blood clot dissolution, cellular migration and cancer metastasis (Vassalli and Saurat, 1996). In mammalian plasma, plasmin degrades blood clot networks to produce soluble products. Plasmin is produced from a precursor, plasminogen, through limited proteolysis by plasminogen activators. Plasminogen activators are used clinically for heart attack sufferers to prevent further blood clotting (Madhani, Movsowitz and Kotler, 1993). In the early 1980s human tissue-type plasminogen activator (t-PA) was purified biochemically from melanoma cells and digested with the protease trypsin. Some of the resulting protein fragments were subjected to amino acid sequencing and one of the peptides produced in this way is shown in Figure 6.2. Using the genetic code (see the Appendix), all possible DNA sequences that could encode this peptide were determined (Figure 6.2). Of the 15 bases within the five codons, only four of them contained potential alternate sequences. Therefore, Pennica *et al.* constructed a degenerate 14-nucleotide antisense DNA probe, which contained a mixture of eight different sequences (Pennica *et al.*, 1983). They used this probe to screen a plasmid cDNA library of 4600 clones prepared from the melanoma cell line. Of these, 12 were scored as positive in the hybridization screen and, after DNA sequencing, one of these was found to contain the DNA that could have encoded the peptide. This process led to the isolation of the full-length cDNA encoding the t-PA protein.

The DNA probe used in a hybridization experiment must be homologous to the sequence it is to detect. There are examples in the literature where the gene

Amino acid sequence:
```
    -Trp-Glu-Tyr-Cys-Asp-
```

Possible DNA sequence:
```
    5'-TGG-GAA-TAT-TGT-GAT-3'
           G    C    C    C
```

Probe:
```
    3'-ACC-CTT-ATA-ACA-CT-5'
           C    G    G
```

Oligonucleotides in probe:
```
    3'-ACC-CTT-ATA-ACA-CT-5'
    3'-ACC-CTC-ATA-ACA-CT-5'
    3'-ACC-CTC-ATG-ACA-CT-5'
    3'-ACC-CTC-ATA-ACG-CT-5'
    3'-ACC-CTC-ATG-ACG-CT-5'
    3'-ACC-CTT-ATG-ACA-CT-5'
    3'-ACC-CTT-ATA-ACG-CT-5'
    3'-ACC-CTT-ATG-ACG-CT-5'
```

Figure 6.2. Probe design for the isolation of the human tissue-type plasminogen activator (Pennica *et al.*, 1983). The sequence of five amino acids of the protein is shown. This sequence (representing amino acids 253–257 of the 527-amino-acid protein), was generated following digestion of the isolated full-length protein with trypsin. Based on the genetic code, a degenerate probe was designed that would bind to all possible DNA sequences that could encode this peptide. The probe sequence outlined in yellow represents that which is precisely complementary to the isolated gene

isolated from one organism has been used as a hybridization probe to detect a homologous gene in a DNA library generated from a different organism. For example, some of the histone genes encoded by the sea urchin have been used to isolate the homologous histone genes from a frog library (Old *et al.*, 1982). As we have seen above, chemically synthesized degenerate oligonucleotide probes, in the range of approximately 14–20 nucleotides in length, can be used to detect DNA sequences encoding particular proteins. In the case of t-PA above, the peptide sequence generated could only be coded by one of a few DNA sequences – i.e. a sequence containing few degeneracies. Other amino acids are encoded by up to six different triplet codons. For example, the amino acid leucine is coded for by the following triplets: CTA, CTC, CTG, CTT, TTA and TTG. A larger number of degeneracies in a probe sequence will result in the binding of the probe to sequences that do not encode the intended target gene, whereas a highly specific probe will bind to relatively few gene sequences. Consequently, peptide sequences containing the amino acids methionine and tryptophan, which are each encoded by a single triplet, are particularly important for designing such probes. As we have previously seen with Southern blotting (Chapter 2), the stringency of washing the probe from the membrane can be used to adjust the number of positive interactions that occur.

Hybridization probes are usually radioactively labelled to aid their easy detection when bound to the membrane. Chemically synthesized oligonucleotide probes can be treated with the enzyme polynucleotide kinase in the presence of γ-^{32}P-ATP so that the radioactive phosphate group from the ATP molecule is transferred onto the 5′-end of the oligonucleotide. Other, non-radioactive alternatives have been developed, e.g. labelling with digoxigenin (McCreery, 1997), and are useful for certain experimental procedures, but the sensitivity and detection power of radioactivity has been difficult to surpass.

Colony screening, as described above, can be used to screen plasmid or cosmid based libraries. However, with only slight modifications it can also be used to screen λ phage libraries (Benton and Davis, 1977). Indeed, the screening of λ plaques is considered more desirable.

- Less DNA from the bacterial host will be transferred to the nitrocellulose membrane when lifting plaques rather than bacterial colonies. This results in a 'cleaner' background (less background probe hybridization) for λ plaque screening.

- Plaques can be lifted several times, so multiple screens can be performed from the same plate.

- Screening can be performed at very high density by screening small plaques. High-density screening has the advantage that a large number of recombinant clones can be screened for the presence of sequences homologous to the probe in a single experiment. Screening in this way, however, means it is unlikely that a single pure recombinant clone will be isolated from one round of screening. As shown in Figure 6.3, screening an agar plate containing 50 000 or more individual plaques with a probe may generate one, or two, spots in the X-ray film corresponding to positions where the probe has bound. In this case, an area of agar corresponding to the site of the spot on the original agar plate was removed and the multiple plaques contained within it were re-plated at a lower density. The screen was then repeated to generate a secondary hybridization pattern that was enriched for the plaque showing as positive in the screen. Two or three rounds of screening are often required before a pure plaque can be isolated.

A logical extension of hybridization screening is, rather than screening a library with a single oligonucleotide to search for homologous sequences, screening by PCR using two primers to amplify homologous portions of genes. The major

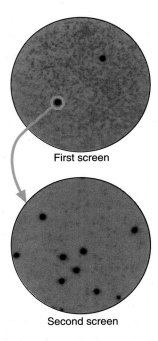

First screen

Second screen

Figure 6.3. Multiple rounds of screening to isolate a single pure recombinant. In a hybridization screen of a λ phage library, positives from the first round of, high-density, screening are re-plated and re-screened at lower density. This allows the isolation of single, pure recombinant plaques. Images courtesy of Michael Bromley and Jayne Brookman, F2G Ltd

advantage of this approach is speed. The library need not be plated prior to screening since the PCR reaction will occur using naked DNA as a template. Degenerate primers can be used to amplify portions of homologous genes from the library (Takumi, 1997). The isolated PCR products usually represent only a small region of the gene. However, this isolated fragment can be used as a highly specific probe in a traditional hybridization screen, or as a starting point to amplify the 5'- and 3'-ends of the gene using various PCR methods (Frohman *et al.*, 1988).

6.2 Immunoscreening

If a DNA fragment library is cloned into an expression vector (see Chapter 3), the gene products encoded within the foreign DNA may be produced within the host cell. Even if the protein is not fully functional, the sequence of the expressed peptide is likely to be unique within the host cell. Therefore, mechanisms to identify these unique polypeptide sequences can be used to screen the library in order to identify particular clones. The detection of polypeptide sequences is usually performed using antibodies. Antibodies are relatively straightforward to produce if a purified, or even partially purified, protein is available. The gene encoding this protein can then be identified using the antibody in screening procedures outlined below. Screening of this type does not rely upon any particular function of the expressed foreign protein, but does require a specific antibody to that protein to be available.

Antibodies are raised when a foreign protein or peptide is injected into an animal. Often, the animal used to raise antibodies for use in the laboratory is the rabbit or mouse, but sheep, goats, pigs and horses have all been used to generate larger amount of antibody (Harlow and Lane, 1999). The presence of the foreign protein (**antigen**) is detected in the animal by surface receptors on B and T lymphocyte cells. Each B cell has many thousands of different receptors on its surface that are able to bind to particular antigens. The binding of the antigen to an individual receptor results, *via* a complex pathway, in the descendants of that B cell secreting vast numbers of the soluble form of that particular receptor. These are the antibodies.

Antibodies are glycoproteins composed of subunits containing two identical light chains (L chains), each containing about 200 amino acids, and two identical heavy chains (H chains), containing about 400 amino acids each (Davies, Padlan and Sheriff, 1990). The amino-terminal 100 or so amino acids of both the H and L chains vary greatly from antibody to antibody – these are termed the variable (V) regions. The amino acid sequence variability in the V regions is especially pronounced at three hypervariable sites (Figure 6.4). The tertiary structure of the antibody brings the three hypervariable regions of both the

(a)

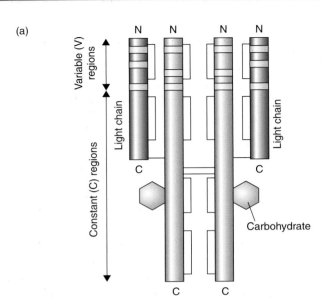

(b)

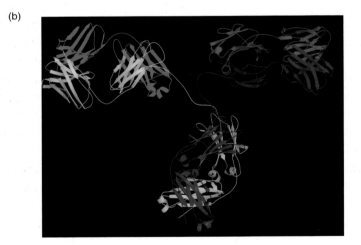

(c)

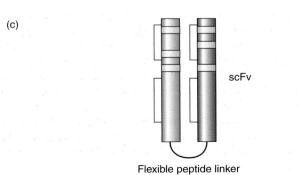

Table 6.1. The five classes of antibodies (Harlow and Lane, 1999)

Class	H chain	L chain	Subunit structure
IgA	α	κ or λ	$(H_2L_2)_2$
IgD	δ	κ or λ	H_2L_2
IgE	ε	κ or λ	H_2L_2
IgG	γ	κ or λ	H_2L_2
IgM	μ	κ or λ	$(H_2L_2)_5$

L and the H chains together to form an antigen binding site. Only a few different amino acid sequences are found at the carboxy-terminal end of H and L chains – the constant (C) regions. Mammals produce two different kinds of C region for their light chains, kappa (κ) L chains and lambda (λ) L chains. Additionally, five different kinds of C region for H chains are produced: α (the heavy chain of IgA antibodies), γ (IgG), δ (IgD), ε (IgE) and μ (IgM). Each type of H chain is able to pair with either λ or κ L chains (Table 6.1). Covalent disulphide linkages hold the pairings of H and L chains together.

Antibody molecules are required to perform two functions – they must recognize and bind to an antigen and then trigger the cellular response to that antigen. The V regions are responsible for antigen recognition, while the C regions are responsible for triggering the cellular response. The five different types of heavy chain provide a mechanism for invoking different cellular responses to an antigen (Wysocki and Gefter, 1989).

Most antibodies in use in the laboratory are described as either **polyclonal** or **monoclonal**. Polyclonal antibodies are isolated from the serum of an immunized

Figure 6.4. The structure of an antibody. (a) The diagrammatic representation of antibody structure. The heavy chains (red) and the light chains (blue) are connected together *via* a series of disulphide bridges (black lines). Both the light and heavy chains possess a series of hypervariable regions (orange) at their amino-terminal ends that provide an immense level of antigen binding site diversity. (b) The X-ray crystal structure of a monoclonal antibody. The heavy chains (yellow and blue) and the light chains (green and red) are shown (Harris *et al.*, 1992). (c) A single-chain antibody variable region fragment (scFv) antibody. The variable regions from the heavy and light chains can be engineered to be expressed as a single polypeptide joined by a 15-amino-acid linker (of the sequence (glycine$_4$serine)$_3$), which has sufficient flexibility to allow the two domains to assemble a functional antigen binding site

animal and contain many different antibodies that recognize different **epitopes** of the same antigen. Monoclonal antibodies are produced from isolated, clonal cells and recognize an individual specific epitope within the antigen. Antibodies that bind to proteins can recognize either continuous (i.e. the primary amino acid sequence of the protein) or discontinuous epitopes. A discontinuous epitope is formed by the folding of the protein to generate a surface area antigen that is composed of different segments of the primary structure. The bulk of naturally occurring epitopes are of the discontinuous type, although antibodies for use in the laboratory are often produced using denatured protein to ensure that the produced antibodies recognize continuous epitopes and can thus be used to detect the denatured protein using western blotting (Chapter 2). Although the use of animals in the production of high-affinity antibodies remains widespread, several protocols are available for the selection and production of specific antibody fragments in bacterial cells (Hexham, 1998; Portner-Taliana *et al.*, 2000; Daugherty *et al.*, 1999).

A generalized scheme for the immunoscreening of a DNA library is shown diagrammatically in Figure 6.5. cDNA is cloned into the expression vector λZAP (see Chapter 3) such that the foreign DNA is placed under the control of the bacterial *lac* promoter. Pooled recombinant λ phages are plated out onto a suitable bacterial host on agar plates. Plates containing the phage library are incubated until small plaques appear. At this point, a nitrocellulose sheet that had previously been soaked in IPTG, the gratuitous inducer of the *lac* promoter, is placed on top of the plaques. The nitrocellulose sheet is left on top of the agar for four hours to both induce the expression of the polypeptides encoded by the cDNA, and bind the proteins that are produced when the *E. coli* cells lyse as a consequence of the phage infection. The nitrocellulose sheet is then peeled off the plate and will contain, adsorbed to its surface, the proteins that were expressed in each individual plaque. The sheet is then incubated with a specific antibody to the protein for which the gene is sought. The antibody should only bind to the nitrocellulose sheet in a position where a plaque expressing that protein was located on the original agar plate. The sheet is then washed to remove any unbound antibody and subsequently incubated with a labelled secondary antibody to detect the presence of the bound primary antibody.

Originally, immunoscreening methods involved the use of radio-labelled primary antibodies to detect antibody binding to the nitrocellulose sheet (Broome and Gilbert, 1978). Such methods have, however, been largely superseded using the antibody sandwiches described above. The secondary antibody recognizes the constant region of the primary antibody and is, additionally, conjugated to an easily assayable enzyme. Such enzymes, for example, horseradish peroxidase or alkaline phosphatase (Mierendorf, Percy and Young, 1987; de Wet *et al.*,

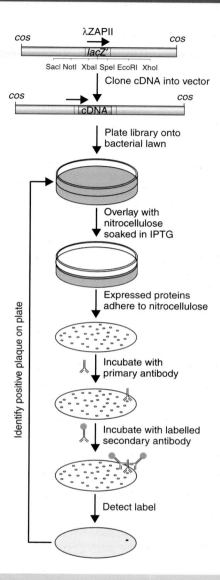

Figure 6.5. Immunoscreening of a λ phage library. *See the text for details*

1984), can be assayed for directly on the nitrocellulose sheet to produce either a colorimetric change, or in reactions linked to the emission of light such that antibody binding can be detected using X-ray film. Antibody sandwiches produce amplified signals since multiple secondary antibodies are able to bind to each primary antibody.

Immunoscreening remains a widely used technique to identify genes from available proteins. For example, Liu *et al.* purified uridine phosphorylase (an enzyme that catalyses the reversible phosphorolysis of uridine to uracil) from a colon tumour (Liu *et al.*, 1998). Antibodies to the purified protein were raised in rabbits and used to screen a human liver cDNA expression library. This resulted in the isolation of a 1.2 kbp clone that contained the entire open reading frame of the human uridine phosphorylase gene.

A particularly successful combination of membrane based protein function screening, sometimes called **south-western screening**, has been used extensively to identify genes that encode DNA binding proteins. Screening of this type is performed as outlined in Figure 6.5, except that, rather than incubating the nitrocellulose membrane with an antibody, radiolabelled double-stranded DNA of known sequence is used instead (Singh *et al.*, 1988). Proteins that are able to bind to this sequence, often originating from the promoter of a gene, will trap the DNA on the membrane, and their position can be visualized using X-ray film. This approach requires that DNA binding can occur on the membrane and also requires that the DNA binding activity is contained with a single polypeptide. Nevertheless, it has been used to isolate numerous DNA binding proteins (Vinson *et al.*, 1988).

6.3   Screening by Function

The screening of cDNA libraries using antibodies described above relies only on the expression of cDNA encoded polypeptide sequences. It does not require the expression of a fully functional protein. Screening can, however, be performed to identify a specific protein function within the host cell. For this type of screening to be successful, a host cell is required that either lacks a biochemical function that can be selected for, or that is specifically disabled in some function that can be compensated for by a protein produced from an expression library. This **functional complementation** is particularly useful for identifying genes from one organism that perform the same role as a defective gene in another organism. For example, *E. coli* cells harbouring a defective copy of the *his*B gene, encoding the enzyme imidazole glycerol phosphate dehydratase that is essential for the biosynthesis of the amino acid histidine, are unable to grow on media lacking histidine. If the *his*B defective *E. coli* cells are transformed with an expression library from yeast and plated onto media lacking histidine, the only cells that will be able to grow will be those producing a functional copy of the yeast enzyme – encoded by the *HIS3* gene (Ratzkin and Carbon, 1977). The yeast *HIS3* and *E. coli his*B genes share little DNA sequence similarity (less than 20% overall identity at the amino acid level), but the encoded proteins, although different in amino acid sequence, perform the same enzymatic function.

Functional cloning has been particularly successful in the isolation of higher-eukaryotic genes as functional homologues of genes found in more experimentally amenable lower-eukaryotic cells. For example, many higher-eukaryotic genes have been isolated by their ability to complement defects in their yeast counterparts. These include

- the genes coding for several human metabolic enzymes, reviewed by Botstein and Fink (1988),

- the *Drosophila* topoisomerase II gene (Wyckoff and Hsieh, 1988),

- a number of human RNA polymerase II transcription factors (Becker *et al.*, 1991) and

- mouse cell cycle control genes (Martegani *et al.*, 1992).

The major drawback with screening of this sort is that an assayable mutation within the host cell must be available that can be compensated for by the gene expressed from the foreign DNA. For many genes, such assays are simply not available. Additionally, mutations may not be fully compensated for by the foreign gene. For example, the foreign gene may be only partially functional within the host cell. A foreign gene may not be expressed within the host cell, or the produced proteins may not be appropriately post-translationally modified to produce the active form. Also, since library construction usually results in the cloning of single genes into vectors, if two or more different foreign gene products are required to produce the active protein, complementation screening is unlikely to succeed.

Genes may be cloned as a consequence of function if the expressed proteins are able to confer a new phenotype upon the host cell into which they are transformed. For example, cellular oncogenes can be isolated from human DNA libraries based on their ability to stimulate cell proliferation in culture (Brady *et al.*, 1985). This 'gain of function screening' may have a limited number of possible uses, but is an extremely powerful way to identify specific genes where it has been applied.

6.4 Screening by Interaction

Most proteins do not exist within cells as single entities. Most interact with a range of other proteins that either regulate their function or assemble them into larger functional complexes (Legrain and Selig, 2000). Therefore, once we have cloned a gene that encodes a protein, we might want to ask with which other proteins it interacts.

6.5   Phage Display

Phage display is a powerful technique for identifying peptides or proteins that bind to other molecules. DNA encoding a specific gene or a library of cDNA fragments is cloned into the M13 bacteriophage genome in frame with gene 3 (see Figure 3.15), encoding the minor coat protein pIII. This results in the expressed peptide or protein being displayed on the surface of the phage particle as a fusion to endogenous pIII (Smith, 1985). Phage display thus creates a physical linkage between a library of random peptide sequences to the DNA encoding each sequence, and allows for the rapid identification of peptide ligands for a variety of target molecules (antibodies, enzymes, cell-surface receptors etc.) by an *in vitro* selection process called panning. In its simplest form, panning is carried out by incubating a library of phage displayed peptides with a plate (or bead) coated with the target, washing away the unbound phage, and eluting the specifically bound phage. The eluted phage is then amplified and taken through additional binding/amplification cycles to enrich the pool in favour of binding sequences. After three or four rounds, individual clones are characterized by DNA sequencing. Phage display has been used to identify peptides that bind to receptors, substrates or inhibitors of enzymes, epitopes, improved antibodies, altered enzymes and cDNA clones (O'Neil and Hoess, 1995). Random peptide libraries displayed on phage have been used in a number of applications, including epitope mapping, mapping protein–protein contacts and identification of peptide mimics of non-peptide ligands. For example, Beck *et al.* used a displayed random peptide library to select for substrates for the HIV-1 protease (Beck *et al.*, 2000). They were able to identify peptides that could be cleaved over 200 times more efficiently than the wild-type sequence, and may provide a basis for designing highly specific inhibitors of the protease itself.

6.6   Two-hybrid Screening

The basis for two-hybrid interaction screening is the modular nature of eukaryotic RNA polymerase II transcriptional activator proteins. They consist of at least two protein domains – a sequence-specific DNA binding domain (DBD) that allows them to bind to the promoters of specific genes, and an activation domain (AD), which serves as a site of recruitment for a number of protein complexes that result in the transcription of the gene. As early as 1985, it had been shown that fusion proteins composed of the DNA binding domain from one protein fused to the activation domain of another could function readily as transcriptional activators (Brent and Ptashne, 1985). This indicated that a precise link between DNA binding and activation functions was not a prerequisite to turn genes on and that the separate domains of the protein can function independently of each other. In many transcriptional activator proteins the DBD and AD functions are found within the same polypeptide but in separate, and separable, parts of the protein. Both of these functions are essential for gene transcription to occur. Other transcriptional activators are composed of two or more subunits where one polypeptide acts as the DBD and another acts as an AD (O'Reilly *et al.*, 1997). In cases such as these, protein–protein interactions hold the DBD and AD subunits together to form the functional activator protein.

Let us consider the structure of a transcriptional activator protein in detail (Figure 6.6). The yeast protein Gal4p is often considered the archetypical activator protein. It activates the genes of the Leloir pathway in response to the yeast cells having to metabolize galactose as their sole carbon source (Reece and Platt, 1997). We will return to this pathway and its importance in genetic engineering in Chapter 11. Gal4p is a single polypeptide comprising 881 amino acids. The protein is, however, divided into distinct regions with the DBD at the amino-terminal end, and the AD at the carboxy-terminal end of the protein (Ma and Ptashne, 1987b).

- Amino acids 1–65 form a domain called a $Zn(II)_2Cys_6$ binuclear cluster (Todd and Andrianopoulos, 1997), in which two zinc atoms are held using six cysteine residues within the protein (Figure 6.7). This region of the protein interacts directly with DNA at a specific sequence called

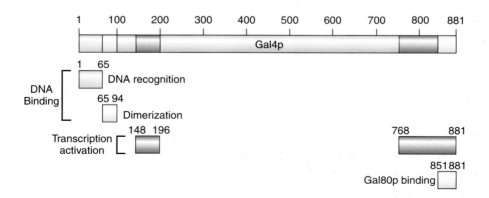

Figure 6.6. The domain structure of the yeast transcriptional activator Gal4p. The protein is composed of two major domains – a DNA binding and dimerization domain located at its amino-terminal end, and an activation domain located at its carboxy-terminal end

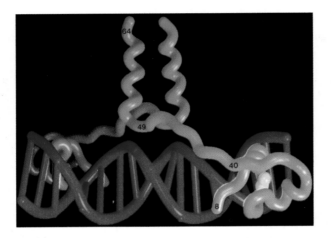

Figure 6.7. The structure of the amino-terminal 65 amino acids of Gal4p bound to DNA as determined by X-ray crystallography (Marmorstein *et al.*, 1992). The DNA (red) is contacted by one-half of the protein dimer on each side of the double helix. The DNA recognition sequences on the protein fold into a $Zn(II)_2Cys_6$ binuclear cluster, which is held by two zinc atoms (yellow). The start of the coiled-coil dimerization surface (residues 49–64) is observable within the crystal structure and extends away from the helix

the UAS_G – upstream activation sequence for galactose. This sequence (5'-$CGGN_{11}CCG$-3', where N can be any nucleotide) is found, often in multiple copies, 50–200 bp upstream of the transcriptional start site of galactose regulated genes (Giniger, Varnum and Ptashne, 1985).

- Amino acids 65–94 form a coiled-coil motif that is responsible for the dimerization of the protein.

- Amino acids 768–881 form the major AD. Like all yeast activation domains, this region is acidic. That is, it contains a preponderance of negatively charged amino acids (Ma and Ptashne, 1987c). Acidity is, however, not the sole determinant defining an activation domain or its relative strength. Activating domains appear to be largely unstructured, with acidic and hydrophobic amino acids contributing to the overall activating potential (Triezenberg, 1995). Gal4p also possesses an additional AD located between amino acids 148 and 196. This region is a weak activator and its relevance to contributing to the overall function of the protein is not understood.

- Amino acids 851–881 also form the binding site for a transcriptional repressor, Gal80p, which functions to modulate the activity of Gal4p (Ma and Ptashne, 1987a; Johnston, Salmeron and Dincher, 1987). The binding of Gal80p to this site appears to maintain the carboxy-terminal end of Gal4p in an orientation that cannot activate transcription.

The basis for interaction screening using transcription factors arises from the observations that the expression of just the DBD or just the AD within a cell is insufficient to elicit transcriptional activation (Ma and Ptashne, 1987b). A general scheme for screening in this way is shown in Figure 6.8. The Gal4p DBD is sufficient for binding to the UAS$_G$, but the lack of an AD means that it will not recruit the RNA polymerase or activate gene expression. Similarly, the Gal4p AD alone will not activate gene expression because the protein cannot be targeted to particular genes. However, if the DBD and AD can be linked in some way, e.g. through the protein–protein interactions of polypeptides fused to them, then a functional activator protein in which the DBD is linked to the AD will be formed and gene activation will occur.

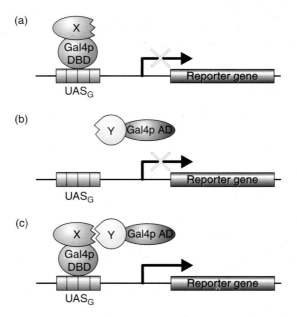

Figure 6.8. The two-hybrid screen. (a) The expression of a hybrid protein composed of the DNA binding domain (DBD) of Gal4p and a bait protein (X) is able to bind to the promoter of the reporter gene but cannot activate it. (b) The expression of another hybrid protein comprising the Gal4p activation domain (AD) and a prey protein (Y) cannot activate transcription because it is unable to bind DNA. (c) The expression of the two hybrid proteins in which the bait and prey interact with each other results in the formation of a functional activator protein and expression of the reporter gene occurs

In 1989 Stanley Fields and Ok-kyu Song first used this split transcriptional activator system to detect protein–protein interactions in yeast cells (Fields and Song, 1989). They constructed yeast expression plasmids in which the gene sequences encoding either the Gal4p DBD or the Gal4p AD were fused to the gene sequences encoding other polypeptides (Figure 6.9). These result in the production of hybrid proteins bearing either the Gal4p DBD or AD. When expressed in cells on their own, neither is able to activate gene expression. Additionally, if both fusion proteins are expressed in the same cell but the polypeptides fused to the DBD and AD do not interact with each other, gene activation will not occur. If, however, the polypeptides fused to the DBD and AD do interact, then the DBD and AD functions will be brought together through this protein–protein interaction and gene activation will occur (Figure 6.8). That is, only if the two hybrid proteins are able to interact with each other will gene expression occur.

Although the activation of the galactose metabolizing genes naturally regulated by the UAS$_G$ could be used to detect fruitful protein–protein interactions (Fields and Song, 1989), it is more convenient to analyse the expression of a more easily detectable gene product. One of the advantages of performing interaction screening in yeast is that a large number of reporters are available for use in yeast – these include those shown in Table 6.2 and a number of

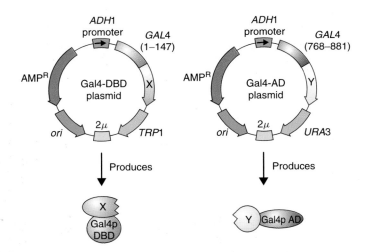

Figure 6.9. Plasmids to generate the DNA binding and activation domain fusions required for the two-hybrid screen. One plasmid produced the DBD fusion hybrid and another produced the AD fusion hybrid. The genes for both hybrid proteins are expressed from a strong constitutive promoter (*ADH1*) and are contained on high-copy-number plasmids bearing the 2μ sequence. Different nutritional genes (*TRP1* and *URA3*) allow for plasmid selection in yeast cells

Table 6.2. Commonly used reporter genes

Gene	Protein	Size (amino acids)	Original source	Detection	Reference
cat	Chroloamphenicol acetyltransferase	219	*E. coli* Tn9 transposon	Acetylation of chloroamphenicol using ^{14}C acetyl-CoA	(Gorman, Moffat and Howard, 1982)
lacZ	β-galactosidase	1024	*E. coli*	Conversion of colourless ONPG to a yellow product; XGal blue–white screening	(Norton and Coffin, 1985)
gusA	β-glucuronidase	603	*E. coli*	Conversion of MUG in a fluorometric assay; XGluc blue–white screening	(Jefferson *et al.*, 1986)
luc	Luciferase	550	*Photinus pyralis* (firefly)	Oxidation of luciferin to produce light	(de Wet *et al.*, 1987)
GFP	Green fluorescent protein	238	*Aequoria victoria* (jellyfish)	Emits green fluorescence when exposed to blue or UV light	(Chalfie *et al.*, 1994)

ONPG – *o*-nitrophenyl-β-D-galactopyranoside.
XGal – 5-bromo-4-chloro-3-indolyl-β-D-galactopyranoside.
MUG – 4-methylumbelliferyl β-D-glucuronide.
XGluc – 5-bromo-4-chloro-3-indolyl-β-D-glucuronide.

nutritional genes that can be selected for in deficient hosts. To make use of the reporter gene, the UAS$_G$ DNA sequence is cloned upstream of a reporter gene so that its expression is dependent on the association of the split DBD and AD. For example, the expression of the *E. coli lacZ* gene in eukaryotic cells, including yeast, produces a functional β-galactosidase enzyme. The production of this enzyme can be detected using methods similar to those we have already discussed for insertional inactivation in cloning vectors (Chapter 3) by, for example, producing a blue colour when grown on plates containing XGal. Yeast cells in which the UAS$_G$ has been placed upstream of the *lacZ* coding sequence will produce β-galactosidase in a galactose-dependent fashion through the action of Gal4p (Yocum *et al.*, 1984).

As originally described, the two-hybrid system was used to detect specific interactions between two known proteins, sometimes called the bait and prey, in yeast (Fields and Song, 1989). The screen was performed in a yeast strain that was deleted for the wild-type copies of the *GAL4* and *GAL80* genes, and the fruitful interaction of the bait and prey was measured using a *lacZ* reporter containing Gal4p binding sites within its promoter. The system was, however, soon adapted to screen cDNA libraries (Chien *et al.*, 1991) and modified to operate in higher-eukaryotic cells (Luo *et al.*, 1997) and in bacteria (Hu, Kornacker and Hochschild, 2000). To screen cDNA libraries, in-frame cDNA fusions are made to the AD. The library is then transformed into yeast cells together with the Gal4p DBD–bait fusion. Interacting partners can be isolated from the library as those that activate the expression of the reporter gene (Figure 6.10). Screening must be performed with the library fused to the AD,

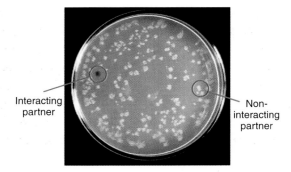

Interacting partner

Non-interacting partner

Figure 6.10. The results of a typical two-hybrid screen. A yeast strain harbouring a UAS$_G$-*lacZ* reporter was transformed with two plasmids – one expressing a Gal4p DBD bait fusion and the other expressing a Gal4p cDNA library fusion. Transformants were selected and the transferred to plates containing XGal. Only one of the Gal4p cDNA library encoded proteins (circled) was able to interact with the bait to produce a blue colony. Non-interacting partners did not elicit *lacZ* transcription and remained white

since cDNA fusion of the Gal4p DBD gives rise to a high number of sequences that are able to activate transcription when tethered to DNA. Experiments have shown that approximately 1 per cent of random *E. coli* genomic DNA fragments will function as transcriptional activators when fused to the Gal4p DBD (Ma and Ptashne, 1987c). This gives an unacceptably high background level of false positives in the screen – that is, many yeast colonies will be able to activate the reporter gene even though they do not contain proteins that interact with each other.

6.6.1 Problems, and Some Solutions, with Two-hybrid Screening

The two-hybrid screen is a very commonly used technique to detect protein–protein interactions (Colas and Brent, 1998). The technique does, however, suffer from a number of deficiencies. Some of these, together with some potential solutions, are listed below.

- *False positives*. Sequences contained within a cDNA library may encode DNA binding proteins that can circumvent the interaction with the DBD–bait protein. If the prey fusion is able to bind fortuitously to the promoter of the reporter gene, then it will promote transcription of the reporter even though it does not interact with the bait. This problem can be overcome by screening for the expression of two, or more, different reporter genes. This approach is essential if screening libraries for interacting partners, and two ways in which this can be achieved are shown in Figure 6.11. Yeast strains harbouring multiple reporter genes, which are different except for the Gal4p binding sites, have been constructed (James, Halladay and Craig, 1996). Colonies are only analysed further if all, and not just a sub-set, of the reporters are activated. An alternative approach is to perform the screen in a yeast strain expressing two different baits and a single prey (Xu, Mendelsohn and Brent, 1997). In this case, activation of only one reporter should occur, and a false positive would be assigned if both reporters become activated (Serebriiskii, Khazak and Golemis, 1999).

- *Weak interactions*. Some protein–protein interactions, although physiologically relevant, are very weak. Weak protein–protein interactions may not be sufficiently strong to recruit the transcriptional machinery to the promoter, and consequently the reporter gene will not be activated. It has been estimated that if the binding affinity between the bait and the prey is weaker than about 10–50 μM, then a transcriptional response will not be elicited (Brent and Finley, 1997).

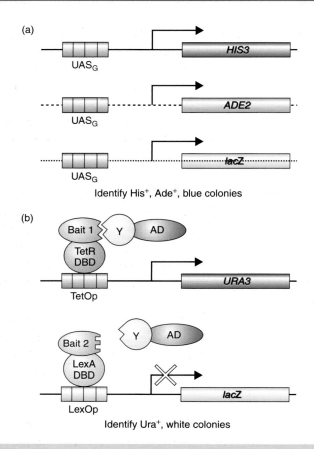

Figure 6.11. Using multiple reporters in the same cell to reduce the number of false positives in a two-hybrid screen. (a) Three different reporters each bearing Gal4p binding sites (UAS$_G$) are screened simultaneously for activity. Interactions between the bait and prey should activate all of the reporters and not a sub-set of them. (b) Screening using two different baits. A specific interaction between one of the bait proteins and the prey will result in the activation of only one of the reporter genes. This type of screening is particularly useful if the two bait proteins are closely related to each other to identify a prey that interacts specifically with only one bait (Serebriiskii, Khazak and Golemis, 1999)

- *Bait protein cannot itself be a transcriptional activator.* If the bait is itself an activator then it cannot be used to find interacting partners using the screens described above. Many proteins, even though they are not normally transcriptional activators, will activate transcription when they are tethered to DNA. Alternative screening systems, such as those based on RNA polymerase III transcription (Marsolier, Prioleau and Sentenac, 1997), or the split-ubiquitin system (Johnsson and Varshavsky, 1994), have been developed to assay baits that fall into this category.

- *Membrane proteins*. The traditional two-hybrid system relies on both the bait and prey proteins entering the eukaryotic nucleus and being able to interact there. Some proteins will not be able to do this. For example, a membrane associated protein will be tethered to the membrane and consequently unable to enter the nucleus. An alternative screening procedure, termed the SOS recruitment system, has been developed, in which the protein–protein interaction screen takes place in the plasma membrane of the yeast cell (Aronheim *et al.*, 1997).

- *Fortuitous interactions*. Some proteins are able to interact within a two-hybrid system even though the interaction is not physiologically relevant. For example, proteins that are normally located in different cellular compartments or at different developmental stages may be capable of interaction with each other when expressed in the same cell, but they would never normally be expressed together in their natural state. It is also possible that the bait and prey do not directly interact with each other, but that their apparent association is mediated by another cellular protein. For example, the HIV encoded protein Rev interacts in a two-hybrid screen with Rip1p, a yeast nucleoporin protein (Stutz, Neville and Rosbash, 1995), but this interaction is in fact mediated *via* another yeast protein, Crm1p, which is involved in nuclear export (Neville *et al.*, 1997).

- *Baits must be tested individually*. A single bait can be screened using a library of interacting prey partners, but the DBD-bait fusions must be produced individually and transformed along with the library into yeast cells. This can be very time consuming if multiple baits need to be screened. Interaction mating is now widely used to screen large numbers of bait proteins (Finley and Brent, 1994). A panel of different baits are constructed and transformed into haploid yeast of one of the two mating types (a or α). A prey library is transformed into a yeast strain of the opposite mating type and the different strains are mated. The resulting diploid yeasts are then screened for reporter gene activity. This method allows for rapid screening of multiple baits in a single experiment (Colas and Brent, 1998).

The two-hybrid system offers excellent opportunities to identify proteins that interact with each other. The caveats described above mean that the technique will not work for every protein but, as we will see in Chapter 10, genome-wide protein interaction maps have been constructed using data collected from two-hybrid screens (Uetz *et al.*, 2000). Many different two-hybrid systems have been described using baits with different DBDs (e.g. those of the bacterial repressor proteins LexA or TetR) or preys with different AD sequences (e.g. the activation domain of the Herpes simplex virus protein VP16 or the artificial activating sequence B42), reviewed by Brent and Finley (1997). The principles described above for the Gal4p based systems hold true for each of these modified versions.

6.7   Other Interaction Screens – Variations on a Theme

The two-hybrid screen has been modified extensively to suit different purposes. The basic objective behind each type of screen that is listed below is to assemble

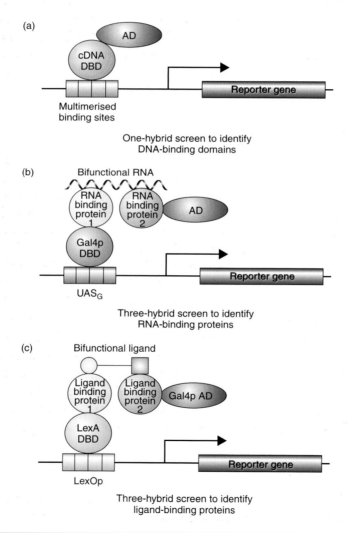

Figure 6.12. Other types of interaction screening. (a) One-hybrid screening. (b), (c) Three-hybrid screening. See the text for details

a functional transcriptional activator protein within the promoter of a reporter gene, whose expression can then be detected (Figure 6.12).

6.7.1   One Hybrid

In a simplification of the two-hybrid system, a cDNA–AD library is introduced into a yeast strain that harbours a reporter gene into which DNA binding sites have been introduced. Proteins encoded by the cDNA that are able to bind to the DNA binding sites will activate the transcription of the reporter gene (Li and Herskowitz, 1993). This approach is useful in the identification of proteins that regulate the promoters of known genes (Wang and Reed, 1993). To perform a screen like this, the DNA binding element is usually multimerized to produce a strong activation element (through the binding of multiple activator proteins) so that expression of the reporter can be observed easily.

6.7.2   Three Hybrid

Any mechanism by which the DBD and AD functions of the transcriptional activator may be brought together can be used to activate the reporter gene. Several screens have been described in which the DBD and AD fusion proteins do not interact directly with each other, but their interaction is mediated through another factor. For example, SenGupta et al. connected two RNA binding proteins (one fused to the DBD and one fused to the AD) through a bifunctional RNA molecule produced within the yeast cells (SenGupta et al., 1996). Screens could then be established to look for RNA binding proteins that bind novel sequences incorporated into the RNA linker. In a similar vein, Licitra and Liu developed a system to detect small-molecule–protein interactions (Licitra and Liu, 1996). In this system, the DBD was fused to a ligand binding protein (a receptor than binds the hormone dexamethasone) and a bifunctional ligand (containing dexamethasone and other groups) was used to screen for proteins in a cDNA–AD fusion library that bound to the other parts of the bifunctional ligand. In both the cases described here a non-protein component holds the DBD and AD together to allow transcription to occur.

6.7.3   Reverse Two Hybrid

The interaction between the bait and the prey is used to drive the expression of a gene whose product is lethal to the cell (Leanna and Hannink, 1996). This is useful to screen for drugs that disrupt the interaction between the proteins and thereby allow the cells to survive through the non-expression of the reporter (Huang and Schreiber, 1997).

7 Creating mutations

Key concepts

◆ Mutations – the altering of one DNA sequence to another – are vital to our understanding of how genes and proteins function

◆ DNA sequences can be altered in a specific, highly directed way

◆ Oligonucleotide binding to complementary DNA sequences can be used to create mutant mis-matches. The mutant oligonucleotide is then used as a primer for DNA synthesis such that the new DNA contains mutations

◆ Strand selection methods distinguish between the newly synthesized mutant DNA and the parental wild-type sequence

◆ PCR can be used to create either specific or random mutations

The sequence of a gene dictates the amino acid sequence of the protein that it encodes. Consequently, mutations within a gene, i.e. alterations to the DNA sequence, may result in changes to the amino acid sequence of the protein. The analysis of mutations is especially useful in the elucidation of protein function. Mutations that either reduce the activity of the protein, or allow the protein to behave in an abnormal way, can be used to ascribe particular functions to individual portions of proteins. The rate of naturally occurring mutations, resulting as a consequence of erroneous DNA replication within genes, is quite low (estimated at a level of one DNA base alteration per 10^6–10^8 bases replicated) and varies widely between individual genes and organisms. Naturally occurring mutations have, however, been used to isolate genes and describe specific functions for their encoded proteins. Prior to the explosion in molecular biology techniques in the 1970s and 1980s, increased mutation rates were usually obtained by treating whole cells with either a physical or a chemical mutagen. For example, the treatment of micro-organisms or

Analysis of Genes and Genomes   Richard J. Reece
© 2004 John Wiley & Sons, Ltd   ISBNs: 0-470-84379-9 (HB); 0-470-84380-2 (PB)

single-cell eukaryotes with X-rays, UV light or chemicals such as ethyl methane sulphonate (EMS) generates DNA base changes throughout the genome. Once produced, these changes will be passed from generation to generation as the cells divide. This type of traditional mutagenesis has, however, a number of drawbacks. For example, the mutations produced are random – they can occur anywhere within a genome and are not restricted to individual genes or parts of genes. Additionally, highly developed and specialized screening procedures are required to identify mutations that have occurred within individual genes. This is relatively straightforward for mutations occurring in genes that encode, for example, one of the enzymes of a metabolic pathway. Mutations that destroy the activity of one member of the pathway are likely to lead to the formation of an organism that is unable to metabolize a particular nutrient. Screens based on growth assays can then be devised to isolate these mutants. Traditional forms of mutagenesis also suffer, since the observed phenotypic change in a screen may not be a result of a mutation within a single gene. Additionally, multiple mutations may be required (perhaps when multiple redundant genes occur within the same cell) before a phenotypic change can be observed.

Mutations within DNA generally fall into one of two categories. In the first, a base or bases within a DNA sequence are changed from one sequence to another, while in the second bases are either inserted into or removed form the gene. Single DNA base pair changes are described as being either **transition** mutations or **transversion** mutations:

- transition mutations – the change of one purine–pyrimidine base pair to a different purine–pyrimidine base pair (e.g. AT → GC, or GC → AT, or TA → CG);

- transversion mutations – the change of a purine–pyrimidine base pair to a pyrimidine–purine base pair (e.g. AT → TA, or GC → CG, or AT → CG, or GC → TA).

Single base changes may result in various alterations to the amino acid sequence of the protein encoded by the gene at the regions of the changed bases. The mutation may be a

- silent mutation – the triplet code is changed, but the amino acid encoded is the same (e.g. the triplets 5′-TCG-3′ and 5′-TCC-3′ both encode the amino acid serine),

- mis-sense mutation – a codon change alters the amino acid encoded (e.g. if the serine codon 5′-TCG-3′ is mutated to 5′-ACG-3′, then the amino acid threonine will be inserted into the encoded polypeptide in place of serine – or

- non-sense mutation – an amino acid codon is changed to produce a stop codon. For example, if the serine codon 5'-TCG-3' is mutated to 5'-TAG-3', then the encoded polypeptide chain will terminate at this point.

The insertion or deletion of a base pair, or base pairs, into the coding sequence of a gene can have drastic implications for the encoded polypeptide. Since the DNA code is read in triplets, the insertion or deletion of bases in multiples other than three will result in a **frame-shift** mutation. If, for example, a single letter in the sentence below is deleted, then the meaning of the sentence beyond the change is radically altered:

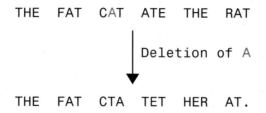

Similarly, the deletion, or insertion, of one or two bases into the coding sequence of a gene will muddle the remainder of the sequence beyond the mutation. Only the deletion or insertion of multiples of three bases will leave the remainder of the encoded polypeptide sequences unaltered, but will remove (or insert) amino acids.

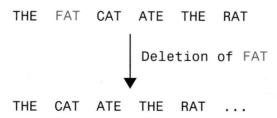

The alteration of a single amino acid to another within a protein can have great consequences on the function of the protein itself. To enable a highly directed approach to the study of the relationship between genes and proteins, mechanisms to create mutations at highly specific regions of DNA are required.

7.1 Creating Specific DNA Changes Using Primer Extension Mutagenesis

As we have already seen in Chapter 3, bacteriophage M13 undergoes a switch during its life cycle during which its single-stranded genome is converted to a double-stranded form. That is, the single-stranded form of the genome serves as a template for the new synthesis of a second DNA strand. This process can be used to our advantage if we want to create mutations within the newly synthesized DNA strand, and is outlined in Figure 7.1. The use of oligonucleotides in creating site-directed mutations was devised in the laboratory of Michael Smith, who shared the 1993 Nobel Prize in Chemistry for his discovery. Smith and his colleagues used single-stranded M13 genomic DNA as a hybridization template for a synthetic oligonucleotide (Zoller and Smith, 1983). The oligonucleotide binds to its complementary sequence within the single stranded genome, and is designed such that one or more mutations (non-complementary base pairings) occur when it binds to the M13 DNA. The binding of the oligonucleotide to the single-stranded DNA is stabilized by the complementary base pairing that occurs elsewhere. In addition to altering individual bases, an oligonucleotide can also introduce base insertions or deletions into a gene. Once bound to its complementary sequence, the oligonucleotide provides a free 3' hydroxyl group as the starting point of DNA synthesis. The hybrid, partially double-stranded, DNA molecule is incubated with a DNA polymerase enzyme in the presence of the four deoxynucleotide triphosphates (dNTP). This will result in the synthesis of a new DNA strand that is entirely complementary to the original DNA strand except at the positions where mutations have been introduced within the oligonucleotide itself. The newly synthesized DNA circle is then completed by the action of DNA ligase, in the presence of ATP, to seal any nicks remaining in the DNA backbone. The naked DNA is unable to infect *E. coli* cells, so it must be introduced into the bacterium where the DNA will be replicated and phage particles produced. When the DNA circles are replicated in bacterial cells, one of two possibilities can arise – either the original wild-type DNA strand or the newly synthesized mutated DNA strand can give rise to progeny M13 bacteriophages. That is, the resulting M13 plaques may either contain the wild-type sequence or the mutated sequence.

Bacteriophages containing either the wild-type or the mutant sequence can be distinguished from each other through hybridization screening (similar to that described in Chapter 6). A radio-labelled version of the synthetic oligonucleotide used to create the mutation will bind preferentially to the mutant sequence when compared with the wild-type sequence (Wallace *et al.*, 1981). Therefore, bacteriophage plaques that are able to bind the oligonucleotide at high stringency should contain the mutant sequence.

The primer extension site-directed mutagenesis procedure became widely adopted in the early 1980s. It suffered, however, from a number of drawbacks as a method for rapidly producing a variety of specific DNA mutations.

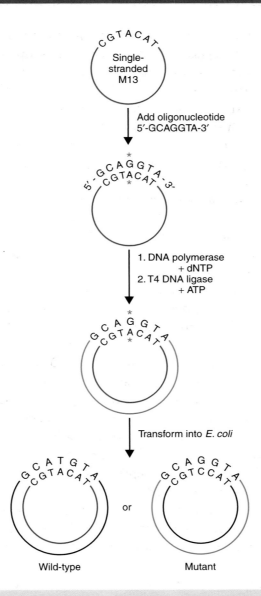

Figure 7.1. Site-direct mutagenesis using a single-stranded DNA template. Single-stranded DNA isolated from a recombinant M13 phage bearing the gene to be mutated is isolated and used as a template for the binding of an oligonucleotide primer. The primer hybridizes to its complementary sequence and introduces a specific mutation(s). The hybrid is then treated with DNA polymerase in the presence of the four deoxynucleotide triphosphates (dNTP) to synthesize a new M13 DNA strand complementary to the original at every base, except for those alterations introduced in the primer. The sugar–phosphate backbone of the new DNA circle is then completed using DNA ligase and the double-stranded DNA is transformed into *E. coli* cells. In *E. coli*, either the newly synthesized mutant DNA strand or the original wild-type DNA strand will be used to create new M13 phages

- The DNA that is to be mutated needs to be cloned into the M13 genome.

- The efficiency of the mutagenesis procedure itself is quite low. In order to create a newly synthesized mutant DNA strand, efficient oligonucleotide binding, DNA replication and DNA ligation are required. Each of these procedures is likely to be less than 100 per cent efficient, and therefore wild-type DNA strands will predominate in the mixture that is transformed into bacteria.

- The newly synthesized DNA will not be methylated as it is produced *in vitro*, while the wild-type M13 genome, isolated from bacterial cultures, will be methylated. This is important because the mismatch repair systems of the *E. coli* favour the repair of non-methylated DNA (Kramer and Fritz, 1984). This will result in the mismatches between wild-type and the mutant DNA strands being repaired in favour of a return to the wild-type sequence.

- The differential screening procedure to identify mutant phages is both slow and cumbersome and often results in the isolation of wild-type rather than mutant phage.

All of these factors result in a mutation efficiency that is low. Typically, mutagenesis frequencies of less than 10 per cent might be obtained for primer extension reactions like those described above. A consequence of the low mutation frequency is that a large number of potential mutants need to be screened to ensure that at least one genuine mutant can be isolated (Figure 7.2). Procedures

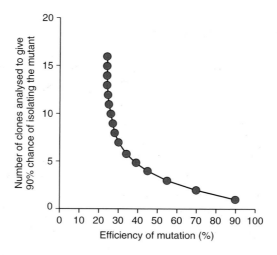

Figure 7.2. Efficient mutagenesis procedures greatly reduced the number of clones that must be screened to a single specific mutant

in which the mutation frequency approaches 100 per cent would mean that only a single bacteriophage plaque (or possibly two) would need to be analysed to ensure that a mutant can be isolated. Lower mutation frequencies result in a greater number of phages that need to be analysed before a mutant is likely to be found, and has consequences for the speed at which specific mutations may be isolated. Methods have been devised to increase the overall efficiency of a mutagenesis experiment by either increasing the efficiency of the mutagenesis reaction itself, or by the use of bacterial strains that are less likely to degrade the newly formed mutant DNA strands. For example, *E. coli* cells that are defective in the *mut*L, *mut*S, *mut*H mis-match repair system can be used for the transformation of the hybrid DNA molecules so that the mutation cannot be repaired back to the wild-type sequence.

7.2 Strand Selection Methods

An extremely effective approach to increasing the mutagenesis efficiency is to devise procedures to select either for the mutant DNA strand, or against the wild-type DNA strand. Here, we will only describe two methods for the latter that remain in use today. In the first the incorporation of nucleotide analogues protects the newly synthesized mutant DNA strand from degradation *in vitro* (Taylor, Ott and Eckstein, 1985), while in the second the wild-type DNA strand is targeted for degradation within *E. coli* cells (Kunkel, 1985).

7.2.1 Phosphorothioate Strand Selection

A phosphorothioate nucleotide contains a phosphorus–sulphur linkage in place of a phosphorus–oxygen group (Figure 7.3). If phosphorothioate deoxynucleotides in which the sulphur is attached to the α-phosphate are used in a DNA synthesis reaction, then the phosphorothioate will be incorporated into the newly synthesized DNA. Certain restriction enzymes are unable to cleave DNA that contains phosphorothioates (Nakamaye and Eckstein, 1986). The mutagenic oligonucleotide is annealed to the single-stranded M13 DNA template as described above, but is extended by DNA polymerase in the presence of three deoxynucleotide triphosphates (dATP, dTTP and dGTP) and a single phosphorothioate nucleotide (dCTPαS). This will result in the formation of the newly synthesized mutant DNA strand, but not the wild-type strand, containing a phosphorothioate at every C residue. The cleavage of the DNA duplex with, for example, the restriction enzyme PstI will result in the nicking of the wild-type DNA strand (no phosphorothioate) but no cleavage of the mutant DNA

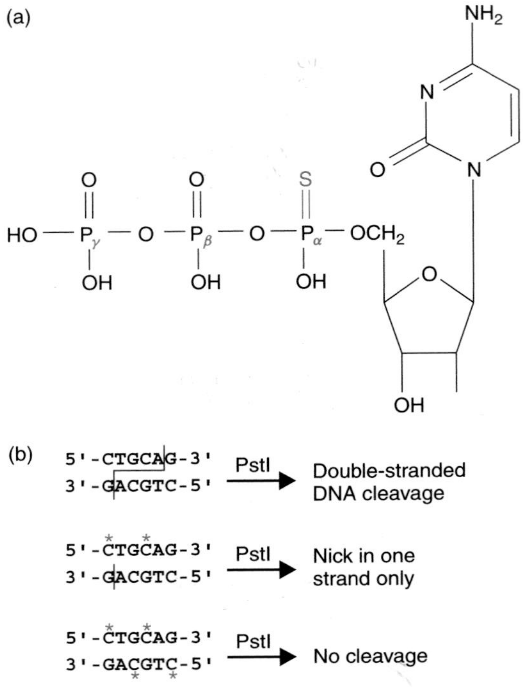

(a)

(b)

```
5'-CTGCA G-3'     PstI      Double-stranded
3'-G ACGTC-5'    ───────▶    DNA cleavage

  *    *
5'-CTGCAG-3'      PstI      Nick in one
3'-G ACGTC-5'    ───────▶    strand only

  *    *
5'-CTGCAG-3'      PstI      No cleavage
3'-GACGTC-5'     ───────▶
     *    *
```

Figure 7.3. DNA containing phosphorothioate linkages are resistant to cleavage by certain restriction enzymes. (a) The chemical structure of dCTPαS. A sulphur replaces an oxygen on the α-phosphate of the nucleotide (shown in red). (b) The effect of phosphorothioate nucleotides on DNA cleavage by the restriction enzyme PstI. PstI cleaves the sequence shown to generate four base overhanging sticky ends. If the sequence contains a phosphorothioate (indicated by the asterisk) at C residues of the recognition sequence in one strand, then the enzyme will nick the other stand only. If both strands contain a phosphorothioate, the enzyme is unable to cleave either strand

strand. The nicked DNA can be removed by treatment with exonuclease III, an enzyme that degrades DNA from its ends. The exonuclease will not degrade the mutant DNA circles since they do not have free ends. The resulting DNA is enriched for mutant circles and is transformed into E. coli. The mutation efficiency using this method can be as high as 40–60 per cent.

7.2.2   dut⁻ ung⁻ (or Kunkel) Strand Selection

In this method, the non-mutated wild-type strand is targeted for degradation. The DNA template for the mutagenesis reaction is obtained from phage grown in E. coli cells that contain mutations in both the *dut* and *ung* genes (Figure 7.4).

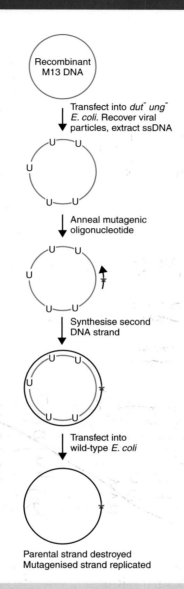

Figure 7.4. The *dut⁻ ung⁻* strand selection method (Kunkel, 1985) to degrade non-mutant sequences during site directed mutagenesis. See the text for details

- The *dut* gene encodes dUTPase whose function is to degrade dUTP within the cell. The *dut* mutation results in an elevated concentration of dUTP accumulating within the cell and results in incorporation of uracil (U) in place thymine (T) at some positions during DNA replication.

- The *ung* gene encodes uracil N-glycosylase, which normally removes uracil from DNA.

Thus, in a double mutant (dut^-, ung^-) uracil is incorporated into DNA and this error is not repaired. U residues have the same base pairing properties as T, so the incorporation of U into DNA in place of T is not mutagenic in itself. M13 phage DNA isolated from a dut^- ung^- E. coli strain will contain approximately 20–30 U residues per ~8000 bases of its genome. The isolated single-stranded M13 DNA is used as a template for the annealing of mutagenic oligonucleotide. The extension reaction is preformed as described above to generate a newly synthesized mutant DNA strand that does not contain uracil residues. Again, after synthesis of the second DNA strand is complete, the ends are covalently joined using DNA ligase. The resulting double-stranded DNA consists of the wild-type strand that contains uracil residues, and the newly synthesized strand that contains the mutant bases present in the oligonucleotide but no uracil residues. The double-stranded DNA is then transformed into ung^+ wild-type E. coli cells, where the uracil N-glycosylase recognizes the uracil residues in the DNA, and excises the uracil base to leave apyrimidinic sites in the template strand. The presence of apyrimidinic sites makes the DNA strand biologically inactive because it cannot be replicated by DNA polymerase. The DNA is cut at the apyrimidinic sites by a specific nuclease (endonuclease IV) within the E. coli cell and degraded. Hence, when the double-stranded DNA is introduced into the ung^+ E. coli, only the mutant strand will be replicated. Using this approach, mutation efficiencies approaching 100 per cent can be obtained.

Although high DNA mutation efficiencies can be obtained using the methods described above, all of these techniques rely on the initial cloning of the gene to be mutated into an M13 phage so that single-stranded DNA can be isolated. The widespread use of double-stranded plasmid DNA as vectors makes the cloning of DNA fragments into an M13 mutagenesis vector, and the subsequent re-cloning of the mutated DNA back into the original plasmid, a time-consuming process.

7.3 Cassette Mutagenesis

Cassette mutagenesis relies on the presence of two restriction enzyme recognition sites flanking the DNA that is to be mutated (Figure 7.5). The plasmid is cut with the enzymes and the large DNA fragment representing the majority of the plasmid is purified from the smaller fragment. The linear plasmid DNA is then ligated to a synthetic double-stranded DNA produced through the annealing of two complementary oligonucleotides (Figure 7.5). The complementary oligonucleotides contain the desired mutation(s) and the required overhanging sequences for the ligation to the restriction enzyme cleavage sites (Wells, Vasser and Powers, 1985). This procedure is highly efficient at producing

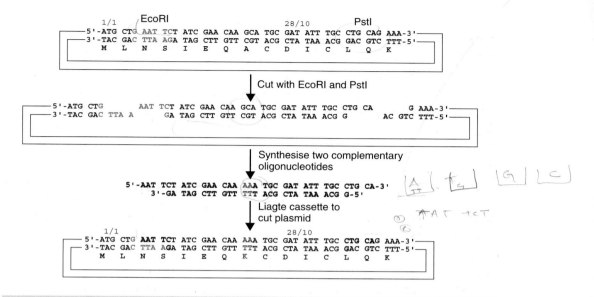

Figure 7.5. Cassette mutagenesis. The DNA flanked by two restriction enzyme recognition sites is cleaved from a plasmid DNA and replaced with a double-stranded oligonucleotide cassette that contains the required mutation. Here, an alanine (A) codon in the protein coding sequence between the EcoRI and PstI cleavage sites is altered to a lysine (K) codon. The single-stranded oligonucleotides are synthesized such that, when hybridized, they will form the required sticky ends for ligation to the cut plasmid DNA

mutations, provided that the small wild-type DNA fragment can be eliminated. The technique does, however, suffer from a major drawback in that it requires two restriction enzyme recognition sequences to flank the DNA that is to be mutated (Worrall, 1994). Oligonucleotides are difficult to synthesize accurately above about 70 nucleotides in length. Although this is not the limit at which that the restriction sites can be separated, since multiple overlapping oligonucleotides can be synthesized to produce larger sequence, it does present a barrier to the physical size of a cassette that can be produced effectively. We will return to cassette mutagenesis again when we look at the production of random mutations in specific genes.

7.4 PCR Based Mutagenesis

We have already seen how, by suitable design of oligonucleotide primers, mutations can be introduced into the ends of PCR products in a way that leads to mutagenesis efficiency of almost 100 per cent (Chapter 4). This method (see Figure 4.7), in which mutations are introduced within the PCR primers themselves, is an immensely powerful tool for introducing DNA alterations into the ends of linear DNA fragments, but is limited to those ends. PCR protocols have, however, been developed to enable the creation of mutation at any point throughout the length of the PCR product (Higuchi, Krummel and Saiki, 1988). This method, often referred to as two-step PCR mutagenesis, requires four oligonucleotide primers and three separate PCR reactions and is outlined in Figure 7.6. Two of the primers (1 and 4) are designed to be complementary

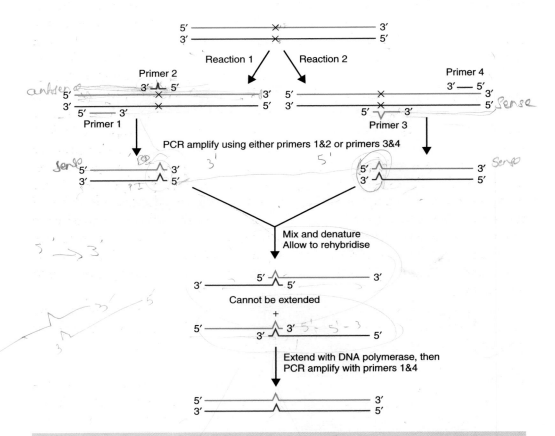

Figure 7.6. Two-step PCR to introduce mutations into the middle of an amplified DNA fragment. Overlapping primers (primer 2 and primer 3) are designed to introduce a mutation onto a newly synthesized antisense or sense strand, respectively. These primers are used in separate PCR experiments to amplify the required DNA fragment in two sections – the DNA containing the mutation and sequences at the 5'-end of the sense strand and, separately, the DNA containing the mutation and sequences at the 3'-end of the sense strand. The two PCR products formed by this process can anneal to each other through their complementary sequences, as dictated by the position of primers 2 and 3. The mixed DNA strands can then be amplified using primers 1 and 4 to generate the intact DNA fragment that now contains the mutation

to the anti-sense strand and the sense strand of the target DNA, respectively. The other two primers (2 and 3) are designed to bind to the different strands of the same DNA sequence and will also introduce the required mutation into each strand. In the first PCR, the 5'-end of the gene is amplified using primers 1 and 2. The resulting product will bear the mutation at its 3'-end. In the second PCR, the 3'-end of the gene is amplified using primers 3 and 4 so that the resulting product will bear the mutation at its 5'-end. Primers 2 and 3 are designed such that they are complementary to each other and overlap with one another. This means that the 3'-end of PCR product 1 will be identical to the 5'-end of PCR product 2. Therefore, if PCR products 1 and 2 are mixed with each other, denatured and allowed to cool, then the individual strands from each reaction can hybridize with each other. Two possible hybrid molecules can form. If the sense strand of PCR 2 binds to the antisense strand of PCR 1, then a molecule is produced that cannot be extended using DNA polymerase (the 3'-ends are not base paired). However, if the sense strand of PCR 1 binds to the antisense strand of PCR 2, then DNA polymerase can produce a double-stranded version of the gene that contains the mutation. In practice, the products of PCR 1 and PCR 2 are mixed in the presence of primers 1 and 4 so that the full-length mutant gene is amplified to yield large quantities of the mutant DNA. In Figure 7.6, primers 1 and 4 are shown as not introducing mutations into the gene. It is often the case, however, that these primers are used to introduce restriction enzyme recognition sites into the PCR product (as illustrated in Figure 4.7) so that the mutant gene may be readily cloned into a plasmid.

The mutagenesis efficiency of PCR methods is very high. The amplification steps ensure that practically no wild-type DNA will be present in the final product. There are, however, several drawbacks to the method.

- The error-prone nature of certain thermostable DNA polymerases means that other, unwanted, mutations may be introduced into the mutant product. It is therefore essential that the entire PCR product be sequenced (see Chapter 9) to ensure that the correct mutation has been made while others have not.

- Large DNA fragments are difficult to amplify using PCR. This may limit the size of the final amplified product that can be successfully produced.

The ability to introduce mutations at will within a segment of DNA has allowed many exceptionally elegant and precise gene analyses to be performed that would have not been previously been possible. Additionally, the initial treatment of the mutagenesis reaction as two separate components means that the technique can readily be adapted to the creation of chimeric DNA sequences. There are a variety of reasons why particular DNA sequences may need to be joined together.

- Deletion analysis – the removal of certain gene sequences can be viewed as the fusion of the remaining sequences to each other.

- Changing the promoter of a gene to express it differently – as we will see in Chapter 8, many genes are expressed in foreign host cells to maximize protein production. So that foreign genes are expressed, they invariably need to be placed under the control of a host promoter sequence.

- Construction of novel genes – new proteins may be produced by the fusion of DNA sequences encoding portions of different genes; e.g. proteins may be tagged with certain sequences to allow their simple purification, or to direct them to certain cellular locations by the addition of signal sequences.

Traditional cloning methods to fuse DNA sequences together rely on the presence of restriction enzyme recognition sites to allow the insertion of foreign DNA. This limits the types of fusion that can be produced and the level of precision to which a particular fusion can be made. Using a PCR based approach can, however, completely alleviate these problems. Suitable PCR primer design, as illustrated in Figure 7.6, will lead to the precise fusion of any two DNA sequences if the ends of the primers contain overlapping sequences. That is, primers (2 and 3 in Figure 7.6) are designed such that they overlap with each other and contain the final fusion junction within their sequence. Thus, the joining of the two initial PCR products together will result in the formation of a precise fusion as dictated by the sequence of the primers in the first reactions.

A specific example, taken from my own work (Reece and Ptashne, 1993), where two-step PCR has been used to great effect to create gene fusions, is shown in Figure 7.7. The yeast *Saccharomyces cerevisiae* contains a large family of sequence related transcription factors called the C_6 zinc cluster proteins. Like the majority of eukaryotic transcription factors, these proteins each have a separate DNA binding domain and an activation domain – whose function is to target the protein to particular genes, and recruit RNA polymerase II, respectively. The DNA binding domains of three of these proteins, Gal4p, Put3p and Ppr1p, is located at the amino-terminal end of each protein, and each contains six highly conserved cysteine residues that chelate two zinc ions to form the functional DNA binding domain (DBD). Each protein binds as a homodimer to its respective DNA binding site. The DNA binding sites of these three proteins are also related to each other (Figure 7.7(b)). That is, each site

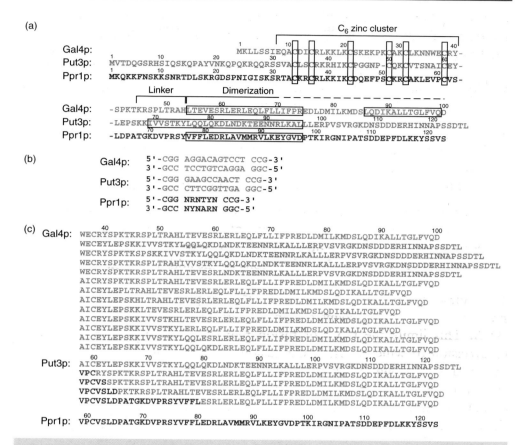

Figure 7.7. The DNA binding domains of three C₆ zinc cluster proteins from yeast. (a) The amino acids sequences of the DNA binding domains of Gal4p, Put3p and Ppr1p. (b) The DNA binding sites for these three proteins are related in that they contain highly conserved inverted 5′-CGG-3′ triplets separated by a different number of base pairs for each protein. (c) Fusion proteins produced by creating chimeric genes between *GAL4*, *PUT3* and *PPR1*. The sequence of each protein is shown from just before the last cysteine residue of the C₆ zinc cluster

contains highly conserved inverted 5′-CGG-3′ triplet nucleotides separated by a more variable spacer. The length of the spacer is, however, different for each protein binding site. In the Gal4p binding site, the CGG triplets are separated by 11 bp, while they are separated by 10 bp in the Put3p binding site and 6 bp for Ppr1p. Since each of these related proteins is able to bind to a related, but different, DNA sequence we wanted to establish precisely which region of each protein is responsible for imparting DNA binding specificity. We therefore wanted to alter the DNA binding specificity of one C₆ zinc cluster protein by replacing some of its own amino acids with the corresponding sequences from a different C_6 zinc cluster protein. In the absence of structural information we had no knowledge of where fusion junctions needed to be made that would allow functional protein production, so we created a large series of fusion proteins from DNA sequences encoding one of the C_6 zinc cluster proteins that were fused to those from another (Figure 7.7(c)). Each fusion was created by two-step PCR. The location of the fusion junction was determined solely by the sequence of the oligonucleotides (2 and 3 in Figure 7.6). This allowed chimeric genes to be constructed to produce, as required, proteins in which the fusion junction could be moved amino acid by amino acid. No restriction enzyme recognition sites were required for cloning, the oligonucleotides themselves provided the overlap so that the genes could be fused. Each chimeric gene was then cloned into an *E. coli* expression vector and the resulting protein was purified and tested for its ability to bind to each of the DNA binding sites.

Each of the wild-type DBD proteins was found to bind to DNA with a particular specificity. Put3p and Ppr1p were found to bind to their own DNA binding site only, while Gal4p bound with high efficiency to its own site and, with approximately tenfold less efficiency, also to the Put3p binding site. Analysis of the chimeras showed that replacing the zinc cluster region (Figure 7.7(a)) of one protein with that of another had no effect on DNA binding specificity. For example, protein 4 in Figure 7.8 has the zinc cluster region of Put3p fused to the carboxy-terminal regions of Gal4p. This protein binds DNA with the specificity of Gal4p rather than Put3p. A similar type of result is noted for proteins 2 and 7. We therefore wanted to know how much sequence to the carboxy-terminal side of the zinc cluster was required to switch the DNA binding specificity to that of the zinc cluster itself. The majority of the fusions produced (Figure 7.7(c)) were unable to bind any DNA sequence, presumably due to the formation of mis-folded or incorrectly aligned protein structures. However, proteins containing the zinc cluster, and an additional 19 amino acids to its carboxy-terminal side, bound DNA with the specificity of the protein from which these sequences were derived (Figure 7.8). Thus, these 19 amino acids to the carboxy-terminal side of the zinc cluster were responsible for determining DNA binding specificity. Subsequent to this work, the structures of the Gal4p–, Put3p– and Ppr1p–DNA complexes have been solved using X-ray crystallography (Marmorstein *et al.*, 1992; Marmorstein and Harrison, 1994; Swaminathan *et al.*, 1997) and are shown in Figure 7.7(c)–(e). Each protein–DNA complex shows the same overall format. Each protein is dimeric and makes specific contacts with the DNA at the 5′-CGG-3′ using the zinc cluster. The zinc cluster forms a compact sub-domain in which two zinc ions (yellow spheres) are bound. Extending away from the DNA is a coiled coil dimerization motif. Joining the zinc cluster to the coiled-coil is a linker region.

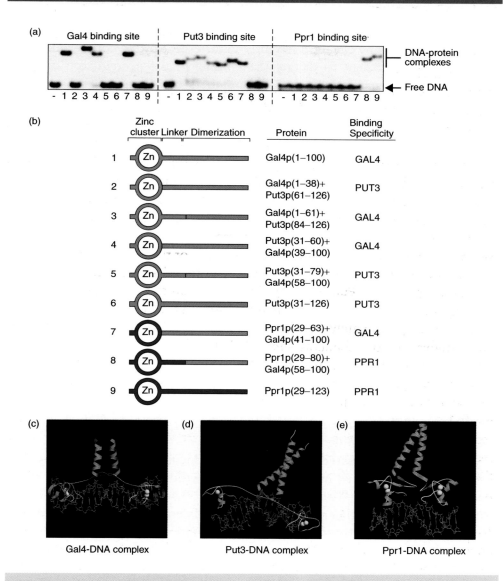

Figure 7.8. DNA binding activity of chimeric Gal4p, Put3p and Ppr1p. (a) Radio-labelled versions of the DNA binding sites for each protein were incubated with purified protein and then subjected to non-denaturing gel electrophoresis. The binding of the protein to the DNA retards its mobility through the gel such that it runs less far through the gel. (b) A summary of the DNA binding activity of each of the functional chimeric proteins. (c)–(e) The structure, as determined by X-ray crystallography, of the Gal4p-DNA, Put3p-DNA and Ppr1p-DNA complexes (Marmorstein *et al.*, 1992; Marmorstein and Harrison, 1994; Swaminathan *et al.*, 1997). In each case the DNA is depicted as a red stick model and the protein is shown as a ribbon diagram (α-helices in purple, β-sheet in blue and other polypeptide chain in white). In each model, yellow spheres represent the locations of the zinc ions

This linker, 19 amino acids in length, forms different structures in each protein that positions the zinc cluster differently with respect to the coiled coil. These differences are sufficient to dictate the binding of the protein to 5'-CGG-3' triplets that are separated by different numbers of base pairs. This kind of precise gene analysis would simply not have been possible using traditional cloning methods.

The power and speed of using PCR techniques to produce mutant DNA molecules, either gene fusions as described above, or point mutants, is unquestionable. Mutagenesis can be performed very quickly. With the availability of suitable oligonucleotides, the two-step PCR strategy can be performed in 3–4 h. The limiting step in the process is the cloning of the mutant linear PCR products into plasmids such that functional analysis may be performed. Procedures in which the double-stranded plasmid DNA can be mutated without the need for additional cloning steps are therefore required.

7.5   QuikChange® Mutagenesis

The PCR based mutagenesis procedures described above require that the linear mutant DNA fragments produced are cloned into plasmid DNAs so that they can be propagated and analysed functionally. A method using the power of PCR to introduce mutations directly into plasmid DNA would alleviate the need for additional cloning steps.

One popular PCR based method for introducing mutations directly into plasmid DNA is outlined in Figure 7.9. This method, often referred to as the QuikChange® method (Wang and Malcolm, 1999), utilizes two oligonucleotide primers. One of the primers is produced so it is complementary to the sense strand of the gene and contains the desired mutation, whilst the other primer is designed to be complementary to the anti-sense strand of the gene, but also contains the mutation. Double-stranded plasmid DNA is used as a PCR template. The plasmid DNA is heated during the course of a normal PCR reaction such that the individual strands become separated (denatured). Cooling the denatured DNA in the presence of the oligonucleotides results in their binding to complementary sequences within the plasmid. Thermocycling is then continued to extend the oligonucleotides to create newly synthesized mutant plasmid DNA. After the PCR reaction is completed, newly synthesized DNA (containing the mutation) comprises two complementary linear DNA

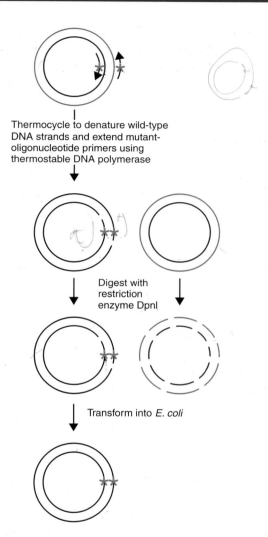

Figure 7.9. QuikChange® mutagenesis. The plasmid to be mutated is mixed with two complementary overlapping oligonucleotide primers, each of which encodes the required mutation. The primers are extended in a PCR reaction to synthesize both plasmid DNA strands, each of which contains the mutation. The DNA is then digested with the restriction enzyme DpnI, which can only cleave methylated DNA. If the parental plasmid was isolated from a *Dam*⁺ *E. coli* strain then the parental DNA strands, but not the unmethylated newly synthesized mutant DNA strands, will be cleaved by the enzyme. Subsequent transformation into *E. coli* will result in the degradation of the cut parental DNA fragments and the repair of the nicks in the newly synthesized DNA. The newly synthesized mutant DNA will then be replicated

molecules that are able to form a double-stranded circle containing staggered DNA nicks (Figure 7.9). The PCR products are then digested with the restriction enzyme DpnI, which can only cleave methylated DNA:

DpnI:

$$
\begin{array}{c}
CH_3 \\
\uparrow \\
5'\text{-}GA\,|\,TC\text{-}3' \\
3'\text{-}CT\,|\,AG\text{-}5' \\
\Downarrow \\
CH_3
\end{array}
$$

The newly synthesized DNA will not be methylated, and consequently will not be cleaved by the restriction enzyme. If the non-mutant parental plasmid DNA, on the other hand, was isolated from an *E. coli* strain that contains the *Dam* methylase (see Chapter 2), then DpnI will cleave at its recognition sequences. Most common laboratory *E. coli* strains are *dam*+, so this method of degradation of the wild-type DNA is applicable to the majority of plasmids available in the laboratory. Transformation of the restriction enzyme products into *E. coli* cells will result in the degradation of the wild-type DNA fragments and the repair of the nicks in the newly synthesized mutant DNA circles, which will then be propagated. This procedure is very rapid (3–4 h) and is highly efficient (~80 per cent) at producing mutant DNA plasmids without the need for additional cloning steps.

7.6 Creating Random Mutations in Specific Genes

The creation of specific directed mutants within genes using oligonucleotides has revolutionized our understanding of protein function. The examples we have discussed so far have, however, been limited to the alteration of specific bases within a gene to other defined bases. This will result in the formation of mutant protein with defined amino acid changes if the alterations are within the coding sequence of the gene. It is not always possible to know which amino acids of a protein should be altered, or what they should be altered to. Some systematic approaches to this problem involve the change of each amino acid coding triplet within a gene to an alanine codon (Cunningham and Wells, 1989). This **alanine scanning mutagenesis** can identify amino acid side chains that are important for protein function with the premise that the presence of alanine will not perturb the overall structure of the protein and will only eliminate amino acid side chain interactions. This type of approach requires that a screen is available for identifying protein function and is especially applicable to small proteins or protein domains owing to the number of individual mutations that must be constructed. An alternative approach is to convert sets of charged amino acid residues that occur consecutively within a linear polypeptide sequence to alanine (Bass, Mulkerrin and Wells, 1991). This **charged to alanine scanning mutagenesis** is based on the observation that most proteins contain a hydrophobic core with charged residues on the outside surface of the protein. Consequently, clusters of charged amino acids in a linear protein sequence are likely located on the surface of the protein and may therefore participate in, for example, protein–protein interactions. Mutation of these charged clusters are more likely to disrupt these protein–protein interactions than mutagenesis of other residues.

Two approaches are commonly used for the creation of random mutations within individual genes, or parts of genes. Again, these methods rely on a screen to analyse mutants with an appropriate phenotype, but do not suffer from limiting mutations types to individual residues or from the types of alteration that can be made.

- *Doped cassette mutagenesis.* An experiment like that already discussed in Figure 7.5 is performed except that a library of oligonucleotides is ligated into the cut plasmid (Figure 7.10). Like conventional cassette mutagenesis, the DNA between two restriction enzyme recognition sites is removed from a plasmid and replaced using a pair of synthetic oligonucleotides. Here, however, the oligonucleotides do not encode a unique sequence. Instead, libraries of oligonucleotides are produced that are based on the same sequence, but contain certain random changes from that sequence. Such **doped oligonucleotides** are synthesized (Figure 4.6) using a mixture of bases. For example, if the next base to be added to an extending oligonucleotide were an A, then rather than chemically adding only the A precursor to the growing oligonucleotide chain a mixture of A and a small quantity of the other nucleotide precursors would be added. Such mixtures might commonly contain 95 per cent of the wild-type nucleotide and 1.7 per cent of each of the other nucleotides. The level of 'doping' gives some control over the level of mutagenesis that will be obtained. In the example shown in Figure 7.10, the sequence between the EcoRI and PstI restriction sites is to be altered. An oligonucleotide is constructed that contains invariant EcoRI and PstI restriction sites that are absolutely required for cloning of the DNA back into the plasmid. The sequences between these sites are doped at a level such that, on average, each oligonucleotide produced will contain a single variation from the wild-type sequence. Two example of oligonucleotides produced are shown in Figure 7.10. By choosing an appropriate level

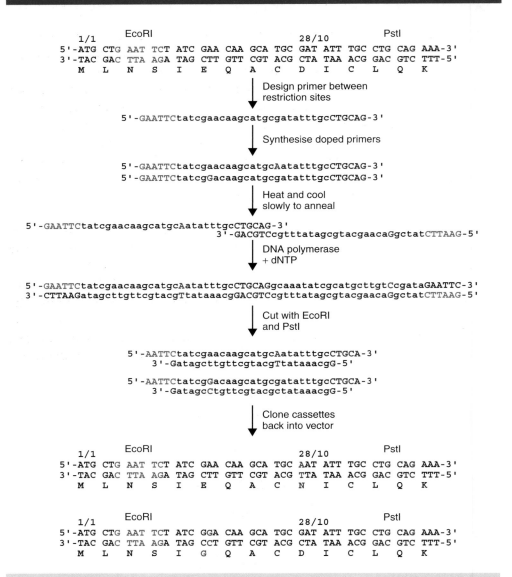

Figure 7.10. Doped cassettes for the introduction of random mutations within a defined segment of a gene. See the text for details

of doping, each nucleotide within this region can be altered to every other nucleotide, but with only one change occurring per oligonucleotide. Oligonucleotides produced in this way are single stranded and therefore cannot be cloned directly into the cut double-stranded plasmid. The cassette can be made double stranded in one of two ways. Either a complementary doped oligonucleotide is synthesized and annealed to form two DNA strands, or the palindromic nature of the restriction enzyme recognition sites at the end of the oligonucleotide are used to create dimeric molecules that can then be cut with the restriction enzymes and cloned into the plasmid. The first method relies complementary mutations existing with the two complementary DNA strands, and suffers from that mutant mis-matches between the oligonucleotide pairs may be tolerated during the formation of double-stranded DNA to increase the number of mutations found in each cassette. The second method (as shown in Figure 7.10) is more desirable since individual mutations within the single-stranded oligonucleotide library will be retained in the double-stranded form.

- *Error-prone PCR.* We have already discussed the error-prone nature of certain DNA polymerases that are used in PCR (Chapter 4). In particular, the lack of a 3′–5′ exonuclease proofreading activity in *Taq* DNA polymerase means that significant mutations may be introduced into PCR products simply as a consequence of the PCR itself (Keohavong and Thilly, 1989). The advantage of this method for introducing random mutations (Figure 7.11) is that only a PCR reaction need be performed. The PCR product can then be cloned and analysed functionally. The error rate of *Taq* DNA polymerase

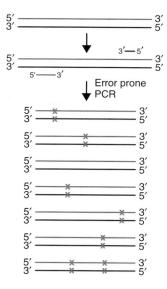

Figure 7.11. Error-prone PCR as a method for mutating a gene. The error-prone nature of certain DNA polymerases using during PCR will result in the creation of mutations within the amplified DNA. The PCR experimental conditions can be altered to increase the error rate so that, on average, each amplified double-stranded product contains a single mutation

may be increased, to increase the mutation frequency obtained, by altering a variety of the PCR reaction conditions. For example, increasing the magnesium concentration in the reaction or adding manganese ions to the reaction will increase the error rate of the polymerase (Lin-Goerke, Robbins and Burczak, 1997). Additionally, changes in the reaction deoxynucleotide concentration, the concentration of the polymerase itself or the length of the extension step of the reaction can each result in an elevated error rate. The ease at which PCR based random mutagenesis can be preformed has made it a popular choice. The main drawback of the technique is the reliance on an enzyme to create random mutations. DNA polymerases have preferences in the mistakes they make. In the case of *Taq* DNA polymerase, transitions are favoured over transversions (Keohavong *et al.*, 1993), so some mutations are difficult to obtain.

7.7   Protein Engineering

Protein engineering can be thought of as the deliberate modification of the sequence of a protein (through the alteration of the DNA sequence encoding it) to impart the protein with a new or novel function. This approach has been used for the creation of enzymes with altered characteristics that may be desirable for particular purposes. The sorts of enzyme characteristics that may be altered include.

- thermal stability

- pH stability

- kinetic properties

- stability in organic solvents

- altered cofactor requirement

- altered substrate binding specificity

- resistance to proteases

- changed allosteric regulation.

Protein engineering has been used to alter the thermal stability of lysozyme in a directed way (Matsumura, Signor and Matthews, 1989). The rationale behind these experiments was that disulphide bonds formed between two cysteine amino acid residues within a protein should be able to lock the protein structure into a conformation that is resistant to heat denaturation. The gene encoding lysozyme from the bacteriophage T4, a disulphide-free enzyme, was engineered by the introduction of cysteine codons in its sequence such that in the resulting protein disulphide bonds were formed to crosslink residues 3–97, 9–164 and 21–142. The mutant protein denatured at 66 °C, compared with 42 °C for its wild-type counterpart (Matsumura, Signor and Matthews, 1989).

Protein engineering can also be used to change the specificity of an enzyme such that it is able to catalyse the reaction of alternative substrates. For example, a single point mutation in the yeast alcohol dehydrogenase I gene, converting aspartic acid 233 to glycine, results in the production of a protein that, rather than solely using NAD^+ as a cofactor for the is reduction of acetaldehyde to ethanol, can using both NAD^+ and $NADP^+$ (Fan, Lorenzen and Plapp, 1991). In a more extreme example, the lactate dehydrogenase from the bacterium *Bacillus stearothermophilus* has been converted, through the mutation of three active site amino acids, into a highly active malate dehydrogenase (Wilks *et al.*, 1988). In both of these cases, the alterations were made in the light of high-resolution structures of the respective proteins and converted the natural enzyme into one with only a slightly altered function. A more difficult problem is to design proteins that have entirely novel functions. Some inroads into this have been achieved by using **directed evolution** – a method in which multiple rounds of random mutagenesis beginning with a gene encoding a known protein function are combined with selection processes to produce a protein with a specific, and new, function. For example, Olsen *et al.* used phage display (Chapter 6) and random mutagenesis to isolate proteases with novel substrate specificity (Olsen *et al.*, 2000). This approach is especially successful at generating altered protein characteristics rather than entirely novel proteins. For example, Williams *et al.* used directed evolution to alter the stereochemical course of a reaction catalysed by tagatose-1,6-bisphosphate aldolase (Williams *et al.*, 2003). After three rounds of mutagenesis and screening, an evolved aldolase was produced, which showed a 100-fold change in stereospecificity toward the non-natural substrate fructose 1,6-bisphosphate. The altered enzyme contains four specific single amino acid changes when compared with the original tagatose-1,6-bisphosphate aldolase, and the changes are spread through the length of the polypeptide. Each of the changes does, however, alter the active site of the protein when it is folded into its three-dimensional form.

8 Protein production and purification

Key concepts

♦ Proteins are over-produced by placing the gene encoding them under the control of a strong promoter

♦ Strong, inducible promoters allow the production of toxic proteins and for proteins to be made under defined conditions

♦ Proteins may be produced in bacterial or eukaryotic cells

♦ DNA encoding a protein purification tag is often added to the expressed gene to aid in the protein purification process

♦ Protein purification tags impart a unique property to the over-produced protein such that it may be purified biochemically

The production and purification of proteins for biochemical and structural analysis have formed the lynchpin of many advances in genetic engineering, drug discovery and medicinal chemistry over recent years. Some proteins are naturally expressed at high levels. For example, actin and certain heat-shock proteins can accumulate at high levels within cells. Many other, potentially biologically important, proteins are expressed at very low levels. For example, many transcription factors involved in turning sets of genes on and off are present at only a few copies per cell. To aid the study of proteins that are produced at a low level, the gene encoding them generally has to be over-expressed. The most straightforward way to achieve this is to fuse the target gene to a strong promoter. The strong promoter, usually derived from a highly expressed gene, will drive the expression of any gene placed under its control through the recruitment of RNA polymerase to that gene.

Much work has gone into the design of vectors for maximizing protein production. The architecture of a typical expression vector is shown in Figure 8.1.

Analysis of Genes and Genomes   Richard J. Reece
© 2004 John Wiley & Sons, Ltd   ISBNs: 0-470-84379-9 (HB); 0-470-84380-2 (PB)

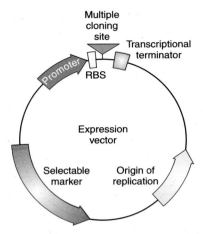

Figure 8.1. The architecture of an expression vector. An expression vector should contain a strong inducible promoter, a multiple cloning site for the insertion of target genes, and a transcriptional terminator. Additionally, a ribosome binding site (RBS) is included to promote efficient translation

Such vectors will often contain a multiple cloning site located between a strong transcriptional promoter and terminator sequence. Additionally, the expression vector, like other plasmids, will contain an origin of replication and a selectable marker such that the vector may be autonomously replicated and maintained within cells.

At high levels, many proteins will be toxic to the host cell in which they are produced. Indeed, some proteins when produced in small amounts will also be toxic to the host. For example, the expression of the poliovirus 3AB gene product is highly toxic to *E. coli* cells, due to the drastic changes it creates in the membrane permeability of the bacteria (Lama and Carrasco, 1992). Therefore, to maximize protein expression it is vital that an inducible expression system be established, so large quantities of the host cells can be grown before the expression of the target protein is initiated. Protein production can then be activated rapidly and the cells harvested soon afterwards prior to the potentially toxic effects of the expressed protein. Here, we will discuss a number of inducible expression systems that are in common use today. Additionally, we will describe the common host–vector systems that are used for protein production in *E. coli*, yeast, insect and mammalian cells.

8.1   Expression in *E. coli*

E. coli remains the host cell of choice for the majority of protein expression experiments. Its rapid doubling time (approximately 30 min) in simple defined (and inexpensive) media, combined with an extensive knowledge of its promoter and terminator sequences, means that many proteins of both prokaryotic and eukaryotic origin can be produced within the organism. Additionally, *E. coli* cells are easily broken for the harvesting of the proteins produced within the cell. Of course, *E. coli* does suffer from the fact that is a prokaryotic organism when it is used to produce eukaryotic proteins. *E. coli* cells are unable to process introns and do not possess the extensive post-translational machinery found in eukaryotic cells that can glycolylate, methylate, phosphorylate or alter the initially produced protein in other ways, such as through extensive disulphide bond formation. The use of cDNA to produce an expression vector overcomes the first of these problems, but if post-translational modifications to the protein are necessary for protein function, then an alternative host must be sought.

Many different promoter sequences have been used to illicit inducible protein production in *E. coli* (Makrides, 1996). Some of these are discussed below.

8.1.1 The lac Promoter

We have already seen (Figure 1.23) that the *E. coli lac* promoter provides a mechanism for inducible gene expression. The *lac* genes are expressed maximally when *E. coli* are grown on lactose. Fusing the *lac* promoter sequences to another gene will result in the lactose- (or IPTG-) dependent expression of that gene. The *lac* promoter suffers, however, from a number of problems that mean that it is rarely used to drive the expression of target genes. First, the *lac* promoter is fairly weak and therefore cannot drive very high levels of protein production, and second the *lac* genes are transcribed to a significant level in the absence of induction (Gronenborn, 1976). The latter problem can be partially overcome by expressing mutant versions of the *lac*I gene that have increased DNA binding (and consequently repressing) ability – for example the *lac*I^q allele results in the overproduction of *Lac*I and consequently results in a reduced level of transcription in the absence of inducer (Müller-Hill, Crapo and Gilbert, 1968).

8.1.2 The tac Promoter

The ease with which the *lac* promoter can be activated (the addition of IPTG to *E. coli* cultures) makes it an attractive system for producing target proteins. However, the relative weakness of the promoter means that the target gene will not be greatly over-produced. Through the analysis of many *E. coli* promoters, consensus sequences for the -35 and -10 regions, to which the RNA polymerase must bind to transcribe the gene, can be determined (Lisser and Margalit, 1993). The *lac* promoter is weak because the -35 region deviates from the consensus (Figure 8.2). The creation of a fusion sequence

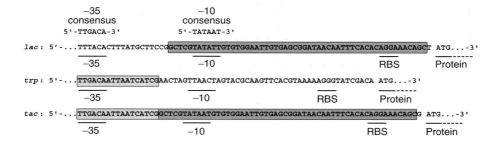

Figure 8.2. DNA sequences of the *lac, trp* and *tac* promoters. The consensus *E. coli* −35 and −10 sequences based on the analysis of naturally occurring promoters are shown above, and the sequences of each of the promoters, extending from the −35 region to the translational start site, are shown. The *tac* promoter is a hybrid of the *trp* and *lac* promoters. The −35 and −10 regions it contains closely resemble the consensus sequences. The *tac* promoter is approximately five times stronger than the *lac* promoter, but is still inducible by lactose or IPTG

containing the −35 region of the *E. coli trp* operon, and the −10 region of the *lac* operon, controlling the expression of the genes responsible for tryptophan biosynthesis and lactose metabolism, respectively, results in the formation of the *tac* promoter, which is five times as strong as the *lac* promoter itself (de Boer, Comstock and Vasser, 1983). The *tac* promoter is able to induce the expression of target genes such that the encoded polypeptide can accumulate at a level of 20–30 per cent of the total cell protein (Amann, Brosius and Ptashne, 1983). Expression vectors that carry the *tac* promoter also carry the *lacO* operator and usually the *lacI* gene encoding the Lac repressor (Stark, 1987). Genes cloned into these vectors are therefore IPTG inducible and can be repressed and induced in a variety of *E. coli* strains (Figure 8.3).

8.1.3 The λP_L Promoter

The λP_L promoter is responsible for transcription of the left-hand side of the λ genome, including *N* and *cIII* (see Chapter 3). λ repressor, the product of the *cI* gene, represses the promoter. Two basic methods are used to activate the λP_L promoter. In the first, a temperature-sensitive mutant of *cI* (*cI857*) is used in conjunction with λP_L for the expression of target genes (Hendrix, 1983). When grown at 30 °C the mutant cI protein is able to bind to the λP_L promoter and repress it. Above 30 °C, however, the mutant cI protein is unable to bind DNA, and the λP_L promoter is activated (Bernard *et al.*, 1979). This method produces high levels of target gene expression, but the heat pulse required to induce protein production can be difficult to control. A second way to induce

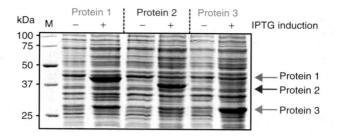

Figure 8.3. The production of three different proteins in *E. coli* whose expression is driven from the *tac* promoter. Bacterial cells, harbouring the appropriate *tac* based expression vector, were grown in liquid cultures and then IPTG was added, as indicated, to half of each culture. Growth was continued for an additional 90 min before the cells were harvested, broken open and subjected to SDS–polyacrylamide gel electrophoresis. The protein content of each culture was observed by staining the gel with Coomassie blue. The locations of the three proteins produced upon IPTG induction are indicated

the λP_L promoter is to transform the expression vector into an *E. coli* strain in which the *cI* gene has been placed under the control of the tightly regulated *trp* promoter. Expression of the target gene can then be induced by the addition of tryptophan to the growth media, which will prevent transcription of the *cI* gene, and consequently activate the strong λP_L promoter. This results in a system that is so tightly controlled that it can be used to express even highly toxic proteins (Wang, Deems and Dennis, 1997; Celis *et al.*, 1998).

8.1.4 The T7 Expression System

This is the RNA polymerase encoded by bacteriophage T7 is different from its *E. coli* counterpart. Unlike the $\alpha_2\beta_2$ subunit structure of the *E. coli* enzyme, T7 RNA polymerase is a single-subunit enzyme that binds to distinct DNA 17 bp promoter sequences (5′-TAATACGACTCACTATA-3′) found upstream of the T7 viral gene it activates. *E. coli* RNA polymerase does not recognize T7 promoter sequences as start sites for transcription. The overall scheme for the production of target proteins using the T7 system is shown in Figure 8.4 (Studier and Moffatt, 1986; Studier *et al.*, 1990). The target gene is cloned into a plasmid expression vector such that it is under the control of the T7 promoter. Propagation of this plasmid in wild-type *E. coli* cells will not result in the expression of the target gene since the T7 RNA polymerase is absent. To elicit target gene expression, the expression plasmid is transformed into an *E. coli* strain that contains a copy of T7 gene 1 that is under the control of the *lac* promoter. Such sequences can be transferred into most *E. coli* strains using a λ lysogen called DE3 that contains the T7 RNA polymerase gene under

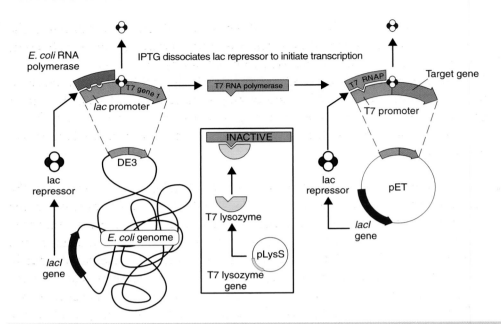

Figure 8.4. The T7 system for the expression of proteins in *E. coli*. The expression vector (pET) contains the target gene under the control of the T7 RNA polymerase promoter. The vector is transformed into an *E. coli* strain that contains, integrated into its genome, a copy of the gene for T7 RNA polymerase (T7 gene 1) under the control of the *lac* promoter. Additionally, the promoters for both the target gene and T7 gene 1 also contain the *lacO* operator sequence and are therefore inhibited by the *lac* repressor (*lacI*). IPTG induction allows the transcription of the T7 RNA polymerase gene whose protein product subsequently activates the expression of the target gene. The presence of an additional plasmid in the *E. coli* cell producing T7 lysozyme inactivates any T7 RNA polymerase that may be produced in the absence of induction. After induction sufficient T7 RNA polymerase is produced to escape this regulation. Reprinted with permission of Novagen, Inc.

the control of the *lacUV5* promoter (Figure 8.4). Therefore, IPTG induction will promote the synthesis of T7 RNA polymerase which will bind to the T7 promoter and drive the expression of the target gene. As we have already noted, the *lac* promoter will express small amounts of the gene it controls even in the absence of inducer. The addition of a *lacO* sequence in between the T7 promoter and the target gene in the expression vector reduces the level of target gene expression (Dubendorff and Studier, 1991). To control the leaky production of T7 RNA polymerase (thereby ensuring that target gene expression is minimized) *E. coli* cells can be co-transformed with an additional plasmid. As shown in Figure 8.4, the plasmid pLysS, which uses a different, but compatible, replication origin to the expression vector, will produce T7 lysozyme, which is a natural inhibitor of T7 RNA polymerase. The production of this inhibitor will inactivate the small levels of polymerase produced in the absence of induction, but will be swamped, and thereby rendered ineffective, by the larger amounts of polymerase produced during induction.

Despite the availability of excellent promoters that will drive high levels of RNA production, many proteins cannot be produced in *E. coli* cells. Promoter strength is not necessarily the determining factor as to levels at which the target protein will accumulate within the cell. Some additional factors are listed below.

- *Expression vector levels.* Naïvely, one would imaging that increasing the copy number of the expression vector would lead to an increase in the accumulation of the protein it encodes. There are, however, documented cases when a very high expression vector copy number (in comparison to the levels obtained for pBR322) did not result in increased protein production (Yansura and Henner, 1990) and others where increased vector levels actually reduce the levels of protein production (Vasquez *et al.*, 1989). Most commercially available expression vectors today contain the replication origin of either pBR322 or pUC (Chapter 3), and altering copy number is not commonly used to modulate protein production, although some systems are available (Wild, Hradecna and Szybalski, 2001).

- *Transcriptional termination.* Although often overlooked in the design of expression vectors, efficient transcription termination is an essential component for achieving high levels of gene expression. Terminators enhance mRNA stability (Hayashi and Hayashi, 1985) and can lead to substantial increases in the levels of accumulated protein (Vasquez *et al.*, 1989). The two tandem transcriptional terminators (T1 and T2) from the *rrnB* rRNA operon of *E. coli* (Brosius *et al.*, 1981) are often present in expression vectors, but other terminators also work well.

- *Codon usage.* The degeneracy of the genetic code means that more than one codon will result in the insertion of an individual amino acid into a growing polypeptide chain. The genes of both prokaryotes and eukaryotes show a non-random usage of alternative codons. Genes containing favourable codons will be translated more efficiently than those containing infrequently used codons. This effect is particularly prevalent in genes that are highly expressed in *E. coli*, where there is a high degree of codon bias. In general, the frequency of use of alternative codons reflects the abundance of their cognate tRNA molecules. For example, the minor arginine tRNA[Arg(AGG/AGA)] has been shown to be a limiting factor in the bacterial production of several mammalian proteins (Brinkmann, Mattes and Buckel,

1989) because the codons AGG and AGA are infrequently used in *E. coli*. The co-expression of the gene coding for tRNA$^{Arg(AGG/AGA)}$ (*dna*Y) can result in high-level production of the target protein whose production is limited in this way. Systems have been established for the expression of other tRNA molecules that occur frequently in mammalian coding sequence but are used rarely in *E. coli*. One such system uses a bacterial strain (called Rosetta^{TM}) that expresses the tRNAs for AGG, AGA, AUA, CUA, CCC and GGA on a plasmid that is compatible with the expression vector. An alternative, although more time consuming, approach to the problem of rare codon occurrence is to mutate the gene that is to be expressed such that the codons it contains are more frequently used by other highly expressed genes in *E. coli*. That is, the DNA sequence of the gene is altered to allow more favourable codons to be used, but the encoded polypeptide remains unchanged. There does not appear to be a simple correlation between the presence of rare codons within a gene and the levels to which protein production can occur. A combination of consecutive rare codons within the target sequence and other factors reduces the overall efficiency of translation.

- *Protein sequence.* The amino acid sequence of the target protein plays an important role in the ability of the protein to accumulate to high levels. First described in the laboratory of Alexander Varshavsky, the 'N-end rule' relates protein stability to the sequences at its amino-terminal end (Bachmair, Finley and Varshavsky, 1986; Varshavsky, 1992). In *E. coli*, an amino-terminal Arg, Lys, Leu, Phe, Tyr and Trp located directly after the initiating methionine results in proteins with a half-life of less than 2 min. Other amino acids at the same location in the same protein confer a half-life of over 10 h (Tobias *et al.*, 1991). Additional amino-acid-sequence-dependent protein stability factors also exist, reviewed by Makrides (1996).

- *Protein degradation.* *E. coli* is often considered as a molecular biology 'bag' for making DNA and proteins. Of course, the organism is highly developed and contains multiple mechanisms for removal of substances that may be toxic to it. For example, *E. coli* contains a large number of proteases located in the cytoplasm and the periplasm and associated with the inner and outer membranes (Chung, 1993; Gottesman, 1996). Proteolysis serves to limit the accumulation of critical regulatory proteins, and also rids the cell of abnormal and mis-folded proteins. Target proteins expressed in *E. coli* may be mis-folded for a variety of reasons, including the exposure of hydrophobic residues that are normally in the core of the protein, the lack of its normal interaction partners and inappropriate or missing post-translational modifications. Some methods used to counteract the effects of proteolysis include the use of protease deficient *E. coli* strains; low-temperature cell growth; expression of the target gene fused to a known stable protein; and the targeting of the produced protein to the periplasm, where fewer proteases exist (Murby, Uhlén and Ståhl, 1996).

Despite the limitations discussed above, *E. coli* remains widely used as the organism of choice for protein production. Some of the expression systems that can be used in the laboratory are not, however, suitable for the production of proteins on a very large scale. For example, IPTG induction of human therapeutic proteins is impractical due to the cost of inducing large cultures and the potential toxicity of IPTG itself (Figge *et al.*, 1988).

8.2 Expression in Yeast

As eukaryotes, yeasts have many of the advantages of higher-eukaryotic cells, such as post-translational modifications, while at the same time being almost as easy to manipulate as *E. coli*. Yeast cell growth is faster, easier and less expensive than other eukaryotic cells, and generally gives higher expression levels. Three main species of yeast are used for the production of recombinant proteins – *Saccharomyces cerevisiae, Pichia pastoris* and *Schizosaccharomyces pombe*.

8.2.1 Saccharomyces cerevisiae

Baker's yeast, *S. cerevisiae*, is a single-celled eukaryote that grows rapidly (a doubling time of approximately 90 min) in simple, defined media similar to those used for *E. coli* cell growth. Proteins produced in *S. cerevisiae* contain many, but not all, of the post-translation modifications found in higher-eukaryotic cells. For example, human α-1-antitrypsin, a 52 kDa serum protein involved in the control of coagulation and fibrinolysis, is normally glycosylated. However, if the protein is produced in *S. cerevisiae*, glycosylation still occurs at the same locations as the human-derived protein, but the glycosylation pattern obtained is very different (Moir and Dumais, 1987).

A number of strong constitutive promoters have been used to drive target gene expression in yeast. For example, the promoters for the genes encoding phosphoglycerate kinase (*PGK*), glyceraldehyde-3-phosphate dehydrogenase (*GPD*) and alcohol dehydrogenase (*ADH1*) have all been used to produce target proteins (Cereghino and Cregg, 1999). However, these suffer similar problems as constitutive *E. coli* expression systems. A variety of systems for the inducible production of target proteins in *S. cerevisiae* have been utilized. Two of these are discussed below.

8.2.1.1   *The GAL System*

In yeast, like almost all other cells, galactose is converted to glucose-6-phosphate by the enzymes of the Leloir pathway. Each of the Leloir pathway structural genes (collectively called the *GAL* genes) are expressed at a high level, representing 0.5–1 per cent of the total cellular mRNA (St John and Davis, 1981), but only when the cells are grown on galactose as the sole carbon source. Each of the *GAL* genes contains within its promoter at least one, and often multiple, binding sites for the transcriptional activator Gal4p. The binding of Gal4p to these sites, and its transcriptional activity when bound, is regulated by the source of carbon available to the cell. When yeast is grown on glucose, its preferred carbon source, transcription from the *GAL4* promoter (regulating the production of Gal4p) is down-regulated so that there is less Gal4p in the cell, and consequently a reduced level of activator binding at the promoters of the *GAL* structural genes (Griggs and Johnston, 1991). In other carbon sources, such as raffinose, Gal4p is produced and binds to the *GAL* structure gene promoters, but a repressor, Gal80p, inhibits its activity. Gal80p binds directly to Gal4p and is thought to mask its activation domain such that it is unable to recruit the transcriptional machinery to the gene (Lue *et al.*, 1987). Only in the presence of galactose is the inhibitory effect of Gal80p alleviated, leading to strong, inducible levels of target gene expression.

To produce a target protein in *S. cerevisiae* using galactose induction, the gene encoding the protein must be cloned so that it is under the control of a *GAL* promoter. The promoter from the *GAL1* gene, encoding galactokinase, is most commonly used, but synthetic promoters containing multiple Gal4p binding sites are also available. Once constructed, the expression vector is transformed into yeast cells and protein production is initiated by switching the cells into a galactose-containing medium. Proteins produced in this way seldom accumulate to the levels of recombinant protein found in *E. coli* cells. It not usually possible to detect protein produced in this way using Coomassie stained gels, such as those in Figure 8.3, and maximum production may represent only 1–5 per cent of the total cell protein. Western blotting, or other methods to detect the target protein, must be used. An additional difficulty is brought about as a consequence of the activator of the *GAL* genes, Gal4p, being normally present in the yeast cell at a very low level. Therefore, if the expression vector, which carries multiple Gal4p binding sites, is a high-copy-number plasmid then there may be insufficient Gal4p to activate the expression of all of the available target genes to a maximum level. To overcome this problem, yeast strains have been constructed in which the coding sequence of the *GAL4* gene has been placed under the *GAL* promoter control (Schultz *et al.*, 1987; Mylin *et al.*, 1990). This results in a feedback loop in which induction by galactose results in the production of Gal4p so that more of the target gene may be expressed (Figure 8.5).

Gal4p produced from its own promoter

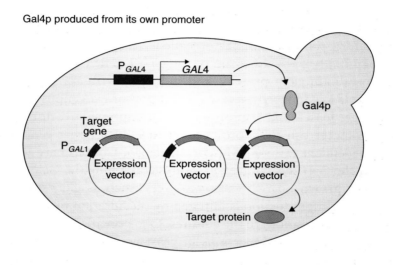

Gal4p produced from the *GAL1* promoter

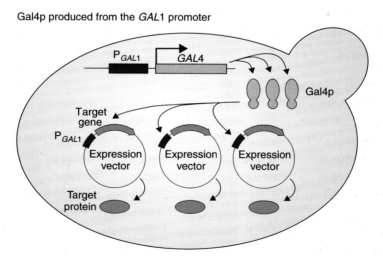

Figure 8.5. Galactose inducible gene expression in yeast. The expression of genes from multicopy vectors under the control of the *GAL1* promoter (P$_{GAL1}$) can be increased substantially if the gene encoding the transcriptional activator of *GAL1*, *GAL4*, is also placed under the control of P$_{GAL1}$. In this case, induction by galactose will produce more Gal4p and consequently more of the target protein

8.2.1.2   The CUP1 System

Copper ions (Cu^{2+} and Cu^+) are essential at appropriate levels, yet toxic at high levels for all living organisms. Cells must therefore maintain a proper cellular level of copper ions that is not too low to cause deficiency and not too high to cause toxicity. In *S. cerevisiae*, copper homeostasis consists of uptake, distribution and detoxification mechanisms (Eide, 1998). At high concentrations, copper ion detoxification is mediated by a copper ion sensing metalloregulatory transcription factor called Ace1p. Upon interaction with copper, Ace1p binds DNA upstream of the *CUP1* gene (Winge, 1998), which encodes a metallothionein protein, and induces its transcription. The transcription of *CUP1* is induced rapidly by addition of exogenous copper to the medium (Winge, Jensen and Srinivasan, 1998). Expression vectors harbouring the *CUP1* promoter can therefore be used to induce target gene expression in a copper-dependent fashion (Mascorrogallardo, Covarrubias and Gaxiola, 1996). Unlike the *GAL* system, yeast cultures containing the *CUP1* expression plasmid can be grown on rich carbon sources, such as glucose, to high cell density, and protein production is initiated by the addition of copper sulphate (0.5 mM final concentration) to the cultures. One potential drawback with this system is the presence of copper ions in yeast growth media, and indeed in water supplies. Therefore, the 'off' state in the absence of added copper may still yield significant levels of protein production.

8.2.2   Pichia Pastoris

Pichia pastoris is a methylotrophic yeast, capable of metabolizing methanol as its sole carbon source. The first step in the metabolism of methanol is the oxidation of methanol to formaldehyde using molecular oxygen (O_2) by the enzyme alcohol oxidase. Alcohol oxidase has a poor affinity for O_2, and *P. pastoris* compensates for this deficiency by generating large amounts of the enzyme (Koutz *et al.*, 1989). The promoter regulating the production of alcohol oxidase (*AOX1*) can be used to drive heterologous protein expression in *P. pastoris* (Tschopp *et al.*, 1987) since it is tightly regulated and induced by methanol to very high levels. *P. pastoris* cells containing the expression vector, which is usually integrated into the genome as single or multiple copies, are grown in glycerol (growth on glucose represses *AOX1* transcription, even in the presence of the methanol) to extremely high cell density prior to the addition of methanol. Once induced, target proteins may accumulate at very high levels, often in the range of 0.5 to tens of grams of protein per litre of yeast culture. For example, the expression of the gene encoding recombinant hepatitis B surface antigen results in the production of more than 1 g of the antigen from 1 L of *P.*

pastoris cells (Hardy *et al.*, 2000). This is much greater than could be achieved in *S. cerevisiae*. Additionally, in comparison to *S. cerevisiae*, *P. pastoris* may have an advantage in the glycosylation of secreted proteins. Glycoproteins generated in *P. pastoris* more closely resemble the glycoprotein structure of those found in higher eukaryotes (Cregg, Vedvick and Raschke, 1993).

8.2.3 *Schizosaccharomyces pombe*

S. pombe is a single-cell eukaryotic organism with many properties similar to those found in higher-eukaryotic organisms. These properties, such as chromosome structure and function, cell-cycle control, RNA splicing and codon usage, suggest that *S. pombe* would make an ideal candidate for the production of eukaryotic proteins (Giga-Hama and Kumagai, 1999). Additionally, eukaryotic proteins expressed in *S. pombe* are more likely to be folded properly, which may reduce protein insolubility associated with the production of many proteins in *E. coli*. Protein production in *S. pombe* is usually controlled by the expression from the *nmt1* (no message in thiamine) promoter (Maundrell, 1993). This promoter is active when the cells are grown in the absence of thiamine, allowing downstream transcription of genes under its control, while in the presence of greater than 0.5 μM thiamine, the promoter is turned off (Maundrell, 1990). Overall protein production levels are similar to those found in *S. cerevisiae*.

8.3 Expression in Insect Cells

Baculoviruses are rod-shaped viruses that infect insects and insect cell lines. They have double-stranded circular DNA genomes in the range of 90–180 kbp (Ayres *et al.*, 1994). Viral infection results in cell lysis, usually 3–5 d after the initial infection, and the subsequent death of the infected insect. The nuclear polyhedrosis viruses are a class of baculoviruses that produce occlusion bodies in the nucleus of infect cells. These occlusion bodies consist primarily of a single protein, polyhedrin, which surrounds the viral particles and protects them from harsh environments. Most viruses of this type need to be eaten by the insect before infection will occur, and the occlusion body protects the viral particles from degradation in the insect gut. The polyhedrin gene is transcribed at very high levels late in the infection process (2–4 d post-infection). In cultured insect cells, the production of inclusion bodies is not essential for viral infection or replication. Consequently, the polyhedrin promoter can be used to drive target gene expression.

The baculovirus *Autographa californica* nuclear polyhedrosis virus (AcNPV) has become a popular tool of the production recombinant proteins in insect cells (Fraser, 1992). It is used in conjunction with insect cell lines derived from the moth *Spodoptera frugiperda*. These cell lines (e.g. Sf9 and Sf21) are readily cultured in the laboratory, and a scheme for constructing baculovirus recombinants is shown in Figure 8.6. The size of the baculoviral genome generally precludes the cloning of target genes directly onto it. Instead, the target gene is cloned downstream of the polyhedrin promoter in a transfer plasmid (Lopez-Ferber, Sisk and Possee, 1995). The transfer plasmid also contains the sequences of baculovirus genomic DNA that flank the polyhedrin gene, both upstream and downstream. To produce recombinant viruses, the recombinant transfer plasmid is co-transfected with linearized baculovirus vector DNA into insect cells. The flanking regions of the transfer plasmid participate in homologous recombination with the viral DNA sequences and introduce the target gene into the baculovirus genome. The recombination process also results in the repair of the circular viral DNA and allows viral replication to proceed through the re-formation of ORF1629 (a viral capsid associated protein that is essential for the production of viral particles). Recombinant viral infection can be observed microscopically by viewing viral plaques on a lawn of insect cells. Plaques containing recombinant virus will be unable to form occlusion bodies due to the lack of a functional polyhedrin protein (Smith, Summers and Fraser, 1983). Screening plaques this way is, however, technically difficult. Therefore, the transfer plasmids also usually contain the *lacZ* gene, or another readily observable reporter gene, which allows for the visual identification of recombinant plaques by their blue appearance after staining with X-Gal (Figure 8.6). Following transfection and plaque purification to remove any contaminating parental virus, a high-titre virus stock is prepared, and used to infect large-scale insect cell culture for protein production. The infected cells undergo a burst of target protein production, after which the cells die and may lyse.

Protein production in baculovirus infected insect cells has the advantage that very high levels of protein can be produced relative to other eukaryotic expression systems, and that the glycosylation pattern obtained is similar, but not identical, to that found in higher eukaryotes (Possee, 1997; Joshi *et al.*, 2000). Baculoviruses also have the advantage that multiple genes can be expressed from a single virus. This allows the production of protein complexes whose individual components may not be stable when expressed on their own (Roy *et al.*, 1997). The main disadvantages of producing proteins in this way is that the construction and purification of recombinant baculovirus vectors for the expression of target genes in insect cells can take as long as 4–6 weeks, and that the cells grow slowly (increasing the risk of contamination) in expensive media. An alternative approach to recombinant viral genome production uses

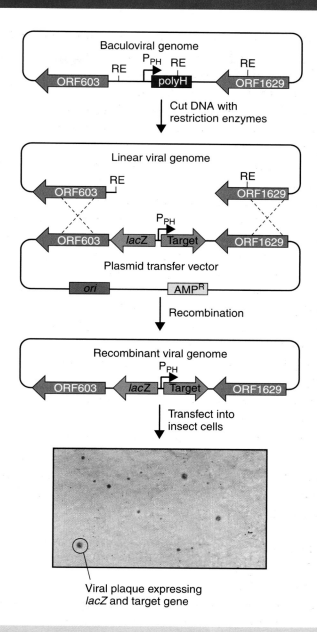

Figure 8.6. The production of a recombinant baculoviral genome for the production of proteins in insect cells. The target gene is cloned under the control of the polyhedrin promoter into a transfer vector that also contains regions of the viral genome that flank the polyhedrin locus. The vector is then co-transfected into insect cells with a viral genome that has been linearized using restriction enzymes (RE) that cut in several places. Homologous recombination between the linear genome and the vector will result in formation of a functional viral genome that is capable of producing viral particles. The inclusion of *lacZ* in the transfer vector allows for visual screening of viral plaques to identify recombinants

site-specific transposition in *E. coli* rather than homologous recombination in insect cells (Luckow *et al.*, 1993). It is based on site-specific transposition of an expression cassette into a baculovirus shuttle vector (**bacmid**) propagated in *E. coli*. The bacmid contains the entire baculovirus genome, a low-copy-number *E. coli* F-plasmid origin of replication and the attachment site for the bacterial transposon Tn7. The bacmid propagates in *E. coli* as a large plasmid. Recombinant bacmids are constructed by transposing a Tn7 element from a donor plasmid, which contains the target gene to be expressed, to the attachment site on the bacmid – a helper plasmid encoding the transposase is required for this function. The recombinant bacmid can be isolated from *E. coli* and transfected directly into insect cells.

8.4   Expression in Higher-Eukaryotic Cells

For the production of mammalian proteins, mammalian cells have an obvious advantage. The major problem with expressing genes in mammalian cells is that expression levels like those we have discussed above are simply not currently available. For many years protein production in mammalian cells has utilized strong constitutive promoters to elicit transcription of target genes. Promoters, such as those derived from the SV40 early promoter, the Rous sarcoma virus (RSV) long terminal repeat promoter and the cytomegalovirus (CMV) immediate early promoter, will all constitutively drive the expression of genes placed under their control (Mulligan and Berg, 1981; Gorman *et al.*, 1982; Boshart *et al.*, 1985). Inducible systems can also be used. For example, heat-shock promoters or glucocorticoid hormone inducible systems have been used to express target genes (Wurm, Gwinn and Kingston, 1986; Hirt *et al.*, 1992). These systems, however suffer from leaky gene expression in the absence of induction and potentially damaging induction conditions. To overcome some of the problems of using endogenous promoters to drive target gene expression, systems have been imported from bacteria to control gene expression in mammalian cells.

8.4.1   *Tet-on/Tet-off System*

As we have discussed in Chapter 1, the control of transcriptional initiation is fundamentally different between eukaryotes and prokaryotes. An activator from prokaryotes is unable to bring about a transcriptional response in eukaryotes and *vice versa*. DNA binding is, however, species independent. The tightly regulated DNA binding properties of prokaryotic activators can be used to direct eukaryotic activation domains to drive the expression of target genes. One such system exploits the DNA properties of the *E. coli* tetracycline repressor.

The *E. coli tet* operon was originally identified as a transposon (Tn10) that confers resistance to the antibiotic tetracycline (Foster *et al.*, 1981). The TetR protein, in a similar fashion to the *lac* repressor protein (LacI) we have already discussed (Chapter 1), binds to the operator of the tetracycline-resistance operon and prevents RNA polymerase from initiating transcription. Activation of the tetracycline-resistance operon occurs when tetracycline itself binds to the repressor and induces a conformational change that inhibits its DNA binding activity. The TetR protein has a very high affinity for the antibiotic (association constant $\sim 3 \times 10^{-9}$ M^{-1}) and will dissociate from its DNA binding site when tetracycline is present at low concentrations (Takahashi, Degenkolb and Hillen, 1991). The regulated DNA binding activity of TetR cannot itself elicit a transcriptional response in eukaryotes, but can if the protein is fused to a eukaryotic transcriptional activator domain. The use of the *tet* system to drive target gene expression in eukaryotes relies on the insertion of two recombinant DNA molecules into the host cell (Figure 8.7).

- *Regulator plasmid* – produces a version of the *E. coli* tetracycline repressor (TetR) that is fused to the transcriptional activation domain of the herpes simplex virus VP16 protein. The fusion protein is constitutively produced in the host cell from the CMV promoter.

- *Response plasmid* – contains the target gene cloned downstream of multimerised copies of the tetracycline operator (*tetO*) DNA sequence that form a tetracycline response element (TRE) cloned into a minimal CMV promoter that is not, on its own, able to support gene activation.

In the absence of tetracycline, the TetR-VP16 fusion protein will bind to the TRE and activate transcription of the target gene. Upon the addition of tetracycline to the cells, however, TetR will dissociate and target gene transcription will be turned off (Gossen and Bujard, 1992). That is, the addition of tetracycline turns target gene expression off. The use of the *tet* system has become more prevalent due to the existence of a mutant version of TetR. The mutant tetracycline repressor contains four amino acid changes (E71K, D95N, L101S and G102D) from the wild-type protein that radically alter its DNA binding properties. Rather than tetracycline inhibiting its DNA binding properties, the mutant protein, called rTetR for reverse tetracycline repressor, will only bind DNA in the presence of tetracycline (Gossen *et al.*, 1995). This means that, with the appropriate TetR fusion to the activation domain of VP16, target gene expression can either be inhibited or activated in the presence of tetracycline.

- *Tet-off* uses the wild-type TetR protein fused to VP16. Target gene expression is active in the absence of tetracycline but not in its presence.

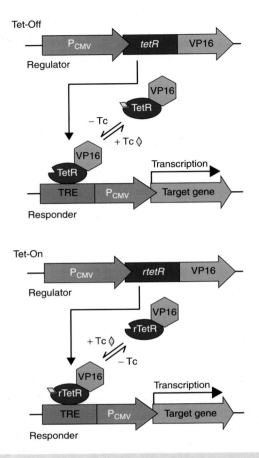

Figure 8.7. Tetracycline regulated gene expression for protein production in mammalian cells. The Tet-off and Tet-on systems differ in their transcriptional response to added tetracycline. The Tet-off system turns transcription of the target gene off in response to tetracycline, whereas the Tet-on system, which contains a mutant version of TetR with altered DNA binding properties, activates gene expression in response to tetracycline addition

- *Tet-on* uses the mutant rTetR proteins fused to VP16. Target gene expression is active in the presence of tetracycline but not in its absence.

The advantage of this on and off switching system is that host cells do not need to be exposed for long times to the antibiotic prior to the induction of either gene expression or gene silencing. Additionally, the control over target gene activation achieved using the Tet system is very tight. For example, transgenic mice have been produced that carry the diptheria toxin A gene under the control for a TRE promoter. Small quantities of the toxin, perhaps as little as a single molecule, will lead to cell death. When fed with water containing tetracycline, mice containing a Tet-off version of the regulator in conjunction with the diptheria toxin responder are healthy until tetracycline is removed, when death ensues as a result of toxin production (Lee *et al.*, 1998).

8.5 Protein Purification

The techniques described above provide mechanisms by which a gene sequence can be efficiently converted into the encoded polypeptide sequence. The next problem facing the researcher is how to separate the target protein from the thousands of naturally occurring host proteins. Many proteins share similar properties. The irreversible denaturation of proteins from *E. coli* cells and their separation by very high-resolution two-dimensional gel electrophoresis, where separation occurs on the basis of both charge and size, reveals a large number of spots corresponding to individual proteins (Figure 8.8). This analysis shows that many proteins are of average size (in the range of 40–80 kDa) and have average charge (isoelectric point in the rage of pH 6–8). Native separation techniques are required for the analysis of functional proteins. Traditional biochemical separation techniques may be employed, such as the separation of proteins on the basis of their size (gel filtration chromatography), their charge (ion-exchange chromatography) or their degree of hydrophobicity (hydrophobic interaction chromatography), but the ability of these techniques to separate similar proteins is severely limited. Consequently, recombinant proteins may often be difficult to purify and will require multiple time-consuming chromatographic steps to be performed before an acceptable level of purity can be achieved. Of course, the required level of purity depends on the use of the protein itself. Many enzymatic

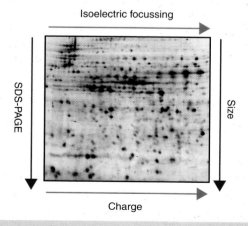

Figure 8.8. The high-resolution separation of proteins by two-dimensional SDS–PAGE. The gel image was kindly supplied by Phil Cash (University of Aberdeen)

reactions will occur in crude cell lysates without the need for protein purification, while other methods, particularly those of structural biology, demand a high degree of protein homogeneity. What is required is the ability to impart the target protein with a unique property that can be used to separate it from all other host proteins. Protein purification tags are protein sequences that possess high-affinity binding properties for particular molecules, and the tag allows the target protein to bind to a solid support, usually in the form of a column matrix, to which very few (if any) other proteins are able to bind. The purification of tagged proteins from host cells consists of four steps: lysis of the host cell, binding of the tagged protein to an affinity column, washing the column to remove untagged proteins, and finally elution of the tagged protein itself. Ideally, the tag should allow binding of the recombinant protein to the column with high affinity and specificity, yet the interaction between the tag and the column needs to be capable of disruption under mild conditions so that the protein is not denatured during the elution process. Additionally, the tag should not interfere with the normal function of the recombinant protein. Some of the commonly used protein purification tags are described below, while others are listed in Table 8.1.

8.5.1 The His-tag

The simplest of all protein purification tags, the his-tag is normally composed of six histidine residues. The DNA encoding these residues is cloned into the target gene such that the produced protein contains, at some point in its polypeptide sequence, six consecutive histidine residues (Hoffman and Roeder, 1991). Cloning is often performed such that the tag is located at either the extreme amino- or extreme carboxy-terminal end of the protein, where it is less likely to impair protein function. The tag may, however, also be placed in the middle of a protein if a central region is already known to be non-essential for function (see, e.g. Zenke *et al.*, 1996). The histidines will bind non-covalently and with high affinity for certain metal ions (Figure 8.9). In a technique called immobilized metal ion affinity chromatography (IMAC), metal ions, e.g. nickel, are bound to a resin matrix and used to capture his-tagged proteins (Yip and Hutchens, 1996). The most commonly used resins for this purpose have nitriloacetic acid (NTA) covalently attached to them. NTA has four coordination sites that bind a single nickel ion very tightly. The charging of NTA with Ni^{2+} leaves two of the six possible coordination sites of the ion free. In solution these will weakly interact with water, but can interact more strongly with the side chain imidazole rings of consecutive histidine residues on a polypeptide chain. At least six histidine residues are required to provide the necessary binding affinity to firmly adhere the tagged protein to the column. The vast majority of host proteins will not be able to bind to a column of this type.

Table 8.1. Other commonly used protein tags

Tag	Sequence	Source	Uses	Notes	Reference
FLAG	DYKDDDDK	Bacteriophage T7 gene 10 leader peptide	Purification, western blots	High-affinity antibodies available	(Einhauer and Jungbauer, 2001)
S-tag	KETAAAKFER QHMDS	Bovine pancreatic ribonuclease A cleaved with subtilisin	Purification, western blots	Binds with high affinity to the ribonuclease S protein	(Raines et al., 2000)
Trx	109-amino-acid thioredoxin protein	E. coli trxA gene	Increases the stability of fusion partners	May stabilize some proteins, and help others fold	(LaVallie et al., 1993a)
NusA	495-amino-acid transcriptional elongation factor	E. coli nusA gene	Increases the stability of fusion partners	Highly soluble protein that confers solubility on fusion proteins	(Davis et al., 1999)
HA	YPYDVPDYA	Influenza haemagglutin	Western blots	High-affinity antibodies available	(Wilson et al., 1984)
Myc	EQKLISEEDL	Human c-myc protein	Western blots	High-affinity antibodies available	(Evan et al., 1985)

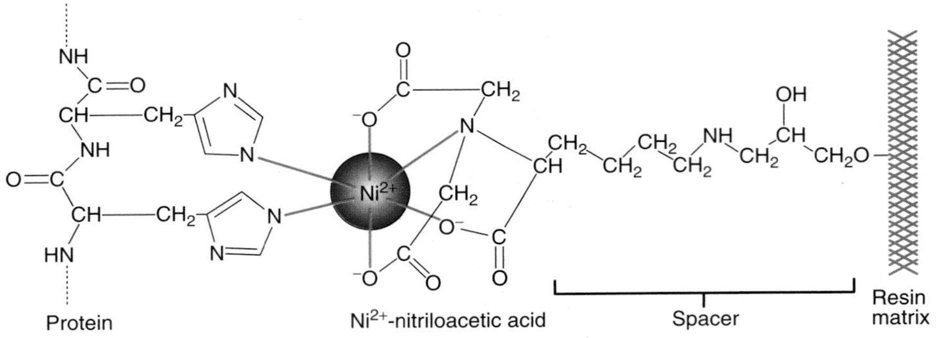

Figure 8.9. The binding of proteins tagged with multiple histidine residues to Ni^{2+}-NTA resin

The purification of a his-tagged protein from *E. coli* cells is shown in Figure 8.10. *E. coli* cells containing an inducible expression vector were grown and induced to produce the tagged target protein. The cells were broken open and insoluble cell debris was removed by centrifugation. The supernatant from this process was applied to a Ni^{2+}-NTA column. The column was washed with a low concentration (20 mM) of imidazole, which will compete with low-affinity histidine–column interactions to remove from the column any, perhaps histidine-rich, proteins that are non-specifically bound. Finally, the tagged protein itself is removed from the column by increasing the concentration of imidazole to a high level (250 mM). This process results in the single-step purification of the tagged protein to yield a very pure, almost homogenous, sample. His-tagged proteins from any expression system including bacteria, yeast, baculovirus, and mammalian cells, can be purified to a high degree of homogeneity using this technique. Alternative elution conditions may also be used. For example, lowering the pH from 8 to 4.5 will alter the protonated state of the histidine residues and results in the dissociation of the protein from the metal complex. The tagged protein can also be removed by adding chelating agents, such as EDTA, to strip the nickel ions from the column and consequently remove the tagged protein.

The small size of the histidine tag means that the tagged recombinant protein often behaves identically to its untagged parent. In some cases, the tagged protein is actually found to be more biologically active than the untagged version of the same protein (Janknecht *et al.*, 1991), although this effect is likely to be due to the speed of the purification process rather than any biological activity of the tag itself. Some proteins have been crystallized in the presence of the his-tag (Kim *et al.*, 1996a). Additionally, the his-tag has extremely low

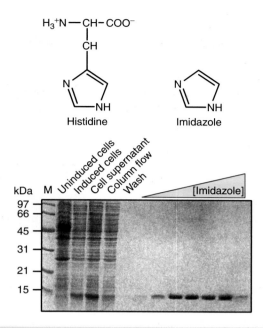

Figure 8.10. The purification of a his-tagged protein. The chemical structures of histidine and imidazole are shown, together with an SDS–polyacrylamide gel of the purification of a his-tagged protein. An *E. coli* cell extract producing a 14 kDa his-tagged protein was applied to a Ni^{2+}-NTA column. The column was washed with a buffer containing a low concentration (20 mM) of the histidine analogue imidazole prior to elution of the tagged protein with an imidazole gradient (20–250 mM). Proteins were visualized after staining the gel with Coomassie blue

immunogenicity and consequently the recombinant protein containing the tag can be used to produce antibodies. There are some reports of the his-tag altering protein function, (see, e.g. Knapp *et al.*, 2000), but, as we will see later, it is more important to remove some other purification tags. An additional advantage of the his-tag is that purification can be performed under denaturing conditions (Reece, Rickles and Ptashne, 1993). The interaction between the histidine residues and the metal ion does not require any special protein structure and will occur even in the presence of strong protein denaturants (e.g. 8M urea). This is particularly important for the purification of proteins that would otherwise be insoluble.

8.5.2 The GST-tag

The glutathione S-transferases (GSTs) are a family of enzymes that are involved in the cellular defense against electrophilic **xenobiotic** chemical compounds.

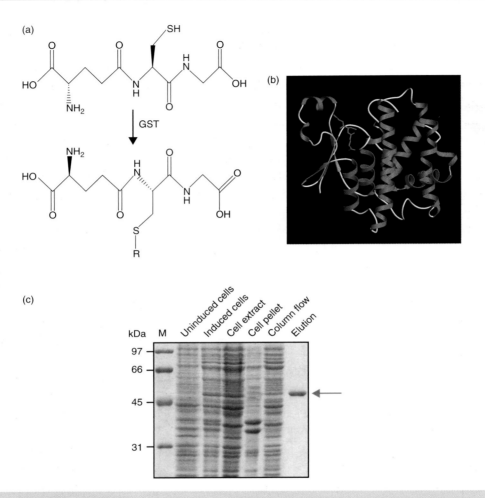

Figure 8.11. The purification of proteins tagged with GST. (a) The chemical structure of the tripeptide glutathione and the action of GST for its addition to electrophilic compounds (R). Glutathione is composed of three amino acids – glutamic acid, cysteine and glycine. Note that the glutamic acid is joined to the Cys–Gly dipeptide through its γ-carboxyl group. (b) The three-dimensional structure of the GST–glutathione complex. The protein is depicted in a ribbon form and the glutathione as a green stick model (Garcia-Sáez *et al.*, 1994). (c) The purification if a GST-tagged protein from *E. coli* cells. The tagged protein was bound to a glutathione-affinity column and eluted using free glutathione itself. The tagged protein is indicated by the arrow

They catalyse the addition of glutathione to these electrophilic substrates, which results in their increased solubility in water and promotes their subsequent enzymatic degradation (Strange, Jones and Fryer, 2000). Glutathione is a tripeptide composed of the amino acids glutamic acid, cysteine and glycine (Figure 8.11(a)). GST binds to glutathione with high affinity (Figure 8.11(b)).

The enzyme from the parasitic flatworm *Schistosoma japonicum* is a 26 kDa dimeric protein (Walker *et al.*, 1993). The gene encoding this protein is fused, in the correct reading frame, to the target gene and a fusion protein is produced from an expression vector. Host cells producing the fusion protein are broken open and soluble proteins are applied to a column to which glutathione is attached (e.g. glutathione-agarose). The specific interaction between GST and glutathione will result in the binding of the fusion protein to the column, while the majority of host proteins are unable to adhere. The bound protein can then be eluted from the column by washing with a high concentration of glutathione (10 mM) to compete for the interaction with the column (Figure 8.11(c)).

Both the large size of GST and its dimeric nature mean that the tag is more likely to influence the biological activity of the target protein than the his-tag. It is therefore desirable to remove the GST portion of the fusion protein to study the activity of the target protein in isolation. This can be achieved by the inclusion, in the expression vector, of DNA coding for the amino acid sequence of a specific protease cleavage site between GST and the target gene. Treatment of the purified fusion protein with the protease will then result in the generation of two polypeptides – the free target protein and GST itself. GST can then be removed from the target protein by applying the mixture back onto a glutathione column. The GST will, again, bind to the column, but the target protein will not. The column flow-through can be collected and will contain the purified target protein.

A variety of specific proteases have been used to cleave purification tags from target fusion proteins (Table 8.2). Unlike restriction enzymes when they cleave DNA (see Chapter 2), many proteases do not have an absolute sequence requirement for their cleavage sites. For example, the protease Factor Xa cleaves after the arginine residue in its preferred cleavage site Ile–Glu–Gly–Arg. However, it will sometimes cleave at other basic residues, depending on the conformation of the protein substrate, and a number of the secondary sites have been sequenced that show cleavage following Gly–Arg dipeptides (Quinlan, Moir and Stewart, 1989). Consequently, the protease may not only cleave the site between the tag and the target protein, but many also cleave the target protein itself. Obviously, this must be avoided to maintain the integrity of the target protein. Other proteases, e.g. the TEV and PreScission proteases, have larger and more specific recognition sequences and are less likely to cleave at alternative sites. The TEV protease has the added advantage that the protease can be produced in a recombinant form from *E. coli* and is therefore not contaminated with other plasma proteases and factors.

Table 8.2. Site-specific proteases. The recognition sequence of each protease is shown, together with the actual site of cleavage, depicted by the arrow

Protease	Recognition and cleavage site	Notes	Reference
Factor Xa	IleGluGlyArg↓	42 kDa protein, composed of two disulphide linked chains, purified from bovine plasma	(Nagai, Perutz and Poyart, 1985)
Enterokinase	AspAspAspAspLys↓	26 kDa light chain of bovine enterokinase produced in and purified from *E. coli*	(LaVallie *et al.*, 1993b)
Thrombin	LeuValProArg↓ GlySer	Purified from bovine plasma	(Chang, 1985)
TEV	GluAsnLeuTyr-PheGln↓Gly	Tobacco etch virus protease	(Dougherty *et al.*, 1989)
PreScission	LeuGluValLeuPhe Gln↓ GlyPro	Protease from the 3C human rhinovirus	(Walker *et al.*, 1994)

8.5.3 The MBP-tag

The target gene is inserted downstream from the *malE* gene of *E. coli*, which encodes maltose binding protein (MBP), in an expression vector that results in the production of an MBP fusion protein (Kellermann and Ferenci, 1982). Maltose is a disaccharide composed of two molecules of glucose (Figure 8.12(a)). MBP is a 40 kDa monomeric protein that forms part of the maltose/maltodextrin system of *E. coli*, which is responsible for the uptake and efficient catabolism of glucose polymers (Boos and Shuman, 1998). The protein undergoes a large conformational change upon binding of maltose, and results in the formation of a stable complex (Figure 8.12(b)). One-step purification of fusion proteins is achieved using the affinity of MBP for cross-linked amylose (starch) (di Guan *et al.*, 1988). Bound proteins can be eluted from amylose by including maltose (10 mM) in the column buffer (Figure 8.12(c)).

8.5.4 IMPACT

Intein mediated purification with an affinity chitin binding tag (IMPACT) is an approach to protein purification that uses the protein self-splicing of

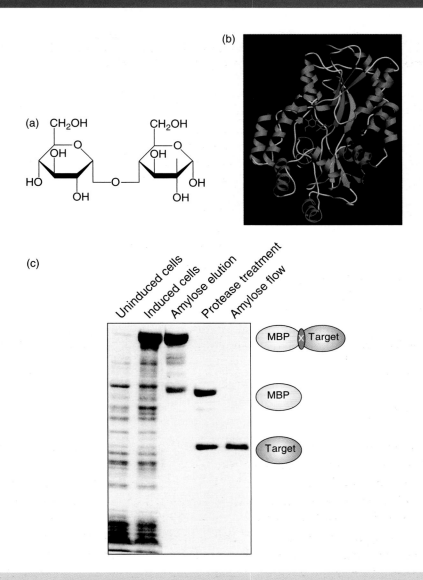

Figure 8.12. The purification of proteins tagged with MBP. (a) The chemical structure of maltose, a glucose disaccharide. (b) The three-dimensional structure of the MBP–maltose complex (Quiocho, Spurlino and Rodseth, 1997). The protein is depicted in a ribbon form with α-helices coloured in purple and β-sheets in blue. Maltose is shown as a green stick model. (c) The purification of an MBP-tagged protein. The tagged protein is bound to an amylose column and eluted with maltose. The MBP–target fusion is then cleaved with a protease at a site indicated by the X, and reapplied to the amylose column. The target protein will not adhere to the column when it is separated from MBP. The gel image is reprinted with permission of New England Biolabs, © 2002/2003

inteins to remove the purification tag and give pure isolated protein in one chromatographic step. Inteins are a class of proteins, found in a wide variety of organisms, that excise themselves from a precursor protein and in the process ligate the flanking protein sequences (exteins) (Cooper and Stevens, 1995). The excised intein is a site-specific DNA endonuclease that catalyses genetic mobility of its own DNA coding sequence. The process of polypeptide cleavage and ligation is dependent on specific chemistry involving thiols and a conserved asparagine residue.

Most inteins have a cysteine residue at their amino-terminal end and an asparagine at their carboxy-terminal end (Figure 8.13(a)). All the information required for the splicing reaction is contained within the intein itself, and if these sequences are placed in the context of a target protein they still splice themselves out. The mechanism of splicing is complex, but the reaction is very efficient. The IMPACT expression system exploits this unusual chemistry by mutation of the C-terminal asparagine to alanine in a yeast intein, VMA1 (Chong and Xu, 1997). This mutation prevents the cleavage reaction occurring at the carboxy-terminal side of the intein and traps the protein in a thioester that can be cleaved by β-mercaptoethanol or dithiothreitol (DTT). The target gene is cloned into an expression vector such that a three-component fusion protein is produced, in which a target protein–intein–chitin binding domain fusion is produced. Chitin is a fibrous insoluble polysaccharide made of β-1,4-N-acetyl-D-glucosamine that is found in the cell walls of fungi and algae and in the exoskeletons of arthropods. Chitinase catalyses the hydrolytic degradation of chitin, and the *Bacillus circulans* enzyme (M_r 74 kDa) is composed of three domains – an amino-terminal catalytic domain (CatD) (417 amino acid residues), a tandem repeat of fibronectin type III-like (FnIII) domains (duplicate 95 residues) and a carboxy-terminal chitin-binding domain (CBD, 45 amino acid residues) (Watanabe *et al.*, 1990). The isolated CBD shows high-affinity binding to chitin.

In the IMPACT system, the fusion protein is made in *E. coli* and passed down the chitin column, where it binds. The protein can be cleaved off the column by using thiol containing compounds, such as DTT, at 4 °C. This is a slow process and requires an overnight incubation to complete, which may prove problematical if the target protein is not stable under these conditions. The final target protein produced by this method is native except for the DTT thioester moiety attached at the carboxy-terminal end. The thioester is, however, unstable and will spontaneously hydrolyse to yield a native protein. Other thiols can also be used to initiate the cleavage process, e.g. β-mercaptoethanol and cysteine. Cysteine induced cleavage results in the insertion of a cysteine amino acid residue at the carboxy-terminal end of the cleaved polypeptide. The cysteine

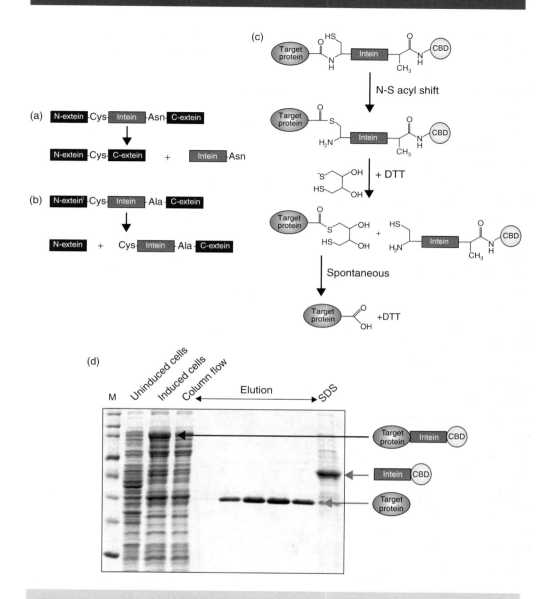

Figure 8.13. The IMPACT system for the purification of tagged proteins and the subsequent removal of the tag. (a) The normal splicing reaction involves the complete removal of the intein and the joining of the polypeptide sequences to its amino- and carboxy-terminal side. (b) A mutant form of the intein, in which an essential asparagine is replaced with an alanine, results in partial cleavage and the release of the amino-terminal side polypeptide only. (c) The chemistry of the splicing reaction used to cleave the target protein from the intein–chitin binding domain (CBD) tag. (d) The purification of *Hha*I methylase using the IMPACT system (Chong *et al.*, 1997). Purified target protein was eluted from the column, while the detergent SDS was used to remove the intein–CBD fusion. The gel image was kindly provided by Ming-Qun Xu (New England Biolabs)

can be radio-labelled, or it can be a site for chemical modification, especially if it is the only cysteine in the protein, since it is a good site to add protein cross-linkers, fluorescent probes, spin labels or other tags.

8.5.5   TAP-tagging

An extension of tagging over-produced proteins for purification is to tag proteins produced at wild-type levels in their native host cells. Protein purification in these circumstances, if performed under suitably mild conditions, can lead to the isolation of naturally occurring protein complexes. Most proteins do not exist as single entities within cells. They are associated, through non-covalent interactions, with a variety of other proteins that may be involved in the regulation of their function. The over-production of a single protein will not result in the over-production of other proteins in the complex. Therefore, to isolate complexes from cells, protein production should be as close to the natural state as possible. The DNA encoding what is termed a tandem affinity purification tag (TAP-tag) is cloned at the 3′-end of a target gene so that little disruption is made to its ability to be transcribed, and the fusion protein should be produced at the same level as the wild-type target protein. The TAP-tag encodes two purification elements – a calmodulin binding peptide and Protein A from *Staphylococcus aureus*. These elements are separated by a TEV protease cleavage site (Puig *et al.*, 2001). Cells containing the tagged protein are gently lysed and then applied to a column containing IgG, which binds with high affinity to Protein A. The fusion protein, and its associated proteins, are removed from the column using TEV protease and then applied directly to a calmodulin bead column, in the presence of Ca^{2+}, and eluted using the chelating agent EDTA. The two-step purification procedure is highly specific and can result in the isolation of contaminant-free protein complexes. The TAP-tag allows the rapid purification of complexes from a relatively small number of cells without prior knowledge of the complex composition, activity or function (Rigaut *et al.*, 1999; Gavin *et al.*, 2002), and, combined with mass spectrometry, the TAP strategy allows for the identification of proteins interacting with a given target protein.

9 Genome sequencing projects

Key concepts

♦ Genetic and physical maps are used to determine the order of genes on a chromosome and their approximate distance apart

♦ DNA sequence determination is performed using dideoxynucleotides that halt replication at a specific base. DNA fragments that differ by a single base can be separated using polyacrylamide gels

♦ Sequencing reactions generate a few hundred bases of sequence

♦ Whole genomes can be sequenced by cloning random small DNA genomic fragments, sequencing them, and then reassembling the genome sequence based on the overlap between the sequenced fragments

♦ Massive computing power is required to assemble the sequenced fragments and determine the locations of genes within the genome

The ultimate goal of all genome sequencing projects is to determine the precise sequence of bases that make up each DNA molecule within the genome. The knowledge of the sequence of individual genes, and the entire genome, is vital if we are to understand not only how genes and proteins work but also how different gene products influence the activity of each other within the context of the whole organism. The sheer amount of DNA contained within the genome of an organism, however, represents a substantial barrier to attaining this level of analysis. Even in the absence of complete sequence knowledge, however, a variety of methods have been used to map the location of genes and other DNA sequences within the genome. On a small scale, mapping DNA fragments is a relatively straightforward process (Figure 9.1). We have already seen (Chapter 2) that restriction enzymes will cleave DNA at specific sequences, termed recognition sites. The

Analysis of Genes and Genomes   Richard J. Reece
© 2004 John Wiley & Sons, Ltd   ISBNs: 0-470-84379-9 (HB); 0-470-84380-2 (PB)

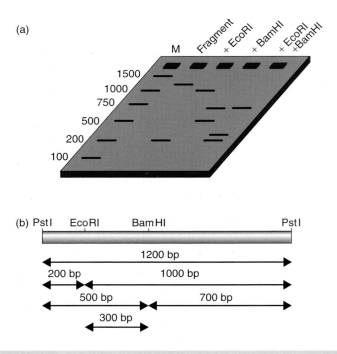

Figure 9.1. Mapping small DNA fragments by restriction digestion. (a) An isolated 1.2 kbp PstI DNA fragment is cleaved with the restriction enzymes EcoRI or BamHI or a mixture of EcoRI and BamHI as indicated. The products are separated on an agarose gel and the sizes of the resulting bands calculated with reference to DNA fragments of known size (M). (b) The deduced restriction map of the DNA fragment. PstI sites must be at either end, and the locations of the EcoRI and BamHI sites are calculated from the sizes of the fragments produced

cleavage sites can be used as map reference points to build up a linear diagram of the order in which the restriction sites occur within a particular DNA molecule and the distance between each site – as determined by the lengths of fragments produced after digestion. On a genome-wide scale, however, analysis of this type is extremely difficult. The massive number of DNA fragments produced upon restriction digestion of genomic DNA makes it almost impossible to order fragments this way. Here, we will discuss a number of genetic and physical methods that have been used to map genomes. Our discussion will concentrate mainly on the mapping and sequencing projects associated with the human genome, although readers should be aware that much of the groundwork for the elucidation of the human genome sequence has come from the analysis of other organisms – both prokaryotic and eukaryotic.

9.1 Genomic Mapping

In eukaryotes the simplest, and most natural, way to split a genome into smaller fragments is to consider the DNA contained within each chromosome individually. Since each is composed of one double-stranded DNA molecule, the chromosome provides the first level of genome mapping. The chromosome content of an organism (its **karyotype**) can be visualized using a microscope. Each chromosome is composed of two arms separated by a centromere. By convention, the shorter arm of each chromosome is designated as p and the longer arm is designated as q. The different chromosomes of an organism are usually different sizes (ranging in the human from 279×10^6 bp for chromosome 1 to 45×10^6 bp for chromosome 21), but most chromosomes are difficult to distinguish based on size alone by microscopy. Distinct chromosome banding patterns can be obtained, however, when they are treated with certain dyes. Approximately 500 different bands can be obtained reproducibly after treating human chromosomes with the stain Giemsa (Figure 9.2). These banding patterns can be used to generate a **cytological map** of each chromosome and provide a low-resolution mechanism to distinguish one portion of a chromosome from another. Some chromosome abnormalities that cause inherited genetic diseases can be observed by karyotype analysis – additional copies of chromosomes can be easily identified, e.g. Down's Syndrome results from an extra copy (trisomy) of all or part of chromosome 21, and sufferers from Klinefelter's Syndrome possess three sex chromosomes (XXY). Additionally, a variety of other chromosome abnormalities, e.g. deletions, inversions, translocations etc., can be detected as alterations in the normal banding pattern. The banding pattern also provides a mechanism for labelling chromosome regions. For example, using some of the techniques described below, the gene mutated in sufferers of cystic fibrosis has been mapped to the long arm of chromosome 7 in banding region 31. The chromosomal location of the gene in the cytological map is therefore designated as 7q31.

Isolated DNA fragments can be plotted onto the cytological map by a variety of methods. For example, fluorescently labelled single-stranded DNA fragments will hybridize to chromosome spreads like those shown in Figure 9.2 to yield the location of the complementary sequence (Taanman *et al.*, 1991). This method of fluorescent *in situ* hybridization (**FISH**) is a powerful way to localize DNA sequences to individual chromosomes and even parts of chromosomes, but is low resolution in that sequences closer than approximately 3 Mbp apart will hybridize indistinguishably from each other. A number of additional genetic and physical maps of chromosomes have also been produced to aid the localization of specific DNA sequences (Figure 9.3), and we will discuss these below.

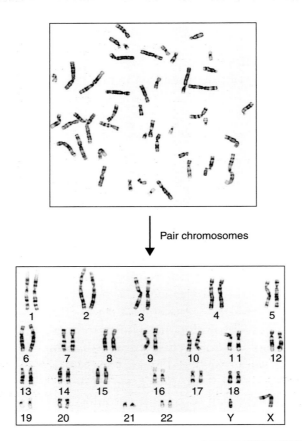

Figure 9.2. The human male karyotype showing the G bands. Metaphase chromosomes from a male were treated with the protease tryspin (to remove protein) and then stained with a mixture of dyes called Giemsa (named after Gustav Giemsa, who first used it) and viewed using a light microscope. Each pair of chromosomes has a similar length and banding pattern that allows them to be aligned. Chromosomes from a female would have two X chromosomes rather than the X and Y shown here

9.2 Genetic Mapping

A genetic map is a representation of the distance between two DNA elements based upon the frequency at which recombination occurs between the two. The first genetic map of a chromosome was constructed by Alfred Sturtevant using data from *Drosophila* mating crosses collected by Thomas Morgan (Morgan, 1910). Sturtevant used the frequency at which particular observable phenotypes were separated from other genes (through recombination events) during meiosis. The information gained from the experimental crosses could be used to plot out the location of genes – tightly linked genes are physically

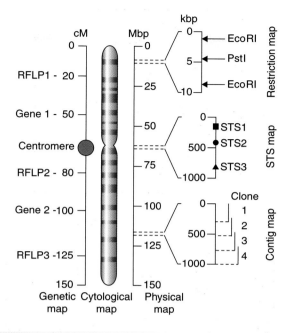

Figure 9.3. The different types of cytological, genetic and physical map of a chromosome. Genetic map distances are based on crossover frequencies and are measured in centiMorgans (cM), while physical distances are measured in megabase pairs (Mbp) or kilobase pairs (kbp)

located close to each other, while those that were only weakly linked are physically further apart. Sturtevant constructed a genetic map of the locations of six genes on the X chromosome of *Drosophila melanogaster* (Sturtevant, 1913). Many other gene traits in a variety of different organisms have been mapped using similar techniques. Genetic maps can be constructed for each chromosome within an organism. Genes on different chromosomes are not linked to each other and are therefore not amenable to this analysis. The major drawbacks with this type of approach are the requirement for a phenotype for the gene that is being mapped and the number of crosses required to generate accurate mapping data. Additionally, a tacit assumption of mapping based on crosses is that the recombination frequency is equal for all part of the chromosome. This is simply not the case, and many recombinational 'hot-spots' and 'cold-spots' have been identified.

In humans, the segregation of naturally occurring mutant alleles in families can be used to estimate map distances, but the relatively low number of previously identified human genes makes this approach difficult. An alternative to genetic mapping using phenotypes is to follow the inheritance of DNA sequence variations between individuals. It is estimated that more than 99 per cent of human DNA sequences are the same across the population. This still allows for huge numbers of variations in DNA sequence between individuals. Several different methods have been used to exploit the inheritance of these variations to map their genomic location.

- *Single-nucleotide polymorphisms*. The most common types of sequence variation between individuals are described as **single-nucleotide polymorphisms** (SNPs), in which a single base pair is different between one individual and another. These differences may occur as frequently as about once every 100–300 bp (Collins *et al.*, 1998). Some of these alterations will be disease causing mutations – they may change the sequence of amino acids within a protein or alter the way in which gene expression occurs to impair the function of the resulting protein. Many SNPs, however, occur in non-coding regions of DNA or, even if they do occur within a coding region, they may not alter the amino acid sequence of the encoded polypeptide due to the degeneracy of the genetic code. Some of the nucleotide differences between individuals will, however, result in the alteration of restriction enzyme recognition sites such that existing sites are destroyed or new sites are created (Figure 9.4). Base changes at these sites results in different length DNA fragments being produced upon restriction digestion. These restriction fragment length polymorphisms (**RFLPs**) are usually detected by Southern blotting (Chapter 2) using a radioactive DNA probe. RFLPs are inherited and segregate in crosses and they can therefore be mapped using linkage analysis like genes (NIH/CEPH Collaborative Mapping Group, 1992).

- *VNTRs*. Another common variation in humans involves short DNA sequences that are present in the genome as tandem repeats. The number of copies of **variable number tandem repeats** (VNTRs) at a specific genomic location can vary widely between individuals, and is described as being highly polymorphic. Restriction fragment sizes (again detected by Southern blotting) using enzymes that cleave the DNA in regions flanking the repeats will be of different sizes depending on the number of repeats present.

- *Microsatellites*. Microsatellites are short, 2–6 bp, tandemly repeated sequences that occur in a seemingly random fashion distributed throughout the genome of all higher organisms. They are generally found in non-coding regions of DNA, and their function (if any) is unknown. The number of repeats found at any particular genomic location is highly individual specific. The repeats are thought to be generated by polymerase 'slippage' during replication (Schlötterer, 2000). In humans, the most common type

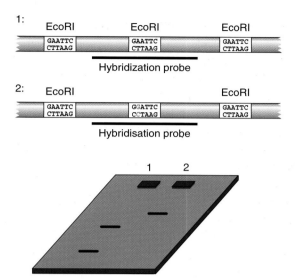

Figure 9.4. Restriction fragment length polymorphisms. (a) A section of DNA that contains three recognition sites for the restriction enzyme EcoRI. A single base change within one of the sites destroys the recognition sequence. (b) Cutting the DNA with EcoRI will generate different sized fragments that will be able to hybridize to the labelled DNA fragment (hybridization probe) shown. In the first case two small fragments will be formed that are capable of binding the probe, while in the second a single, larger fragment will bind. The restriction fragments are separated on an agarose gel and subjected to Southern blotting (see Figure 2.21) to identify sequences that are complementary to the probe

of microsatellite is 5′-AC-3′ and several thousand different AC arrays may occur throughout the genome. Dinucleotide microsatellites in mammals typically vary in repeat number from about 10 to 30 repeats. The microsatellite DNA is subjected to PCR amplification using primers that flank the repeated region. The size of the PCR product obtained will therefore depend on the number of repeats. Microsatellites are inherited from one generation to the next and can thus be used for mapping by linkage analysis (Dib *et al.*, 1996).

9.3 Physical Mapping

The information held within genetic maps provides vital clues as to the order and approximate distance between particular DNA sequences within a chromosome. The map, although not providing sequence information itself, yields a framework onto which subsequently obtained sequence information can be applied. The physical map of a genome is a map of **genetic markers** made by analysing a genomic DNA sequence directly, rather than analysing recombination events. As with genetic maps, physical maps for each chromosome within the genome can be constructed. Again, a variety of different techniques have been used to construct physical maps in the absence of complete sequence information.

- *Restriction maps.* The digestion of genomic DNA, or even isolated chromosomes, with restriction enzymes produces a large number of fragments that appear to run as a continuous smear, rather than as discrete bands, on agarose gels after electrophoresis. However, certain restriction enzymes, *e.g.* NotI, have a comparatively large recognition sequence (5′-GCGGCCGC-3′) that is rarely found in human DNA sequences. The recognition site for NotI would be expected to occur, by chance, every $4^8 = 65\,536$ bp. Experimentally, NotI cleaves human DNA on average once every 10 Mbp. The discrepancy between these two numbers arises from the fact that the DNA sequence within the genome is not random. For example, the sequence 5′-CG-3′, occurs comparatively rarely in the human genome and clusters of this dinucleotide tend to accumulate only at the 5′-end of actively transcribed genes (Cross and Bird, 1995). The recognition sequence for the NotI restriction enzyme contains two of these dinucleotide repeats and explains why the enzyme cuts human DNA so infrequently. Even using rare cutting restriction enzymes such as NotI, the construction of genomic restriction maps like those generated for small DNA fragments (Figure 9.1), is extremely difficult. Restriction mapping does provide highly reliable fragment ordering and distance estimation, but has only been completed for a few human chromosomes (Ichikawa *et al.*, 1993; Hosoda *et al.*, 1997).

- *Radiation hybrid maps.* A radiation hybrid is, usually, a hamster cell line that carries a relatively small DNA fragment from the genome of another organism, e.g. human. Irradiating human cells with X-rays causes random breaks within the DNA and produces fragments. The size of the fragments produced decreases as the dose of X-rays increases. The radiation levels used are sufficient to kill the human cells, but the chromosome fragments can be rescued by fusing the irradiated cells with a hamster cell *in vitro*. Typically, the human DNA fragments in the hybrid are a few Mbp long. The human DNA within the hybrid cell line is then analysed for the genetic markers it carries, either by hybridization, or by PCR. The closer the two markers are, the greater the probability those markers will be on the same DNA fragment and therefore end up in the same radiation hybrid.

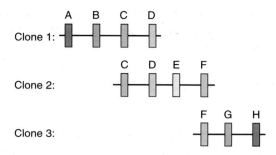

Figure 9.5. Aligning clones by STS mapping. Each clone contains several STSs. Clone 1 has four (A, B, C and D). Clone 2 also contains STSs C and D. Therefore clones 1 and 2 overlap with each other

- *STS maps.* A sequence tagged site (STS) is a DNA fragment, typically 100–200 bp in length, generated by PCR using primers based on already known DNA sequences. The genomic site for the sequence in question can be 'tagged' by its ability to hybridize with that sequence. STSs can be generated from previously cloned genes, or from other random non-gene sequences. Genomic DNA fragments that have been cloned into a library can then be ordered on the basis of the STSs they contain (Figure 9.5). This technique has been used to order inserts from individual human chromosomes in a YAC library (Foote *et al.*, 1992), but fell foul when it was discovered that some YACs contained DNA from more than one human genome location. An STS map of the human genome has, however, been constructed using a series of radiation hybrids (Hudson *et al.*, 1995).

The physical maps, although not aligning DNA base sequences themselves, have proved immensely useful in producing ordered library clones. The final stage of any sequencing project is then to determine the individual base sequence of each clone. Before we look at how the human genome sequence was attained and assembled, we needed to understand how the DNA sequence information itself is obtained.

9.4 Nucleotide Sequencing

The uniformity of the DNA molecule and the seemingly monotonous repetition of the nucleotide bases may seem like impenetrable barriers to determining the precise sequence order of the bases within nucleic acid. In 1966, Robert Holley published the results of a 7 year project to sequence the alanine tRNA from yeast (Holley, 1966). At 80 nucleotides in length, tRNAs are relatively small molecules in comparison to complete genes, or even complete genomes. The first DNA molecule to be sequenced was that of the bacteriophage λ cohesive (*cos*) ends (Wu and Taylor, 1971). These sequences, which are only 12 bases long, were obtained after the synthesis of a complementary RNA molecule and the subsequent use of RNA sequencing procedures. The methods used were, however, impractical for DNA sequencing on a large scale. In 1975, Fred Sanger and Alan Coulson devised a method of direct DNA sequencing referred to as the plus–minus method (Sanger and Coulson, 1975). This method utilized a DNA polymerase, primed by synthetic radio-labelled oligonucleotides, to generate fragments of DNA that could be analysed following electrophoresis and autoradiography. This technique was used to determine the entire 5386 bp sequence of the bacteriophage øX174 genome (Sanger *et al.*, 1977).

9.4.1 *Manual DNA Sequencing*

Two alternative, and improved, sequencing methods were described in 1977. Allan Maxam and Walter Gilbert devised a chemical method for cleaving the sugar–phosphate backbone of a radio-labelled DNA fragment at specific bases (Maxam and Gilbert, 1977). They used specific chemicals to modify individual DNA bases (e.g. the modification of T residues with potassium permanganate) or sets of bases (e.g. the modification of both A and G residues with formic acid) prior to cleavage of the sugar–phosphate backbone with piperidine at the modified bases (Maxam and Gilbert, 1980). The separation of the cleaved products using high-resolution polyacrylamide gel electrophoresis allowed unequivocal assignment of individual bases within a DNA sequence. Their method was, however, limited in the length of the DNA that can be sequenced during a single reaction (approximately 100 bases) and by the use of harsh chemicals required to modify and cleave the DNA.

Fred Sanger and his colleagues devised an alternative sequencing approach based upon the faithful replication of DNA using a DNA polymerase (Sanger, Nicklen and Coulson, 1977b). They relied on the incorporation of 2′, 3′ dideoxynucleotides into a newly replicated DNA chain to generate DNA fragments that ended at a specific base (Figure 9.6). The dideoxynucleotide lacks a 3′ hydroxyl group and, consequently, when it is incorporated into an extending DNA chain, DNA replication cannot continue as the 3′ hydroxyl group is not available for the addition of further nucleotides. Thus, the growing DNA chain is terminated after the addition of the dideoxynucleotide. As originally described by Sanger, DNA replication was initiated by the binding of a complementary oligonucleotide to the DNA sequence and subsequent

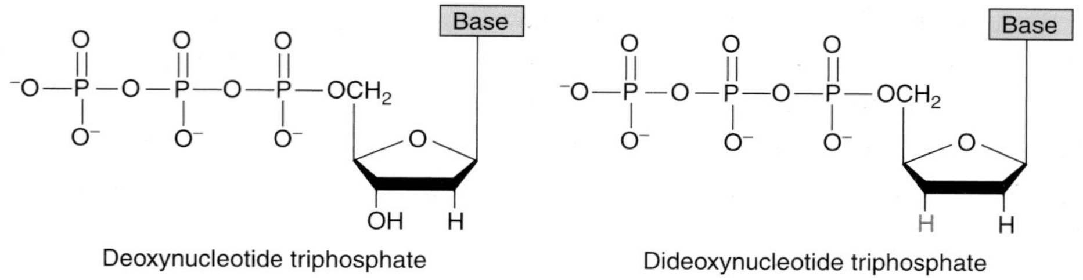

Figure 9.6. The structure of a deoxynucleotide triphosphate and its dideoxy derivative

incubation with DNA polymerase. The newly synthesized DNA will thus be complementary to the strand of DNA to which the oligonucleotide binds. The sequencing reaction was then split into four separate parts. To each was added a mixture of the four nucleotide triphosphates (dNTPs) required for the synthesis of new DNA. One of these was radio-labelled so that the newly synthesized DNA could be easily detected. Additionally, a single dideoxynucleotide triphosphate (either ddATP, ddGTP, ddCTP or ddTTP) was included in each reaction at a concentration of approximately 1/10 of its deoxynucleotide counterparts. Therefore, in the reaction containing ddATP, for example, when a T residue occurs on the template strand, in most cases a dATP will be inserted into the newly synthesized chain. However, at a relatively low frequency the dideoxy form of the nucleotide will be incorporated and the chain will terminate at this point. Since many DNA molecules are produced at the same time, this process results in the formation of a population of partially synthesized radioactive DNA molecules each having a common 5′-end, but each varying in length to a specific base at the 3′-end (Figure 9.7). These products can be separated using polyacrylamide gel electrophoresis and the sequence of the newly synthesized DNA can be read. The gel used to separate the newly synthesized DNA fragments usually contains high concentrations of urea (7 M) and is run at a high power level to heat the gel to about 70 °C. Both of these have denaturing effects on DNA fragments and help reduce secondary structure that could occur in the single-stranded molecules that may make them run anomalously through the gel.

The use of DNA replication as a tool for sequencing has several advantages.

- DNA synthesis can be initiated at any known point in a DNA sequence through the design of an oligonucleotide. This does mean that some knowledge of the DNA sequence is required before sequencing can commence. Many popular cloning vectors (Chapter 3) contain common oligonucleotide binding sequences flanking cloning sites so that unknown DNA cloned into them may be sequenced.

- Unlike the Maxam–Gilbert technique, the DNA strand that is being sequenced does not need to be radio-labelled. Labelling is required so that the extended and chain terminated products can be detected after gel electrophoresis. A radio-label (e.g. ^{32}P, ^{35}S, or ^{33}P in the form of an α-modified deoxynucleotide) can be incorporated into the newly extended chain as part of the replication process itself.

- The DNA molecule to be sequenced does not necessarily have to be single stranded. The original Sanger method was used to sequence linear double-stranded restriction digestion products, but was not directly applicable to the sequencing of double-stranded plasmids. This led to the widespread use of M13 vectors to produce single-stranded templates for sequencing (see Chapter 3). The single-stranded DNA produced from M13 generally yielded very clean readable sequence. The Sanger technique was subsequently adapted to allow for the denaturation of plasmid DNA using alkali that was suitable for sequencing (Yie, Wei and Tien, 1993).

Many modifications have been made to the chain-terminating sequencing protocol since its inception, but the basic chain-termination method devised by Sanger has remained the cornerstone of almost all sequencing projects. The Klenow fragment of *E. coli* DNA polymerase I was originally used as the replicating enzyme, but this was superseded by a modified form of the DNA polymerase from the bacteriophage T7 (also known as Sequenase™), which proved to be a more processive enzyme that allowed more sequence to be read from a single reaction (Griffin and Griffin, 1993). It is essential that an enzyme lacking a 5′–3′ exonuclease activity is used for sequencing to ensure the integrity of the newly synthesized DNA fragments. Nowadays, most sequencing is performed using Taq DNA polymerase. The high temperatures at which the thermostable enzyme can replicate DNA ensure that secondary structure is kept to a minimum so that cleaner, more readable sequences can be obtained. The use of Taq DNA polymerase is often combined with thermocycling to amplify a single DNA strand of a duplex in a linear manner from a single primer (Murray, 1989). This eliminates the requirements for separate double-stranded DNA denaturation and primer annealing steps. The method enables sequencing from very small amounts of double-stranded DNA, and also allows direct genomic DNA sequencing from bacterial colonies or phage plaques, thereby bypassing the requirement for cloning entirely (Slatko, 1996).

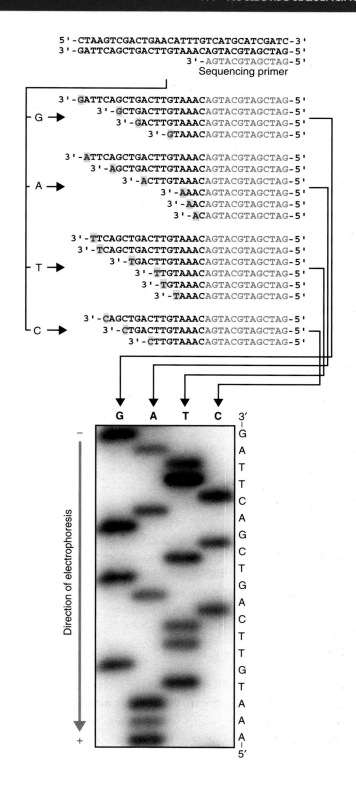

9.4.2 Automated DNA Sequencing

Sequencing using the Sanger technique can lead to clean and unambiguous assignment of about 300 bases per reaction. The method is, however, quite labour intensive. For example, multiple pipetting steps are required to set up each reaction and then the reactions must be loaded onto four lanes of a gel to separate the products. Additionally, the manual reading of sequencing gels (Figure 9.7) can be both time consuming and error prone. To tackle the sequence of the human genome ($\sim$3.2 $\times$ 10^9 bp), more automated and rapid methods of sequence collection were required.

A straightforward way to increase the throughput of DNA sequencing would be to combine the four individual sequencing reactions (each containing a different ddNTP) into a single reaction that could be analysed on a single lane of a gel. This is not possible using radioactivity since each band (Figure 9.7) is distinguishable only by the position in which it runs on the gel. Therefore, combining all four lanes would merely result in a series of bands differing in size by a single base (Figure 9.8). However, if the terminal base of each DNA fragment can be identified specifically then, since each band on the gel is a different size, the DNA sequence can be unambiguously assigned from a single gel lane. A set of dideoxynucleotides has been developed that are labelled with fluorescent dyes precisely for this purpose (Glazer and Mathies, 1997). The dideoxynucleotide can still be incorporated into DNA opposite its complementary base, which again results in the termination of DNA synthesis. The dye structures attached to the dideoxynucleotide contain a fluorescein donor dye linked to a dichlororhodamine (dRhodamine) acceptor dye *via* an aminobenzoic acid linker and are called BigDye™ terminators. An argon ion laser is able to excite the fluorescein donor dye that efficiently transfers the energy to one of the four acceptor dyes, each of which has a distinctive emission spectrum (Figure 9.9). Each dideoxynucleotide is labelled with a different acceptor dye so that DNA fragments ending in a different ddNTP

Figure 9.7. DNA sequencing using dideoxynucleotide chain terminators. DNA replication is initiated from an oligonucleotide primer and four individual sequencing reactions are performed each of which contains all the dNTPs and a single ddNTP (either G, A, T or C, as indicated). DNA replication is terminated when the ddNTP (highlighted in yellow) is incorporated to generate a series of different length DNA molecules that can be separated using a polyacrylamide gel. The sequence of the newly synthesized DNA can be read in a 5′ to 3′ direction from the bottom of the gel to the top. In the example shown, the primer produces a new 'bottom' DNA strand. The sequence of the 'top' strand can be obtained from this

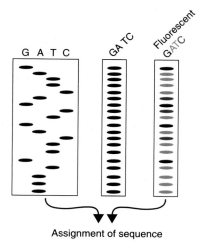

Assignment of sequence

Figure 9.8. Sequencing in a single gel lane is only possible if a mechanism exists to identify each individual band on the gel as being terminated by a particular dideoxynucleotide

will fluoresce at a different wavelength. Sequencing reactions can therefore be performed in a single tube (or single well of a microtitre dish) and the products separated either on a single lane of a gel, or using a capillary tube containing a gel matrix (Karger *et al.*, 1991). The intensity and wavelength of the fluorescent emission is measured as the DNA fragments move past a laser and fluorescence detector located at the bottom of the gel. This information is fed directly into a computer so that the resulting sequence can be automatically assigned and stored.

Sophisticated base calling software is available to convert the fluorescent patterns obtained into a sequence of DNA bases (Figure 9.10). Sequencing in this way has massive speed advantages over manual sequencing methods. As many as 1000 bases can be read automatically from a single reaction, although the sequence obtained from within 500 bp of the primer is generally more reliable than that further away. Additionally, the detection methods used during automated sequencing are far more reliable than sequence interpretation from an autoradiograph. Even so, automated DNA sequencing is not infallible. For example, long continuous runs of the same nucleotide can become compressed together as they travel though a gel. This may result in multiple, overlapping peaks on the fluorescent trace that need to be deconvoluted manually.

The main advantage of sequencing in this way is the ability to automate almost all parts of the process. Sequencing reactions can be set up robotically in the wells of microtitre dishes and subjected to thermocycle sequencing. The products can then be purified from the plates, loaded onto capillary columns,

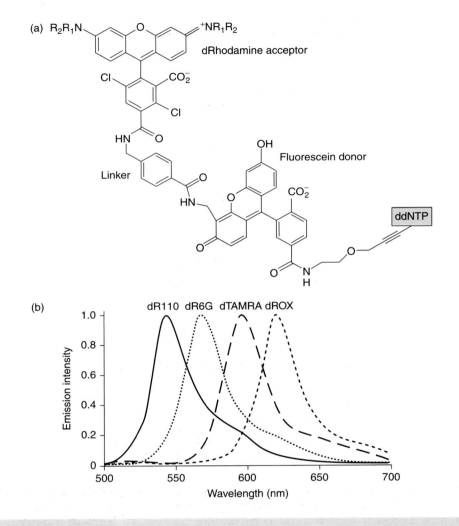

Figure 9.9. Dideoxynucleotide terminator dyes used for DNA sequencing. (a) The general structure of BigDye™ terminators. The different terminators have different emission properties depending on the nature of the R groups. (b) The emission spectra of the four BigDye™ terminators

and subjected to electrophoresis without the need for human intervention. A single DNA sequencing machine working like this is capable of generating between 1 and 2 million bases of DNA sequence per day with a single-run accuracy of between 98 and 99 per cent. This level of accuracy may sound impressive, but if one base in every 100 is incorrectly assigned, then virtually all genes whose sequence is obtained in this way will contain errors. It is therefore

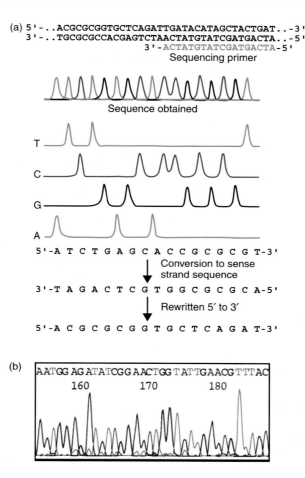

Figure 9.10. Automated DNA sequencing. (a) A sequencing primer is used to initiate DNA synthesis that is terminated using a fluorescently labelled dideoxynucleotide. The series of peaks obtained has been separated into the individual fluorescent components and the sequence assembled based on the data obtained. (b) The printout of an actual sequence obtained from an ABI Prism 377 DNA sequencing machine

necessary to sequence the same section of DNA, preferably from multiple clones containing different lengths of the same DNA sequence on multiple occasions, to ensure that accuracy levels are as high as possible.

9.5 Genome Sequencing

As we have already seen (Chapter 3), genomes have to be split into sections of a suitable size such that they can be maintained within a vector. Although sequencing can be performed directly on genomic DNA, this is generally impractical on a large scale due to the large number of different oligonucleotide primers that would need to be synthesized to initiate the sequencing reactions. Genomic DNA fragments are therefore cloned into a vector and each fragment is subsequently sequenced using the automated methods described above. The problem then is how to reconstruct the original genome sequence based on the small fragments that are cloned into individual vectors. Several basic approaches have been used.

- *Clone contigs*. The simplest way to generate overlapping DNA sequence is to isolate and sequence one clone, from a library, then identify (by hybridization) a second clone, whose insert overlaps with the first. The second clone is then sequenced and the information used to identify a third clone, whose insert overlaps with the second clone, and so on. This is the basis of **chromosome walking** and was first used to build up large continuous DNA sequences (**contigs**) from small fragments cloned into vectors. This method is, however, laborious. A single clone has be isolated and sequenced before the next overlapping clone can be sought. Additionally, repetitive sequences within the genome can give rise to incorrect contig assignment.

- *Whole genome shotgun*. The fragments of the genome, which have been randomly generated, are cloned into a vector and each insert is sequenced. The sequence is then examined for overlaps (sequences that occur in more than one clone) and the genome is reconstructed by assembling the overlapping sequences together (Figure 9.11). This approach was first used to sequence the genome of the bacterium *Haemophilus influenzae*. The entire genome of the organism was randomly fragmented using sonication and then small fragments (in the range of 1.5–2 kbp) were cloned into a vector (pUC18). The resulting library consisted of approximately 20 000 individual clones. Each of these was then sequenced to generate approximately 12 million base pairs of sequence information (Fleischmann *et al.*, 1995). This corresponds to six times the length of the *H. influenzae* genome. The sequence obtained from each clone was then assembled into contigs based on the overlaps between the individual clones. Any sequence gaps were filled in subsequently by identifying additional clones (from a different library) that contained sequences close to the gap-point. The main advantage of the shotgun approach is that no prior knowledge of the sequence of the genome is required. The approach is, however, limited by the ability to identify overlapping sequences. Every sequence obtained must be compared with every other sequence in order to identify the overlaps. This can be a time-consuming process and requires large amounts of computational

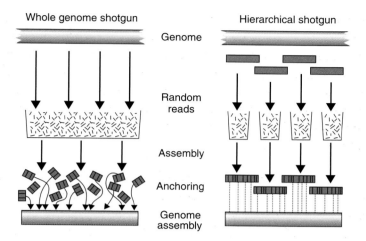

Figure 9.11. Assembling genomic data using the hierarchical and whole genome shotgun approaches. Adapted from Waterston, Lander and Sulston (2002), with permission

power (Myers *et al.*, 2002). The search for overlaps is also hampered by the presence of repetitive DNA sequences in the genome that could lead to the incorrect assignment of contigs (Waterston, Lander and Sulston, 2002).

- *Hierarchical shotgun.* The shotgun approach to contig assembly has proved immensely successful in sequencing comparatively small genomes. The majority of bacterial genomes can be sequenced by this method and contigs assembled within a matter of weeks using the automated DNA sequencing technologies described above. For larger genomes, however, it is not the sequencing itself that is the rate limiting step, but the assembly of contigs. Contig assembly can, however, be greatly simplified if the genomic DNA is first broken up into a series of overlapping large clones such as those produced by cloning into BACs (Chapter 3). A library of smaller clones is then produced from each BAC and subjected to shotgun sequencing as described above. The hierarchical approach provides a mechanism to relatively easily construct assembled contigs since sequence assembly is local and anchored to a particular part of the genome – the sequence contained within the BAC.

9.6 The Human Genome Project

Officially begun in 1990, the Human Genome Project (HGP) was established with the following goals:

- determine the DNA sequence of the entire human genome

- store this information in databases

- identify all of the genes in human DNA

- improve tools for data analysis.

The project was planned to last for 15 years, but technological advances have accelerated the expected completion date to 2003. On 26 June 2000, in a joint press conference, President Bill Clinton (USA) and Prime Minister Tony Blair (UK) announced that the first rough draft of the human genome sequence had been completed. In generating the draft sequence, the order of bases in each chromosome was determined at least four or five times to ensure data accuracy and to help with contig assembly. This repeated sequencing is known as genome 'depth of coverage'. Draft sequence data is mostly in the form of ~10 000 bp contigs whose approximate chromosomal locations are known. The human genome sequence does not represent an exact match for any single person. DNA samples, from both male sperm and female blood, from several different anonymous individuals comprise the genomic libraries that were used for sequencing. A large number of people have donated samples for the construction of these libraries, but only a few samples have been used and neither the donors nor those responsible for the sequencing know whose DNA is actually in them. It is likely that DNA from 10–20 individuals has been used to produce the draft genome sequence. The initial analysis of the draft sequence data was published in February 2001 by two competing groups – the HGP (Lander *et al.*, 2001) and the biotechnology company Celera Genomics (Venter *et al.*, 2001). HGP is a consortium of 16 academic research laboratories and institutes that used a hierarchical shotgun approach to sequencing the genome. As a publicly funded group, the data obtained from their sequencing efforts is made freely available. The **Bermuda Principle** (determined during a 1997 conference in Bermuda) states that sequence assemblies of 2 kbp or larger should be automatically released to public databases within 24 hours of their generation. Celera Genomics undertook a whole genome shotgun approach to sequencing the human genome. Using large numbers of DNA sequencing machines, Celera completed the sequencing phase of their project within 9 months and then employed massive computing power to reconstruct the genome based on their data.

To generate a high-quality final sequence of the human genome, additional sequencing is needed to close gaps, reduce ambiguities and allow for only a single error every 10 000 bases, giving an accuracy of 99.9999 per cent – the agreed standard for the finished sequence. The finished version will provide an estimated eight- to nine-fold depth of coverage for each chromosome. To date, finished sequences have been generated for the three smallest human chromosomes −20, 21 and 22. Fully sequenced genomes are, in fact, not usually complete. For example, the 56 Mbp sequence of human chromosome 22 was declared essentially complete in 1999, yet only 33.5 Mbp were sequenced (Dunham *et al.*, 1999). Also, the 180 Mbp genome of the fruit fly *Drosophila* was announced as completed, although just 120 Mbp were sequenced (Adams *et al.*, 2000). Higher-eukaryotic genomes have large regions of DNA that currently cannot be cloned or assembled. These regions include telomeres and centromeres (the ends and middle of chromosomes), as well as many chromosomal areas rich in other sequence repeats. Most of the regions that cannot be sequenced contain **heterochromatic** DNA, which has few genes and many repeated regions. Current efforts are aimed at sequencing **euchromatic** DNA – defined as gene-rich areas including both exons and introns. In the case of human chromosome 22, the sequenced 60 per cent represents 97 per cent of euchromatic DNA. Similarly, nearly all the euchromatic regions were sequenced in *Drosophila*. Although the goal of the human genome project is to have a complete sequence for each chromosome, obtaining a full sequence still presents a great challenge.

9.7 Finding Genes

Once the DNA sequence of the genome has been obtained, various methods can be employed to identify and locate the genes that reside within it.

- *Sequence inspection.* Genes that code for proteins comprise an ORF consisting of a series of triplet codons that specify the amino acid sequence of the protein. The ORF begins with an initiation codon (usually ATG) and ends with a termination codon (either TAA, TAG or TGA). The average length of an ORF varies between species. In *E. coli*, the average ORF length is 317 codons, while in yeast the average length is 483 codons. Most ORFs are, however, more than 50 codons in length. The search for a gene can therefore be thought of as a scan for an initiation and termination codon that are separated by, say, at least 100 codons. The approach works well for compact prokaryotic genomes, where genes are separated by only small regions of non-coding DNA, but is difficult to apply to the genomes of higher eukaryotes, where ORFs are split into introns and exons. Several computer programmes are available for the identification of ORFs (Fickett, 1996). In general these programmes are species specific; e.g. GRAIL (Roberts, 1991) and GeneMark (Isono, McIninch and Borodovsky, 1994) are designed to identify ORFs within human DNA and other individual species, respectively. These programmes assign ORFs not only on the basis of initiator and terminator codons, but also using codon bias to identify likely coding regions, and the identification of both intron–exon boundaries and transcriptional control elements (e.g. the TATA box). Unfortunately, these latter sequences can be quite variable, and precise gene identification remains problematical. An alternative approach to gene identification is to use previously identified genes as a guide. Is the gene we are trying to assign similar (homologous) to any existing genes? If so, it is likely that assignment has been made correctly. One danger with this type of approach is that **pseudogenes** (generally non-transcribed genomic DNA with a high degree of sequence similarity to a real gene) may be assigned as real genes.

- *cDNA comparison.* The simplest way to identify a gene within a segment of genomic DNA is compare the sequence to a copy of the corresponding cDNA. Readers will remember that cDNA (Chapter 5) is produced from mRNA and contains just the exon sequences of the ORF joined together. This can be achieved either through the hybridization of genomic DNA fragments to mRNA separated on an agarose gel (northern blotting, see Chapter 2) or through the comparison with databases of sequenced cDNA fragments. **Expressed sequence tags** (**ESTs**) are small pieces of cDNA sequence (usually 200 to 500 bases long) that are generated by sequencing either one or both ends of an expressed gene. Random cDNA clones are sequenced to generate sections of sequence that represent genes expressed in certain cells, tissues or organs from different organisms. These tags can then be used to identify the gene encoding them from genomic DNA by sequence comparison. Because ESTs represent a copy of just the interesting part of a genome – that which is expressed – they are powerful tools in the hunt for genes. ESTs also have a number of practical advantages – the sequences can be generated rapidly and inexpensively; only one sequencing experiment is needed for each cDNA generated; they do not have to be checked for sequencing errors as mistakes do not prevent identification of the gene from which the EST was derived using similarity searches. Databases of EST sequences are publicly available, e.g. dbEST (http://www.ncbi.nlm.nih.gov/dbEST/), which contains over 12 million sequences from different organisms including 4.5 million human sequences. Many of these sequences are, of course, repetitious (Banfi, Guffanti and Borsani, 1998), with highly expressed genes being represented many times. Additionally, it should be noted that genes that are expressed at a low level, or those whose expression pattern is highly tissue or developmental stage specific, might not be present within an EST database.

9.8 Gene Assignment

Genome sequencing projects have thrown up some interesting, and somewhat unexpected, results. For example, even though *E. coli* and *S. cerevisiae* had been studied extensively in the laboratory for many decades, when their genomic sequence became available, it was realized that only between 30 and 40 per cent of the genes they contained had been previously characterized. In less experimentally amenable organisms, especially in humans, comparatively few genes were known before large-scale sequencing projects were undertaken. There are, however, several methods there are currently used to assign the function of a gene based only on its sequence.

- *Similarity searches.* Just as computational methods play an important role in defining those portions of the genome that may encode genes, the availability of large databases of known gene sequences can also be used to assign function to unknown ones. Similarity searches like this are usually performed using amino acid sequences because the comparisons of the four DNA bases will often yield similar sequences even through the encoded proteins are very different. Many genes that encode proteins with the same function in different organism will be similar. For example, almost all organisms have the ability to convert the sugar galactose into glucose-6-phosphate so that it can be fed into the glycolytic cycle. The first step of this pathway is the conversion of galactose into galactose-1-phosphate – a reaction that is catalysed by the enzyme galactokinase. All organisms possess their own galactokinase enzyme, and the galactokinases from different organisms each have their own unique sequence. However, most likely as a result of having to perform the same chemical reaction, galactokinase enzymes are related to each other. That is, the amino acid sequence of the galactokinase from one organism shares similarity to the galactokinase from another organism (Figure 9.12). Although only 30 per cent of the approximately 6000 yeast genes had previously ascribed function (see above), the function of an additional 30 per cent could be ascribed based on similarity searches. This still leaves 40 per cent of the identified yeast genes having no known function. Of course, some of these genes may not be real, perhaps being incorrectly assigned as genes, but many will need to have their function assigned by other mechanisms.

- *Experimental gene assignment.* In experimental organisms, such as *E. coli* or yeast, one of the most popular ways of ascribing a function into an unknown gene is to make a gene knockout. As we will see later chapters,

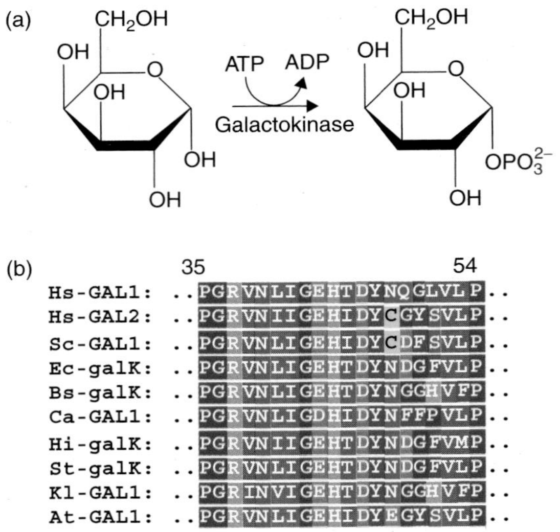

Figure 9.12. Sequence comparison of galactokinases from different species. (a) The galactokinase reaction. (b) Comparison of a region of human galactokinase (amino acids 35–54 of the 392 amino acid protein). Key: Hs – *Homo sapiens*, Sc – *Saccharomyces cerevisiae*, Ec – *Escherichia coli*, Bs – *Bacillus subtilis*, Ca – *Candida albicans*, Hi – *Haemophilus influenzae*, St – *Salmonella typhimurium*, Kl – *Kluyveromyces lactis*, At – *Arabidopsis thaliana*. Amino acids have been coloured according to their properties. Blue indicates positively charged amino acids (H, K, R), red indicates negatively charged residues (D, E), green indicates polar neutral residues (S, T, N, Q), grey indicates non-polar aliphatics (A, V, L, I, M) and purple indicates non-polar aromatic residues (F, Y, W). Brown is used to indicate proline and glycine, while yellow indicates cysteine

homologous recombination is used in both yeast and in higher-eukaryotic cells to disrupt the functional copy of a gene within a genome. The phenotype of the disrupted mutant can then be assessed in order to attempt to identify the natural function of the wild-type gene. This approach works well for many genes. For example, the previously uncharacterized yeast gene *SNU17* shows little similarity to other proteins when compared using database searches. A yeast strain knocked out for *SNU17*, however, shows a slow-growth phenotype and is defective in pre-mRNA splicing (Gottschalk *et al.*, 2001), indicating that the protein is involved in the splicing process. The difficulty with this approach is that, often, the deleted strain is either non-viable or is indistinguishable from the wild-type. Neither of these outcomes makes functional assignment possible – the non-viable state suggests that the protein may be playing a vital role in the cell, but may not yield any further clues to that role. An alternative approach to gene assignment is to overproduce a protein, by carrying the gene on a high-copy-number plasmid, to attempt to observe a phenotype.

Despite the availability of the techniques described above, much of the assignment of gene function must be performed on an individual gene basis. This remains a large task in an experimentally tractable organism for the 2000 or so unidentified yeast genes, but the complete identification of the 30 000 or so human genes seems daunting.

9.9 Bioinformatics

The availability of huge amounts of sequence information from an increasingly large number of fully characterized genomes has led to problems in the way in which the data is stored and accessed. Bioinformatics is the study of this biological information. It brings together the avalanche of biological data (genome sequence and other experiments) with the analytical theory and practical tools of mathematics and computer science. Bioinformatics aims to

- develop new algorithms and statistics with which to assess the relationships among members of large data sets,

- analyse and interpret various types of data including DNA and amino acid sequences, protein domains and protein structures and

- develop and implement tools that enable efficient access and management of different types of information.

Table 9.1. Curated genome sequencing projects

Organism (type)	Web site(s)
Escherichia coli (bacterium)	www.genome.wisc.edu
Bacillus subtilis (bacterium)	genolist.pasteur.fr/SubtiList
Saccharomyces cerevisiae (yeast)	genome-www.stanford.edu/Saccharomyces
Caenorhabditis elegans (nematode worm)	www.wormbase.org
Drosophila melanogaster (fruit fly)	flybase.bio.indiana.edu
Arabidopsis thaliana (plant)	www.arabidopsis.org
Mus musculus (mouse)	www.informatics.jax.org
Homo sapiens (human)	www.ncbi.nlm.nih.gov/genome/guide/human/

We have already touched upon the use of computers to align DNA sequences to form contigs and in the search for similar genes, but their role does not stop there. Raw sequence information, e.g. the entire sequence of a chromosome, deposited into a database is important for the analysis of gene and gene function. Perhaps more important, and certainly more useful to the majority of researchers, is to have an integrated collection of genes, proteins and experimental evidence relating to the function of both. Curated databases (Table 9.1) attempt to collate the available information and present it in a format that is more user friendly than a list of DNA sequences. These databases generally allow users to search for gene or protein names or sequences and will often also guide users to published literature relating to their search topic. As we will see in Chapter 10, the analysis of the relationship between gene products under a variety of experimental conditions provides another layer of complexity to understanding gene function. The ability to integrate and analyse this data is vital if we are to gain real benefits in a post-genome age.

10 Post-genome analysis

> ## Key concepts
>
> ♦ The availability of genomic sequence has allowed experiments to be designed to investigate each gene or gene product within an organism
>
> ♦ Changes in the number and magnitude of genes expressed by cells in different conditions can give vital clues to the cellular response
>
> ♦ The analysis of genomes, transcriptomes and proteomes have been made possible through increases in automation and computational power

With the arrival of a fully sequenced genome, readers may be forgiven for thinking that further gene analysis is not required. What more information could possibly be gained after the sequence of every gene is known? Rather than representing the end, however, the knowledge of the genome sequence of an organism has initiated a whole new series of experiments that could not have even been envisaged previously. In recent years the term '–ome' has been used as a suffix attached to almost any phenomenon that can be described on a cell-wide level (Fields and Johnston, 2002). The genome is easy to understand as the total DNA content that is usually invariant for an individual cell. Some of the other 'omes', however, are more difficult to define and in many cases are subject to change depending upon the conditions in which the cell is grown.

- *Transcriptome* – the mRNA content of a cell. This will change as different genes are expressed at different developmental stages, or in response to external factors or nutrient availability. The transcripts expressed within a cell are an indication of the proteins it is producing, but do not reflect protein function and may not necessarily reflect protein abundance. For example, a transcript that contains many seldom used codons may be translated at

Analysis of Genes and Genomes   Richard J. Reece
© 2004 John Wiley & Sons, Ltd   ISBNs: 0-470-84379-9 (HB); 0-470-84380-2 (PB)

a low frequency to produce much less protein than the same amount of transcript containing commonly used codons.

- *Proteome* – the protein content of a cell. This could be thought of as the translated component of the genome, but in many cases the protein products produced by a cell may differ from those predicted from the transcriptome. Post-translational modifications can radically alter the function of many proteins.

- *Metabolome* – the small molecule metabolites present within the cell. The quantity and identity of primary and secondary metabolites in a cell will vary greatly depending upon its physiological state. Changes in the metabolome may, however, reflect the function of the proteins required for metabolism.

The ability to measure changes in each of the above can help to define the precise cellular processes that occur under particular circumstances. For example, what changes occur within a cell during its conversion from a normal to a cancerous state? What genes are turned on or off? What proteins are made, and how do these differ from the normal complement of proteins? A number of techniques have been devised to address these issues. Many of the experiments designed to address the global effects on gene function have been performed using the yeast *Saccharomyces cerevisiae* as a model eukaryotic organism. Yeast has the advantage of a relatively small genome (~6300 genes) with compact intergenic regions and few introns. This, combined with the ability to perform rapid and powerful genetic analyses, makes it an ideal system to study the interactions between genes and gene products. Most of the experiments described below were performed first on yeast prior to moving to the larger genomes of more complex higher eukaryotes.

10.1   Global Changes in Gene Expression

The expression levels of individual genes can be modulated in response to a variety of extra-cellular and intra-cellular signals. The complement of genes that are expressed within a cell at a particular time gives a 'snap-shot' of the proteins that it is currently producing. For example, the treatment of human cells with a particular drug may induce changes in expression of genes required for the response to that drug; e.g. those proteins required for drug metabolism may be produced at a higher level. We have already discussed a number of techniques that are aimed at monitoring changes in gene expression, e.g. Northern blotting (Chapter 2) and RT-PCR (Chapter 4). These methods, however, require that alterations in the expression levels of specific genes be observed. This approach both limits the number of genes that can be analysed and biases the results obtained just to the genes whose expression pattern is observed. The researcher performing the experiment will have to make a call as to the expression of which genes are likely to be altered by a particular treatment. A far more systematic approach is to test the expression levels of all genes within the genome and to see how these levels are altered under particular circumstances. This allows for 'unexpected' gene expression alterations to be observed. A number of approaches have been designed to monitor gene expression changes on a genome-wide level.

10.1.1 Differential Display

Although not requiring prior knowledge of the sequence of the genome, differential display is a method for monitoring global changes in gene expression levels based on the systematic amplification of the 3'-ends of mRNA molecules (Liang and Pardee, 1992). As we have seen previously, the 3'-ends of most mRNA molecules contain a poly(A) tail. Anchored primers are designed to bind to the 5' boundary of the polyA tail and act as starting points for a reverse transcription reaction (Figure 10.1). The single cDNA strands produced are then PCR amplified using the anchored primer and an upstream primer of arbitrary, but known, sequence. Different arbitrary primers are used to amplify different sets of cDNA molecules derived from a population of mRNA molecules. The population of PCR products produced by this method are then separated using by denaturing polyacrylamide electrophoresis – like the DNA sequencing gels we saw in Chapter 9. The amount of PCR produced from each mRNA molecule in the original sample should be proportional to the amount of RNA from which it was derived. Consequently, the relative abundance of individual mRNA molecules can be compared directly in different RNA samples from related sources. Using multiple primer combinations, differential display is able to visualize all the expressed genes in a cell in a systematic and sequence-dependent manner. One of the first reported uses of differential display was to compare the mRNAs from normal and tumour derived human mammary epithelial cells, cultured under the same conditions. The identification of genes specifically expressed in tumour cells but not in normal cells (potential **oncogenes**), or those expressed in normal cells only (potential **tumour suppressor genes**), is important for understanding the molecular basis of cancer (Liang *et al.*, 1992).

Sequences that are identified by differential display as being either up- or down-regulated under particular conditions may be excised from the gel in which they are separated, re-amplified by PCR, cloned and sequenced.

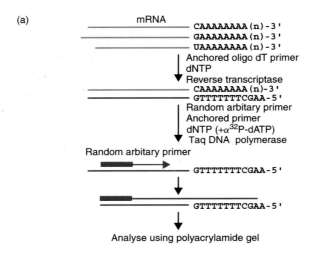

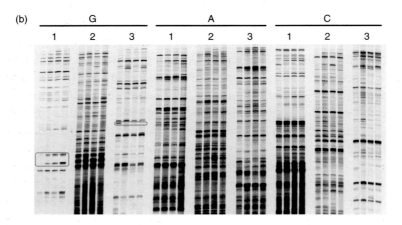

Figure 10.1. Differential display to detect changes in gene expression. (a) Conversion to cDNA is achieved by first using an anchored oligo-dT primer to create a single cDNA strand corresponding to the 3'-end of the mRNA. Each of the three anchored primers (whose 3'-end is either G, A or C) will produce a population of single-stranded cDNA molecules based upon the presence and abundance of individual mRNA molecules within the sample. Second-strand synthesis is then performed using a set of arbitrary primers – of known sequence. Different combinations of arbitrary primers and anchored oligo-dT primers will amplify all possible permutations of the first cDNA strand. The PCR fragments produced are separated using a polyacrylamide gel and differences in expression of genes within the tissue samples can be detected through the analysis of the intensity and individual bands. (b) Differential display of four RNA samples (one normal and three cancerous) using three different anchored primers (G, A and C) in combination with three arbitrary primers (1, 2 and 3). The red box indicates some of the gene products that are highly expressed in the cancer cells and not in the normal cells, and the green box indicates genes expressed only in normal cells. Reproduced, with permission, from GenHunter Corporation, www.genhunter.com

The fragments made in this way are biased toward the 3′-end of genes and are therefore unlikely to represent full-length cDNA clones. The differentially expressed sequences can, however, be used as probes to isolate full-length cDNA and genomic DNA – either through library screening or computer searches. Differential display is a powerful technique for analysing gene expression changes. It does, however, suffer from the problem that even seemingly modest changes in cellular conditions can be accompanied by alterations in the levels of massive numbers of genes. Additionally, multiple primer combinations (>300) are required to analyse effectively all potential mRNA molecules that may be produced within a cell (Crawford *et al.*, 2002).

10.1.2 Microarrays

We have already seen that the pattern of genes expressed within a cell is characteristic of its current state. The realization that the genomes may not contain as many genes as was once thought – for example 6000 in yeast and perhaps as few as 30 000 in humans – opened the possibility of individually analysing the expression of all genes within an organism. 30 000 individual experiments is still a huge number, but advances in automation and sample processing means that this is now achievable. Virtually all changes in cell state or type can be correlated with alterations in the mRNA levels of genes. In some cases, alterations in massive numbers of genes occur. For example, in yeast, the process of sporulation is associated with a change in the expression of at least 1000 different genes – representing almost 20 per cent of the total number of genes (Chu *et al.*, 1998). In other cases, changes in cellular environment may only alter the expression of a small subset of genes – e.g. treatment of yeast cells with copper sulphate significantly alters the expression of only five genes (Gross *et al.*, 2000). Knowledge of the expression patterns of many previously uncharacterized genes may also provide vital clues to their function. The analysis of changes in the expression of all of the thousands or tens of thousands of genes within a genome is essential if we are to understand the interplay between genes and gene products.

DNA microarrays have been developed as a method for rapidly analysing the expression of all genes within a genome (Shalon, Smith and Brown, 1996). They work by providing a fixed single strand of DNA to which labelled cDNA fragments can bind (Figure 10.2). The DNA fragments are physically attached to an inert support (called a chip). Several different technologies are currently used to perform microarray experiments. These differ in the way in which the DNA sequences are attached to the chip, and the length of the DNA sequence itself. The two most commonly used systems are that

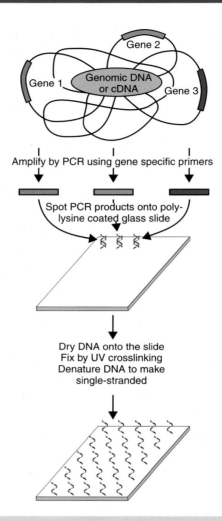

Figure 10.2. The construction of a DNA microarray. PCR products representing individual genes, parts of genes or other DNA are spotted onto a glass slide, and then fixed into position. The resulting single-stranded DNA can then act as a binding point for complementary, labelled, DNA sequences

developed in the laboratory of Patrick Brown at Stanford University, and that sold commercially by the company Affymetrix®. The Affymetrix chips contain chemically synthesized single-stranded DNA oligonucleotides (~25 bases in length) that are attached to a quartz surface using photolabile agents and photolithographic techniques similar to those used in the production of silicon microchips. This process results can result in as many as 500 000 individual DNA molecules being placed precisely within a 1.28 cm² chip. The Stanford chips are constructed by robotic spotting of DNA fragments, derived from the PCR amplification of entire genes, onto precise points of a glass slide. In the case of the yeast genome 6116 separate PCR reactions are performed to amplify every individual gene. These PCR products are then spotted onto a glass slide measuring approximately 2 cm^2. Each PCR product is spotted onto a precise location. Consequently, the sequence of the DNA at any individual spot is known. Once spotted onto the slide, the DNA is dried by heating to 100 °C for 2 s, fixed by UV cross-linking, and then denatured (95 °C, 2 min) to make the DNA single stranded.

The prepared chips are then used as templates for the binding of labelled cDNA fragments. cDNA fragments are prepared as indicated in Figure 10.3. RNA samples from two related sets of cells are isolated. The cells generally have to be related to each other so that the majority of genes that they express will be similar. The cells have, however, been either grown under different conditions, or derived from, say, a normal and cancerous version of the same tissue. Single strands of cDNA are produced using reverse transcriptase from an oligo(dT) primer. The synthesis of cDNA is carried out in the presence of three normal deoxynucleotide triphosphates and a single fluorescently labelled deoxynucleotide triphosphate. One of the cDNA samples is produced using a green fluorescent label (nucleotide labelled with Cy3), while the other is produced using a red fluorescent dye (nucleotide labelled with Cy5). Consequently, the cDNA isolated from each different cell types will bear a different fluorescent marker. The cDNA samples are then mixed in equal quantities and allowed to hybridize to the microarray chip. The colour and intensity of each spot on the microarray is then monitored using a fluorescent scanning confocal microscope. The labels are excited using a laser, and the fluorescence at each spot detected with the microscope to give an indication of the relative amount of each RNA species within the original samples. For example, as shown in Figure 10.3, if the cells grown under condition 1 express a gene that is not expressed under condition 2, then the mixed cDNA will only contain the Cy3 labelled version of the corresponding cDNA and the complementary spot on the microarray will be labelled green. Similarly, a gene only expressed in growth condition 2 will yield a red spot on the array. Genes expressed at equal levels in both samples will give rise to both green and red labelled cDNAs that can both bind to the complementary sequence on the array. In this case a yellow spot (mixture of green and red) will be generated.

It is important that each microarray is calibrated so that the levels of mRNA from each of the samples can be accurately assessed (Bilban et al., 2002). The ratio between the green and red fluorescent signals is calculated for several array spots that contain total genomic DNA. The fluorescence detector can

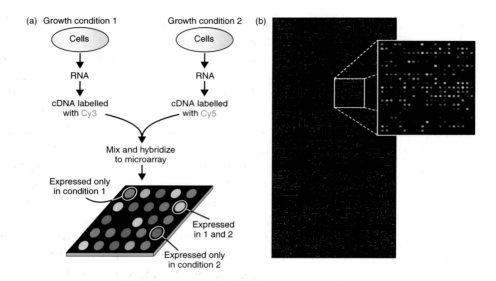

Figure 10.3. Identifying genes whose expression differs between two RNA samples using microarrays. (a) The RNA from cells grown in condition 1 is isolated and converted to cDNA using an oligo-dT primer that is itself labelled with a green fluorescent dye (Cy3). Similarly, RNA from cells grown in condition 2 is isolated and converted to cDNA using an oligo-dT primer labelled with a red fluorescent dye (Cy5). The cDNA samples are mixed in equal proportions and hybridized to the microarray. Viewing the microarray with a fluorescent microscope reveals a series of coloured spots. A green spot indicates that the corresponding gene is only expressed in condition 1, while a red spot indicates that a gene is only expressed in condition 2. If equal quantities of cDNA derived from cells 1 and 2 bind to the single-stranded DNA on the array, then equal amounts of the red and green dyes will bind to yield a yellow spot. (b) A microarray derived from mouse gene sequences used to screen different RNA samples for the expression of genes, courtesy of Yanxia Hao and Christopher Barker, The J. David Gladstone Institutes Genomics Core Laboratory, San Francisco

then be calibrated so these elements have a measured relative intensity ratio of 1.0. The relative intensity of green and red signals can then be used as a measure of the relative abundance of a specific mRNA in each sample.

There has been an explosion in the use of DNA microarrays to analyse gene expression on a genome-wide scale. Here we will discuss several specific examples where **transcript profiling** has been used to address biological problems, but the literature contains numerous others (Lockhart and Winzeler, 2000; Shoemaker and Linsley, 2002).

- *Gene expression associated with metabolic changes.* DNA microarrays have been used to measure the relative expression of all of genes in yeast during the diauxic shift (the change from fermentative to respirative growth that occurs as glucose is depleted from the growth media) (DeRisi, Iyer and Brown, 1997). Yeast cells were harvested from a culture every two hours over a period of time and used to compare the genes being expressed in them to the initial expression pattern. From the isolated mRNA, fluorescently labelled cDNA was prepared using Cy5, while cells harvested at the first time-point (a reference sample) were labelled with Cy3. The cDNA from each of the time-points was then mixed with the reference cDNA and hybridized to a microarray chip containing each of the ~6000 yeast genes. As glucose was depleted from the media, marked changes in the global pattern of gene expression were observed (Figure 10.4). Approximately 30 per cent of the genes within the genome showed altered expression levels. These genes are either up- or down-regulated by at least a factor of two compared to the starting material. The metabolic shift during which glucose is depleted was correlated with widespread changes in the

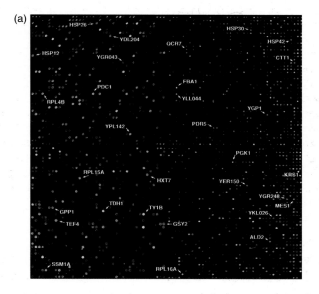

Figure 10.4. Microarrays to identify the flux through metabolic pathways. (a) A yeast genome microarray in which RNA from cells grown in glucose-rich medium was labelled with Cy3 (green) and RNA form cells grown in glucose-depleted medium was labelled with Cy5 (red). (b) Metabolic reprogramming during the diauxic shift. Genes encoding the enzymes shown in red increase by the factor shown during the diauxic shift, while those in green decrease. The red arrows indicate steps in the metabolic pathway whose genes are strongly induced upon glucose depletion. These figures were kindly provided by Joe DeRisi (UC San Francisco) and are reprinted with permission from DeRisi, Iyer and Brown (1997). Copyright 1997 American Association for the Advancement of Science

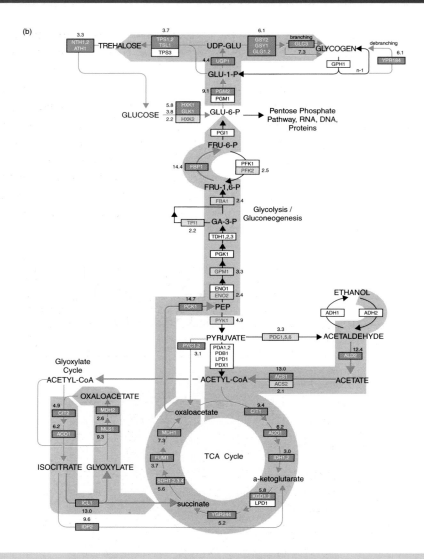

Figure 10.4. (continued)

expression of genes involved in many fundamental cellular processes, e.g. carbon metabolism, protein synthesis and carbohydrate storage. However, ~50 per cent of differentially expressed genes identified by the microarray analysis had no previously characterized function.

- *Expression differences between normal and cancerous cells.* DNA microarrays have been used extensively to characterize genes that are differentially expressed in cancer cells. For example, the analysis of ~5500 genes expressed in normal human breast epithelial cells were compared to those expressed in breast cancer tissue (Perou *et al.*, 1999). This has allowed sets of genes to be identified that may be involved in the process of cellular proliferation during cancer growth. Another important use of microarrays in cancer treatment is the classification of cancerous cells based upon the genes they express. For example, the assignment of previously unidentified subtypes to malignant melanoma cells can be made through their gene expression pattern (Bittner *et al.*, 2000). These subtypes (Figure 10.5) can be used to predict the clinical outcome of individual melanomas that would not be otherwise possible. A similar analysis to compare gene expression patterns with clinical outcomes has also been undertaken for a variety of breast cancers (Sorlie *et al.*, 2001).

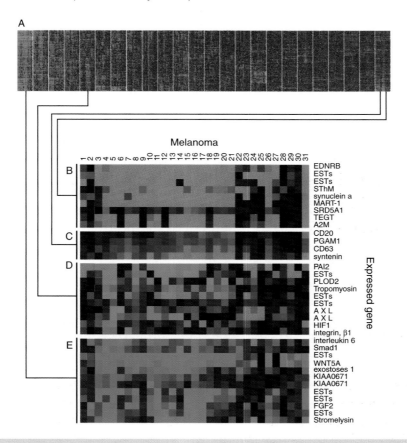

Figure 10.5. The characterization of malignant melanomas based on their gene expression profile. See the text for details. Figure courtesy of Mark Bittner (The National Human Genome Research Institute, NIH, USA), reprinted by permission from *Nature* (Bittner *et al.*, 2000) copyright (2000) Macmillan Publishers Ltd

- *Defining cell type.* Stem cells, which we will discuss in greater detail in Chapter 13, differ from most other cells within mammals in that they have the capacity to both self-renew and can, given the appropriate signals, differentiate into a variety of distinct cell types. Three of the best charac-terized types of stem cell have been isolated from embryonic, neural and hematopoietic tissue (Blau, Brazelton and Weimann, 2001). Using microar-rays, transcript profiling of these different types of stem cell has revealed the presence of approximately 200 genes that are expressed within different stem cells that are not expressed within differentiated cells (Ivanova *et al.*, 2002; Ramalho-Santos *et al.*, 2002). These genes may describe a unique genetic programme that allows the cells to function as stem cells.

DNA microarrays offer the opportunity to analyse the expression of many thousands of genes in a single experiment. They do, however, suffer from a number of deficiencies. For example, like all the experiments we have discussed that rely on nucleic acid hybridization, the cross-hybridization of highly similar sequences can prove problematical. The data obtained from the microarray experiment must therefore be confirmed by more traditional gene expression analysis experiments. Additionally, the overall sensitivity of the microarray experiment is not high. Changes in the expression of a gene can only usually be detected if changes by a factor of two or more occur. Subtle changes in expression, which may have vital physiological roles, may go undetected during microarray analysis. Of course, the production of some proteins is not regulated at the level of transcription. For example, the gene encoding the yeast transcriptional activator Gcn4p is transcribed constitutively, but the production of protein is controlled at the level of translation in response to amino acid starvation (Hinnebusch, 1997). Finally, the sheer amount of data generated by microarrays has led to difficulties in how this information is analysed, and stored in an accessible format. The efforts of many bioinformaticians are currently aimed at presenting such data in a format that is usable by biologists. Databases of publicly available raw and normalized microarray data (e.g. http://www.dnachip.org) allow the inspection of microarrays to see how your favourite gene is regulated under particular conditions.

10.1.3   ChIPs with Everything

A natural extension of seeing which genes are expressed under particular conditions is to ask what are the transcription factors that control the expression of the regulated genes. Traditional biochemical methods (e.g. DNA footprinting and gel retardation) have been used to identify the regions of DNA to which a transcription factor can bind. Inferences can then be made as to which genes are controlled by the transcription factor by looking for particular DNA binding sites within the promoters of genes. This approach has been very successful in identifying potential target genes, but suffers from the fact that many transcription factors bind to relatively ill defined DNA sites that occur far more frequently within the genome than the number of genes actually regulated by the protein. What is required is a way of identifying genuine DNA binding sites for particular transcription factors based on the sequences that they actually bind within the cell. **Chromatin immunoprecipitation** (ChIP) is a method designed to do just that (Strahl-Bolsinger *et al.*, 1997). Growing cells are treated with formaldehyde, a cross-linking agent, to attach covalently proteins and DNA that are in close physical proximity with each other (Figure 10.6). The cells are then broken open and the DNA sheared by sonication to generate fragments of about 500 bp in length. This process will not disrupt the formaldehyde cross-links, so proteins should remain attached to their cognate DNA binding sites. DNA fragments bound by individual proteins are then separated from the rest through the interaction of the protein with a specific antibody raised against it. The antibody–antigen complexes may be precipitated using beads that will specifically bind to the antibodies. Heating the immunoprecipitated samples to 65 °C for 12–18 h results in the reversal of the formaldehyde cross-links and consequent release of the DNA from the DNA–protein complexes. The presence of individual promoter sequences in the immunoprecipitated fraction can then be assayed using PCR. Specific primers to the promoters of individual genes are then used to detect the presence of that particular DNA sequence in the antibody associated DNA fraction (Hecht and Grunstein, 1999). A positive PCR reaction is an indication that one or more binding sites for the protein against which the antibody was raised are present within the promoter being tested. Through careful design of the PCR, the relative occupancy of individual DNA binding sites within a promoter can be determined. Additionally, changes in promoter occupancy associated with altered cellular conditions can be observed as changes in the intensity of the PCR product.

ChIP analysis is a powerful way to identify DNA binding sites for proteins at known promoters. However, the technique is even more useful when combined with microarray chips to identify protein binding sites on a genome-wide scale. As shown in Figure 10.6, DNA fragments cross-linked to a protein of interest are enriched by immunoprecipitation with a specific antibody. After the reversal of the cross-links, the enriched DNA is amplified and labelled with a fluorescent dye (Cy5). A sample of DNA not enriched by immunoprecipitation is labelled with a different fluorophore (Cy3), and a mixture of the enriched and un-enriched pools of labelled DNA are hybridized to a microarray containing

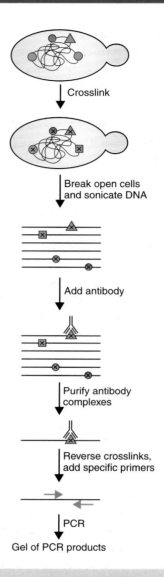

Figure 10.6. Chromatin immunoprecipitation (ChIP) to identify the occupancy of specific DNA binding sites within genomic DNA. See the text for details

all intergenic and promoter DNA sequences. This was first used to identify the genomic binding sites of two yeast transcription factors, Gal4p and Ste12p (Ren *et al.*, 2000). Three previously unidentified gene targets for the exceptionally well characterized Gal4p were identified using this ChIP-on-chip analysis. Subsequently, the binding sites for (and by inference the genes regulated by) 106 yeast transcriptional activators have been identified (Lee *et al.*, 2002). A number of additional experiments have extended the analysis of transcription factor binding sites to insect (Berman *et al.*, 2002) and human cells (Weinmann *et al.*, 2002). Perhaps unexpectedly, analysis of this type has shown that many genes in higher eukaryotes that are co-ordinately expressed from the same transcription factor binding sites are located in clusters next to each other on the genome (Roy *et al.*, 2002). The significance of this finding is not completely understood, but may suggest a higher order of gene organization within the genome that may be involved in the control of their expression.

10.2 Protein Function on a Genome-wide Scale

Knowing what genes are expressed in a cell at any particular time is informative, but it does not give information as to the function of particular gene products. Methods have therefore been devised to analyse protein function at a genome-wide level. These have relied on either disrupting individual genes so that individual proteins are eliminated from the cell, or mapping the interactions between sets of proteins.

10.3 Knock-out Analysis

Despite decades of extensive genetic analysis, approximately one-third of the ~6000 genes in the yeast genome have no ascribed function. The elimination of a single gene product from the genome can yield important clues as to the function of that gene through the phenotypic analysis of the resulting mutant. For many years, researchers have been able to knock out individual genes in yeast cells using homologous recombination (Rothstein, 1983). This process, which occurs at high frequency in yeast, replaces a target gene with one possessing a selectable phenotype (Figure 10.7). A linear DNA fragment bearing the selectable gene is constructed such that it is surrounded at its 5'- and 3'-ends by at least 50 bp of sequence derived from the gene to be disrupted. Transformation of yeast with this DNA fragment, which cannot independently replicate, results in its integration into the genome at the precise location of the homologous sequences. Consequently, the target gene, originally located between the homologous sequences, will be eliminated and replaced by the selectable gene. The elucidation of the entire yeast genome sequence has led to the systematic disruption of every gene (Winzeler *et al.*, 1999). The method used involves a PCR based gene strategy to generate deletion mutants of each gene from its start codon to its stop codon. During the deletion process, each target gene is replaced with a KanMX cassette (Wach *et al.*, 1994). This

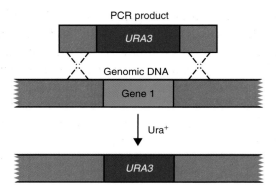

Figure 10.7. Single-step gene knock-outs in yeast. The transformation of yeast with a linear DNA fragment containing a selectable gene (*URA3* forms part of the uracil biosynthetic pathway) flanked by genomic DNA sequences results in its insertion at homologous regions of the genome (indicated by the dashed lines). The gene between the homologous sequences is eliminated and replaced by the selectable gene

cassette contains the KAN^R ORF of the *E. coli* Tn903transposon fused to transcriptional and translational control sequences of the *TEF* gene of the filamentous fungus *Ashbya gossypii*. The insertion of the cassette into the yeast genome permits efficient selection of transformants resistant to the antibiotic geneticin (G418).

The disruption of a gene is likely to lead to one of the following phenotypes.

- *Lethal.* The deleted cell is unable to grow. The gene is essential for some part of the growth process. Essential genes in yeast can be distinguished by the way in which they segregate during sporulation.

- *Wild type.* There is no detectable difference between the mutant and wild-type cells. If no differences can be detected under a variety of conditions, it may suggest that another gene is able to compensate for the loss, or that particular conditions have not been found where the loss will show as a phenotypic change.

- *Growth difference.* The deleted strain may grow slower (or faster) than the wild-type parent under particular conditions. A battery of different growth tests, each under a variety of conditions, may then be used to identify potential roles of the protein encoded by the gene.

The analysis of yeast disruption mutants showed that almost 20 per cent of yeast genes within the genome were essential for growth on glucose-rich media and, in other screens, 15 per cent of the disruptions had an effect on overall cell size or morphology (Giaever *et al.*, 2002). Screening like this is a fairly crude measurement of the function of individual genes, but the analysis of different sets of genes required for growth under different conditions can be informative. Additionally, deletion analysis can be combined with gene expression profiling (microarrays, see above) to compare expression patterns between the wild-type and the mutant strains to give further clues to gene function. For example, the deletion of the *TUP1* transcriptional repressor in yeast results in increased transcription of three per cent of yeast genes (DeRisi, Iyer and Brown, 1997). These genes are likely to be repressed the TUP1 protein in wild-type cells.

10.4 Antisense and RNA Interference (RNAi)

The construction of gene knockouts in higher eukaryotic organisms is not as straightforward as that that described above for yeast, either as a result of low rates of homologous recombination, or the lack of suitable genetics. For some time, however, it has been known that the expression of antisense RNA (representing the sequence of the opposite sense to mRNA) can inhibit gene expression (Green, Pines and Inouye, 1986). Antisense sequences can be produced within cells by inverting the coding sequence of a gene with respect to its promoter such that the complementary strand is transcribed. The formation of a base-paired RNA duplex between the mRNA and the antisense RNA is thought to interfere with either RNA processing or translation, and the encoded protein is produced at much lower levels. For example, tomato plants expressing an antisense version of the polygalacturonase gene, the product of which is involved in softening and over-ripening, produce approximately five per cent of the normal polygalacturonase protein levels and have longer shelf-life and increased resistance to bruising (Smith *et al.*, 1988). The expression of antisense RNA within cells, or the introduction of antisense oligonucleotides into cells, can have dramatic effects on the production of the protein encoded by the corresponding mRNA. Introducing an inducible antisense expression vector into cells may bring about conditional gene silencing. For example, the expression of antisense RNA from a tetracycline inducible promoter (see Chapter 8) may allow specific inhibition of protein production in the presence of tetracycline only (Handler and Iozzo, 2001). The efficiency of antisense inhibition can vary widely between species and individual genes, and specific antisense sequences may lead to a non-specific inhibition of protein production (Cohen, 1991).

It has also been discovered that the expression of individual genes in a variety of eukaryotes can be reduced dramatically through the introduction of specific double-stranded RNA molecules into cells. This phenomenon, termed

RNA interference (RNAi), was first noticed during RNA injection experiments into the nematode *Caenorhabditis elegans*. The injection of either the sense or the antisense RNA strands of a particular gene into the organism caused little reduction in the expression of the gene. However, co-injection of both the sense and antisense RNA strands caused a massive reduction in the expression of the gene (Fire *et al*., 1998). A number of other experimental observations were made.

- The injection of double-stranded RNA for specific genes into *C. elegans* caused a specific disappearance of the corresponding gene product from both the somatic cells and the F1 progeny.

- Double-stranded RNA was able to inhibit gene function at a distance from the site of injection and appeared to be able to cross cell boundaries, suggesting that a small diffusible molecule may be responsible for the repressing effect.

- Only double-stranded RNA sequences from exons had any effect on protein production with sequences from introns having no effect.

- Relatively small double-stranded RNA sequences – considerably less than the full gene sequence – effectively turn off the production of the protein encoded by the corresponding mRNA.

- Very small amounts of double-stranded RNA, representing only a few molecules of double-stranded RNA per cell, were required to repress protein production, suggesting that a catalytic or amplification process was occurring.

RNAi can even be induced in *C. elegans* through the ingestion of double-stranded RNA. *E. coli* cells expressing specific double-stranded RNAs were fed to worms, and 40–80 per cent of the worms showed the specific phenotype associated with silencing (Timmons, Court and Fire, 2001). RNA silencing has been used extensively to analyse the function of genes in *C. elegans*. The ease with which silencing can be achieved and its overall effectiveness has made it a powerful tool in gene analysis. *C. elegans* has six chromosomes, and virtually all of the genes on two of these have been knocked out using RNAi (Gönczy *et al*., 2000; Fraser *et al*., 2000). Knock-outs of a number of the genes result in sterile or embryonic lethal phenotypes that make additional analysis difficult, but many other genes can be assigned distinct functions based on this type of analysis (Kamath *et al*., 2003).

As with the introduction of antisense RNA, double-stranded RNA molecules can also be produced inside cells rather than being added to them. For example, in trypanosomes, integrative plasmids have been constructed in which a trypanosome gene is under tetracycline (Tet) control and transcribes a head to head gene fragment with a spacer between the fragments to produce a double-stranded hair-pin RNA that induces RNAi (Shi *et al.*, 2000). Initially, it was thought that RNAi might be limited in its scope, since the expression of double-stranded RNA in most mammalian cells results in a general down-regulation of protein synthesis rather than a gene-specific effect (Hope, 2001). However, it was found subsequently that the expression of 21-nucleotide RNAs, called small inhibiting RNAs (siRNAs), paired so that they have a two-nucleotide 3'-overlap, are able to down-regulate the expression of specific genes in mammalian cells (Elbashir *et al.*, 2001). The repressive effects of RNAi in mammalian cells are not as great as that observed in flies or worms, but RNAi has great potential for determining gene function in a variety of previously genetically intractable organisms.

Box 10.1. The mechanism of RNAi

The molecular mechanism of RNA interference is not fully understood at present. RNAi has been shown to be a post-transcriptional phenomenon (Montgomery, Xu and Fire, 1998), but how does it work? The expression of double-stranded RNA induces the specific degradation of the mRNA to which it is complementary (Ngo *et al.*, 1998). For example, in *Drosophila*, the expression of a 540-nucleotide double-stranded RNA corresponding to the cyclin E gene (a critical component of cell cycle control) induces cell cycle arrest and the production of 21–25-nucleotide RNA fragments that are homologous to the cyclin E gene – to both the sense and antisense sequences (Hammond *et al.*, 2000). RNA degradation is ATP dependent, does not require the presence of the corresponding mRNA and is catalysed by a sequence-specific nuclease (Zamore *et al.*, 2000). Through genome sequence comparison analysis, the *Drosophila* and *C. elegans* genomes are found to contain only three types of double-strand-specific RNase enzyme (Bernstein *et al.*, 2001). One of these, named RNase III or **Dicer**, has been shown to be responsible for RNAi. The depletion of Dicer activity from cells results in the loss of the ability to silence genes by RNAi (Hutvágner *et al.*, 2001).

The short interfering RNA molecules (siRNA) generated from double-stranded RNA serve as primers to transform the target mRNA into double-stranded RNA, which can then be degraded to generate new siRNAs (Lipardi, Wei and Paterson, 2001). An RNA-dependent RNA polymerase enzyme may play a role in amplifying the double-stranded RNA, but this has yet to be

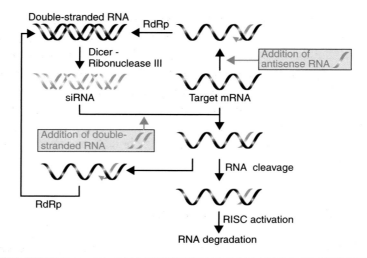

Figure 10.8. The mechanisms of antisense inhibition and RNA interference. The addition or expression of either antisense RNA (shown in green) or double-stranded RNA (pink) within cells can cause specific mRNA degradation. See Box 10.1 for details

shown experimentally. A model for the molecular mechanism of RNAi is shown in Figure 10.8 (Nishikura, 2001). The introduction of the antisense RNA sequence (shown in green) to a specific mRNA results in the formation of a double-stranded RNA hybrid between the two. The action of an RNA-dependent RNA polymerase (RdRp) produces a larger double-stranded RNA molecule that is degraded into the 21–23-nucleotide fragments by the Dicer RNase. These siRNAs are also capable of binding to the specific mRNA from which they were originally derived. When bound to the mRNA, some of the siRNAs will serve as templates for the RdRp to produce more siRNA (Hutvágner and Zamore, 2002). The net result of this process is that the target mRNA is efficiently degraded, either through participation in siRNA production, or though the activation of an RNA induced silencing complex (RISC) that will degrade the mRNA–siRNA duplexes (Martinez *et al.*, 2002).

The mechanism of RNAi is complex (see Box 10.1), but its natural role seems to have evolved to protect cells against transposons and viruses that may produce double-stranded RNA during their replication process. Of course, there are many examples of double-stranded RNA that occur naturally within cells, for example as part of the spliceosome. These do not, however, induce

RNAi, perhaps because this double-stranded RNA is also associated with cellular proteins. In plants, RNA silencing may be used as defence mechanism against viral infections. Viruses containing RNA genomes are strong inducers of RNA silencing since double-stranded RNA is formed during replication. Additionally, RNAi may confer immunity against closely related viruses. As discussed above, a mobile silencing signal (possibly double-stranded RNA) can spread from cell to cell in the plant to provide viral immunity. Some viruses are thought to circumvent RNA silencing by spreading rapidly throughout the plant (Voinnet, 2001).

10.5 Genome-wide Two-hybrid Screens

We have already discussed the use of two-hybrid screens to detect specific protein–protein interaction and to clone potential interacting partners from cDNA libraries (Chapter 6). To perform this type of analysis on a genome-wide scale requires that every possible 'bait' (Figure 6.8) be tested against every potential 'prey' from an organism so that a complete protein–protein interaction map for the organism can be deduced. Moreover, if every gene within the genome has been identified through sequence analysis, the need to construct and screen a complex cDNA library is negated. A far more systematic approach is to clone and analyse individually each protein–coding gene within the genome. So, in the case of a two-hybrid screen, every potential protein–coding gene must be fused, in the correct reading frame, to the sequence encoding a transcriptional activation domain. In the case of yeast, approximately 6000 individual plasmids must be constructed so that all baits may be tested. This is by no means a trivial task. Cloning on this scale can be achieved, however, through the PCR amplification of the ~6000 yeast genes from genomic DNA using a specific primer set for each gene (Figure 10.9). The primers were constructed such that each of the forward primers had a specific common 5'-tail of 20 nucleotides, and each of the reverse primers had a different common tail. The common tails could then serve as priming sites for a second round of PCR using a single primer set for all ~6000 PCR products (Uetz et al., 2000). During the second round of PCR, each of the amplified products has a common 50 bp sequence added to their 5'-ends and a different common 50 bp sequence attached to their 3'-ends. The PCR products are then mixed with a linearized plasmid that contains these common sequences at its ends and transformed into yeast cells. Within the yeast, the common sequences in the PCR product undergo homologous recombination with the linear plasmid to reform a circular plasmid that can subsequently be replicated. This process results in the insertion of each PCR product, and encoded gene, into a yeast

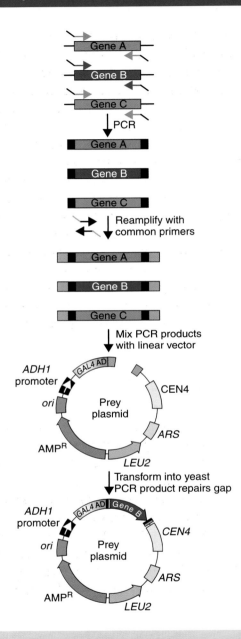

Figure 10.9. The construction of plasmids producing different prey proteins for use in a genome-wide two-hybrid screen. Genes are amplified using primers that contain common sequences at their 5′-ends. This allows their re-amplification using a second pair of primers such that all genes are tagged at their 5′- and 3′-ends (Hudson *et al.*, 1997). The tagged genes are then mixed with a linearized plasmid and transformed into yeast. Homologous recombination between the plasmid and the PCR product occurs within the yeast cell through complementation between the tag and the ends of the linear plasmid. This results in the insertion of the PCR product into the plasmid and the subsequent production of the prey protein

plasmid without the need for additional DNA manipulation. Similar **ligation-independent cloning** methods have also been devised for plasmid construction in *E. coli* cells (Donahue, Turczyk and Jarrell, 2002).

Once constructed, the ~6000 yeast strains each producing a different transcriptional activation domain fusion (prey) can then be mated to a single yeast strain expressing a unique bait (fused to a DNA binding domain) to create a series of diploid yeast cells that all produce a single bait and each produce a different prey. The diploid cells can then be grown in small cultures in the wells of a micro-titre dish under conditions in which only an interaction between the bait and the prey will permit cell growth. Thus all of the potential interacting partners of a single bait can be identified in one experiment (Figure 10.10). The data obtained by repeating the screen using different baits can then be used to build up a comprehensive protein–protein interaction map (Uetz *et al.*, 2000). To date, maps of this kind have only been produced for the protein interactions that occur inside yeast cells, but important information about the number of distinct protein complexes that exist within the cell has emerged. Many of the problems of two-hybrid screening that we discussed in Chapter 6 are also applicable here.

10.6 Protein Detection Arrays

Rather than relying on the presence of an RNA transcript to infer the presence or absence of a protein within a particular cell, a better approach would be to detect the presence of the protein directly. Perhaps an even more stringent approach would be to detect the activity of individual proteins produced by a cell (Kodadek, 2002). The development of protein assays, akin to their DNA microarray counterparts, is still in its infancy. Some protein recognition chips are, however, available (Fung *et al.*, 2001). These are composed of ligands, e.g. antibodies or small molecules, embedded in a surface such that they are immobilized, but still able to bind specifically to proteins. Cell extracts are then washed over the surface of the chip and bound proteins can be detected by mass spectrometry analysis.

10.7 Structural Genomics

The importance of structural biology in advancing the understanding of molecular processes cannot be over-stated. The ability to visualize protein molecules in three dimensions at high resolution yields tremendous insights into their mechanism of action that could not have otherwise been obtained.

Non-selective growth:

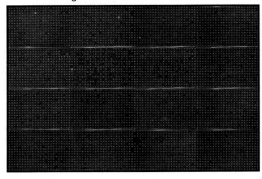

Selective growth:

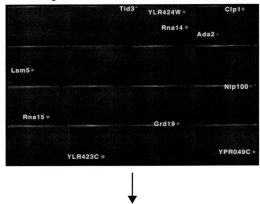

Identify interacting partners

Figure 10.10. A genome-wide two-hybrid screen to identify protein–protein interactions. Yeast cells, each producing a single bait protein (in this case Pcf11p fused to the DNA binding domain of Gal4p), were mated with one of 6000 each producing a different prey (fused to the transcriptional activation domain of Gal4p) and grown in 384-well micro-titre plates. Yeast growth will only occur under selective conditions if an interaction between the bait and the prey occurs to activate the transcription of a reporter gene. The genes shown in the bottom panel are those prey fusions that potentially interact with Pcf11p. Images courtesy of Stan Fields (University of Washington), reprinted by permission from *Nature* (Uetz *et al.*, 2000) copyright 2000 Macmillan Publishers Ltd

High-resolution structures are usually obtained using one of two methods – X-ray crystallography or nuclear magnetic resonance (NMR). In both cases, structure determination can be both time consuming and labourious. The first protein structures to be solved by X-ray crystallography were those of myoglobin and haemoglobin. Max Perutz began working on the structure of haemoglobin (molecular weight 67 kDa) in 1936 and finally solved the structure in 1959 and published in 1960 (Perutz *et al.*, 1960). These days, protein structure determination by X-ray crystallography is a great deal faster, primarily due to advances in computational power, but still relies on many of the techniques that Perutz and his colleagues pioneered. In general, both X-ray crystallography and NMR are dependent on the availability of large quantities of highly purified protein. Traditionally, protein structures have been solved on a piecemeal basis. Someone working on a biologically interesting protein finds that they are able to produce large quantities of it and then attempts structural analysis. The availability of genome sequences, however, provides an alternative route to solving protein structures – based solely on genomic DNA sequences (Figure 10.11). This approach is currently being attempted on a variety of completely sequenced organisms. Analysis of this type has only been made possible through the automation of almost all parts of the structure determination scheme shown in Figure 10.11. For example, 1376 of the predicted 1877 genes (73 per cent) of the thermophilic bacterium *Thermotoga maritime* have been cloned into an *E. coli* expression vector such that the produced protein bears a poly-histidine tag (Chapter 8). 542 of these clones were able to produce sufficient purified protein to attempt crystallization, and successful crystallization conditions were identified for 432 proteins, representing 23 per cent of the *T. maritime* proteome (Lesley *et al.*, 2002). It is likely that not all of these crystals will yield protein structures, resulting in further attrition. The data above shows that the major stumbling block to successful structure determination is the availability of purified protein. Many proteins produced using general methodologies like this will be insoluble and therefore not amenable to structural analysis. Membrane bound proteins and others that might be deleterious to *E. coli* cell growth may be difficult to produce. An alternative approach is to make the proteins for structural analysis *in vitro* where cell related problems may be overcome. *In vitro* transcription/translation systems have been used for many years. In general they operate using plasmid DNA in which the gene to be expressed is cloned downstream of an RNA polymerase promoter binding site, e.g. T7 or SP6. The plasmid is then mixed with a recombinant form of the polymerase to produce RNA. The RNA is translated *in vitro* using cell lysates – derived from either *E. coli* or rabbit reticulocytes (Turner and Foster, 1998). Such systems are, however, limited in the amount of protein that can be produced and will not usually yield sufficient for structural analysis. Coupled transcription/translation systems have recently been developed, where the reaction occurs in a chamber separated from the substrates and energy components needed for a sustained reaction *via* a semi-permeable membrane. Transcription and translation can take place simultaneously in the reaction chamber, while inhibitory reaction by-products

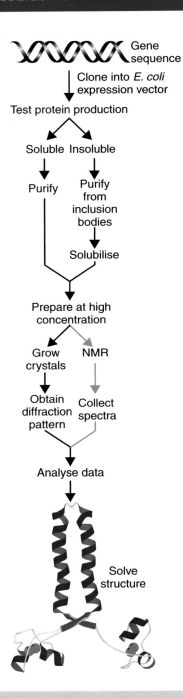

Figure 10.11. Structural genomics to solve protein structures based only on the knowledge of gene sequence. Most structural genomic strategies rely on either X-ray crystallography or NMR and not both

are diluted by diffusion through the membrane into the feeding chamber. This can result in high levels of protein expression by allowing protein synthesis to continue for long periods of up to 24 h (Martin *et al.*, 2001).

Structural genomics promises to increase greatly our knowledge of the types of protein fold that polypeptides may adopt. The danger is that only those proteins that either are easy to crystallize, or are suitably small to be studied by NMR, will have their structures solved. Efforts will still be required to analyse biologically interesting, but structurally difficult, proteins.

11 Engineering plants

Key concepts

- Plants are attractive targets for genetic modification, given their commercial importance

- Genes can be inserted into a variety of dicotyledonous plants by modifying the Ti plasmid from *Agrobacterium tumefaciens*. Genes become randomly inserted into the plant genome

- Transformation of the chloroplast genome allows the insertion of foreign genes at specific loci and does not result in the transfer of the foreign gene to pollen

- A vigorous debate over the necessity and potential dangers of genetically modified crops is currently raging

In the preceding chapters, we have primarily discussed the manipulation of DNA in prokaryotes, such as *E. coli*, or in lower single-celled eukaryotes, such as yeast. We have, however, seen that DNA from practically any source can be introduced into other organisms and is capable, providing certain criteria are met, of functioning within the foreign host. The manipulation of genes in higher eukaryotes offers the opportunity to alter gene expression within a whole organism.

11.1 Cloning in Plants

The engineering of plants offers many attractive potential outcomes. For example, the altering of the protein composition of plants to make them more nutritious or able to grow in difficult circumstances or to impart them with properties to make them more desirable to shoppers has potentially enormous

Analysis of Genes and Genomes   Richard J. Reece
© 2004 John Wiley & Sons, Ltd   ISBNs: 0-470-84379-9 (HB); 0-470-84380-2 (PB)

economic benefits. We will discuss some of these issues at the end of this chapter. First, however, we need to understand the differences between animal and plant cells and the mechanisms that can be employed to get foreign DNA – called a **transgene** – into the latter. One of the major differences between plant and animal cells is that plants retain a high degree of plasticity. That is, an isolated plant cell under appropriate culturing conditions can have the capability of regenerating an entire new plant. This ability has proved vital for the production of engineered plants since individual cells, either as cell cultures or other forms, can be manipulated and then used to reform an intact plant. Recently, a series of *in planta* transformation techniques have been developed in which plant embryos, seeds or even pollen have been treated to take up foreign DNA (Touraev *et al.*, 1997), but culture and regeneration remains widely used.

11.1.1 *Agrobacterium Tumefaciens*

Plants do not contain any naturally occurring plasmid DNA molecules. However, a bacterial plasmid, the tumour inducing (**Ti**) plasmid of the soil microorganism *Agrobacterium tumefaciens*, has been used extensively to introduce genes into plant cells. *A. tumefaciens* is responsible for crown gall disease in a variety of **dicotyledonous** plants – such as tomato, tobacco, potato, peas, beans etc. (Figure 11.1). A wound on the stem of the plant allows the bacteria to invade and cause a cancerous proliferation of the stem tissue. A plasmid carried within the bacterium is responsible for its ability to cause crown gall disease (Zaenen *et al.*, 1974; Van Larebeke *et al.*, 1974). This tumour inducing (**Ti**) plasmid is large (~200 kbp) and carries a number of genes that are required for the infection process (Figure 11.2) (Suzuki *et al.*, 2000). After infection, part of the Ti plasmid, called the **T-DNA**, becomes integrated into the plant genome at an apparently random position through non-homologous recombination. T-DNA, approximately 23 kbp in size, contains not only the genes responsible for the cancerous properties of the transformed cells (e.g. those controlling the production of the plant hormones auxin and cytokinin to stimulate cell division and growth) but also those responsible for the synthesis of **opines**, which are amino acid derivatives (Figure 11.2). Different Ti plasmids direct the synthesis of different opines. The opines produced by the infected plant, e.g. nopaline, which is formed through the reductive condensation of arginine and α-keto-glutarate, can be used by *Agrobacterium* cells as their sole source of carbon and energy. Other soil bacteria are unable to metabolize opines, and thus these molecules serve to promote the growth of more *Agrobacterium* cells, and may provide a selective advantage to *Agrobacterium* over competition from other microorganisms present in the soil. Within the Ti plasmid, the

Figure 11.1. Crown galls formed by the invasion of *Agrobacterium tumefaciens* into wounds on the stem of a potato plant. These images were kindly provided by Professor Paul Hooykaas (Leiden University, The Netherlands)

T-DNA is flaked by two 25 bp imperfect direct repeats, known as the left and right **border sequences** (Figure 11.3). These sequences are necessary for the integration of T-DNA into the plant genome (Yadav *et al.*, 1982), and any DNA sequence located between them will be integrated into the plant genome. Once integrated into the plant, the T-DNA is stably maintained and is passed on to daughter cells.

In addition to the T-DNA, the Ti plasmid also carries the genes needed for opine utilization, a series of approximately 35 genes required for virulence (*vir*) and an *Agrobacterium* origin of replication (*ori*) (Figure 11.3). Since the T-DNA fragment becomes integrated into the plant genome, the Ti plasmid offers an excellent mechanism for the introduction of novel or manipulated genes into the plants. First, the T-DNA needs to be 'disarmed' so that it is unable to promote the formation of cancerous growths that would be disruptive to gene cloning experiments. To do this, the genes encoding the proteins for the production of auxin and cytokinin are simply removed from the T-DNA fragment (Zambryski *et al.*, 1983). New DNA can then be inserted between the left and right border repeats and will be integrated without concurrent tumour

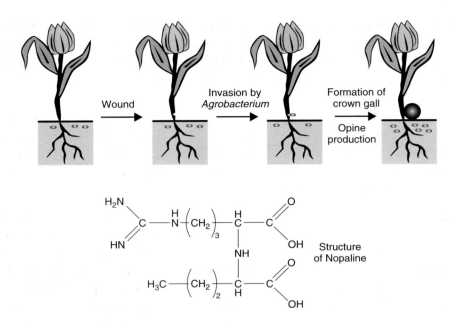

Structure of Nopaline

Figure 11.2. The mechanism of crown gall formation. The wounding of a healthy plant allows soil-borne *Agrobacterium* to invade. Part of a plasmid DNA molecule carried by the bacterium becomes inserted in the genome of infected cells and results in the formation of the cancerous crown gall growth. Additionally, the crown gall produces opines – modified amino acids, like nopaline shown below – which can be used by *Agrobacterium* as a source of metabolites

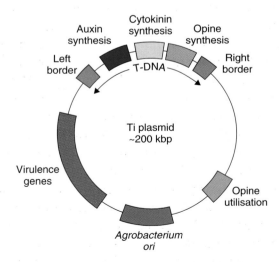

Figure 11.3. The Ti plasmid carried by *Agrobacterium*. The DNA between the left and right borders becomes integrated into the plant genome

formation. We have seen previously, however, that large DNA molecules are difficult to manipulate *in vitro* (Chapter 3) and the size of the Ti plasmid precludes simple cloning procedures. Two basic strategies have been devised for inserting new sequences into the T-DNA region.

- *Co-integration vectors.* The insertion of new DNA into a Ti plasmid results from the recombination of a small vector plasmid, for example an *E. coli* vector, and a Ti plasmid harboured in *Agrobacterium*. The recombination takes place through a homologous region present in both of the plasmids (Matzke and Chilton, 1981).

- *Binary vectors.* The T-DNA does not need to be physically associated with the virulence genes in order to become integrated into the plant genome (Bevan, 1984). Therefore the disarmed T-DNA can be cloned into a small plasmid and manipulated appropriately. The plasmid can then be transformed into *Agrobacterium* that possesses a Ti plasmid that still contains all of the virulence genes, but from which the T-DNA has been removed. The virulence proteins provide all the function required to integrate the T-DNA into the plant genome.

The transformation of plants using modified *Agrobacterium* Ti plasmids, like the transformation of *E. coli* (Chapter 2), requires selection to identify the transformed cells from their untransformed counterparts. Disarmed T-DNA still contains the genes responsible for opine production, and screening for transformants on the basis of this new biochemical activity is possible, but technically difficult. Most vectors used today contain a selectable marker gene between the left and right border repeats for transformant identification. This is commonly the kanamycin-resistance gene (see Chapter 3), but may also be a gene required for resistance to other antibiotics or herbicides. For example, the *csr1-1* gene from *Arabidopsis*, encoding a form of the enzyme acetolactate synthase, renders plants resistant to sulphonylurea herbicides such as chlorsulphuron (Haughn *et al.*, 1988). As we will discuss later, however, the introduction of resistance genes into plants has caused widespread concern about their dissemination into the environment.

Plant transformation using *Agrobacterium* is most commonly performed using leaf tissue (Horsch *et al.*, 1985). Small discs are taken from the leaf, using a paper punch, and incubated with the recombinant *Agrobacterium*. Bacterial invasion occurs at the 'wounded' edges of the disk. The infected leaf disc is transferred to a medium containing high levels of cytokinins to induce shoot formation prior to incubation with a mixture of the antibiotics kanamycin and ampicillin – to both select for the transformed plant cells and kill the bacteria.

Cells surviving this treatment are then used to regenerate plants from tissue culture using plant hormones.

Agrobacterium has proved to be an incredibly useful tool for the insertion of genes into plants (Hooykaas, 1989), but its scope is limited to plants that can be infected by the bacteria. **Monocotyledonous** plants, such as wheat, barley, rice and maize, are attractive targets for genetic manipulation, but are generally resistant to *Agrobacterium* infection. The reason for this is not entirely understood. It is possible that the induction of cell division, and consequent synthesis of DNA, that occurs upon wounding in dicotyledonous plants aids the integration process. Wounding of monocotyledonous plants tends to result in lignification. *Agrobacterium* mediated transformation of rice can occur at high frequency if acteosyringone, a phenolic compound produced by wounded dicotyledonous plant tissues, is included in the transformation reactions (Hiei *et al.*, 1994). A variety of other monocotyledonous plants have also been transformed successfully using *Agrobacterium* either in the presence of chemicals, or using plant embryos, in which cell division may occur sufficiently rapidly to allow integration of T-DNA (Cheng *et al.*, 1997). Another difficulty with *Agrobacterium* mediated transformation is the requirement to regenerate intact plants from cell cultures. The ease with which this can be achieved is species dependent and can form a major stumbling block to the generation of transgenic plants.

If the foreign DNA that has been inserted into the plant genome is to be expressed, it must be under the control of a plant promoter. The promoters from the *Agrobacterium* opine synthesis genes have been widely used to generate high levels of transgene expression that is inducible by wounding (An, Costa and Ha, 1990). In dicotyledonous plants, the promoter from the cauliflower mosaic virus (CaMV) 35S RNA gene is capable of driving high levels of transgene expression (Rathus, Bower and Birch, 1993), but the promoter is not active in monocotyledonous plants. Additionally, a number of promoters have been used to direct the expression of genes in particular tissues. For example, the promoters from several seed storage proteins can be used to target gene expression in the seeds only (Schubert *et al.*, 1994). Other factors, in addition to promoter strength, also play a role in dictating the levels to which transgenes can be expressed. The expression of the transgene will generally occur to higher levels if 5′ untranslated sequences and an intron are included within the transgene, and a polyadenylation site is located at its 3′-end. However, the expression of transgenes has been found to be quite variable. Much of this appears to be a result of the location into which the transgene has integrated. The non-homologous recombination that is required to insert a foreign gene into the plant genome means that the foreign gene is inserted randomly.

At some locations in the genome, the expression of a particular transgene may be extremely high, while at other locations the same transgene may be barely expressed at all. **Positional effects** such as this are particularly noticeable if the foreign DNA is inserted near telomeres. Additionally, the random nature of the integration event often results in multiple copies of the transgene being inserted into the plant genome – either in the form of tandem repeats at a single locus, or scattered throughout the genome of the plant. Rather than increasing the levels at which the foreign gene is expressed, this can lead to silencing of the transgene so that it is not expressed at all (Plasterk and Ketting, 2000). **Transgene silencing** is a complex process. In an excellent demonstration of the process, Garrick *et al.* (1998) constructed a transgene that contained multiple repeated copies of a gene. Each of the genes also contained a *loxP* site (see Figure 3.7). The multiple repeats were expressed very poorly, but if the Cre recombinase was also expressed in the cell (to recombine the *loxP* sites and eliminate all but one of the gene sequences) then the transgene was highly expressed. In addition, the reduction in gene copy number was also associated with a decrease in chromatin compaction and a decrease in DNA methylation at the transgene locus. These appear to be critical factors for repressing gene activity. Attempts have been made to develop a system for targeted transgene insertion into plants through the introduction of elements of a homologous recombination system, e.g. the Cre/*lox* system (Kumar and Fladung, 2001). To date, however, these have not yielded consistent results.

11.1.2 Direct Nuclear Transformation

The relatively limited spectrum of *Agrobacterium* infection led to the search for alternative methods for introducing DNA into plant cells. These generally fall into two categories.

(a) *Protoplast transformation.* A protoplast is a cell that has had its cell wall removed. In plants, this can be achieved by treating tissue with pectinase and cellulase enzymes to, respectively, break up cell aggregates and digest away the cell wall. Protoplasts will synthesize a new cell wall within 10 days, after which they are capable of undergoing cell division and, under appropriate conditions, the regeneration of new plants. In the protoplastic state, cells can be persuaded to take up DNA through either chemical (Negrutiu *et al.*, 1987), electroporetic (Shillito *et al.*, 1985) or liposomal (Hansen and Wright, 1999) means. The DNA inserted into cells in this way is not capable of independent replication. However, it may become randomly integrated into one of the plant chromosomes through a process of non-homologous recombination (Paszkowski et al., 1984). If the DNA also contains a marker gene, e.g. that encoding kanamycin resistance, then transformants may be selected.

(b)   *Biolistics.* Biolistics describes the firing of small metal particles, usually tungsten or gold, coated with DNA, into target cells (Finer, Finer and Ponappa, 1999). Particle acceleration may be achieved using gun-powder, electrical discharges or pressurized gas. This method is applicable to the introduction of DNA into a wide range of cells. The fate of the DNA after it has entered the plant cell is complex. The foreign DNA is often inserted as multiple repeated copies at a particular random location within the plant genome (Pawlowski and Somers, 1998).

11.1.3   *Viral Vectors*

Plants have many natural viral pathogens. We have seen previously how bacterial viruses have been exploited to introduce genes into bacteria (Chapter 3). Two classes of plant DNA viruses have been used as vectors to introduce foreign DNA into plants.

- *Caulimovirus.* Cauliflower mosaic virus (CaMV) contains an ~8000 bp double-stranded circular genome that encodes eight genes (Hull and Shepherd, 1977). The genome is packaged into coat proteins, which, as for bacteriophage λ, introduces restraints on the size of recombinant genomes that can be produced that retain the ability to be infective (Figure 11.4(a)). The maximum packageable size of the viral genome has been found to be 8.3 kbp (Daubert, Shepherd and Gardner, 1983). With the removal of two non-essential viral genes, the maximum insert size is still less than 1 kbp. This greatly restricts the use of such vectors. The replication of caulimoviruses proceeds through RNA intermediates, and is consequently relatively error prone (Hull, Covey and Maule, 1987). Additionally, CaMV has a narrow host range (e.g. cauliflower, turnip and cabbage) so they can be used to introduce foreign DNA into relatively few plant species. Despite these limitations, some small genes and gene fragments have been introduced into plants using CaMV vectors (De Zoeten et al., 1989).

- *Geminivirus.* Geminiviruses are characterized by twin particles that encapsulate single-stranded DNA whose replication occurs in the nucleus *via* a double-stranded DNA intermediate, which remains infectious when cloned (Figure 11.4(b)). Monopartite geminiviruses, such as maize streak virus

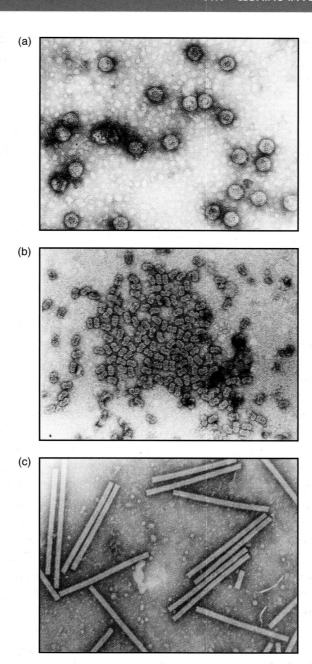

Figure 11.4. Plant viruses. (a) Cauliflower mosaic virus. (b) Maize streak virus – a member of the geminivirus family. (c) Tobacco mosaic virus. Each image is at 300 000 × magnification, and is reproduced from the following web site: http://www.iacr.bbsrc.ac.uk/ppi/links/pplinks/virusems/index.html

(MSV), contain a single genomic DNA, whereas bipartite geminiviruses, such as tomato golden mosaic virus (TGMV), have a segmented genome (Scholthof, Scholthof and Jackson, 1996). Geminivirus replication involves DNA-dependent DNA replication, rather than the relatively error-prone reverse transcription process employed by caulimoviruses. This led to the speculation that geminiviruses could serve as relatively stable transient gene expression vectors – which is particularly important since they infect many crop plants (Stanley, 1993). Geminiviruses, however, tend to undergo high levels of gene rearrangement and deletion during their infective cycle, which is highly undesirable for a cloning vector.

The vast majority of plant viruses possess RNA genomes and are therefore difficult to manipulate using standard techniques. However, the isolation of a full-length cDNA clone corresponding to entire RNA genomes (Ahlquist and Janda, 1984) has permitted the manipulation of the genome. Subsequent transcription of the genome *in vitro* using bacterial RNA polymerases allows for the reintroduction of the modified RNA genome into plant cells. For example, tobacco mosaic virus (Figure 11.4(c)) and potato virus X have both been modified to allow the introduction of foreign genes into plants (Zaitlin and Palukaitis, 2000).

11.1.4 Chloroplast Transformation

In recent years, chloroplast transformation has emerged as a serious alternative to nuclear transformation (Figure 11.5). The chloroplast, like the mitochondrion, possesses its own genome. The chloroplast genome is a single circular double-stranded DNA molecule, which, in the model dicotyledonous plant *Arabidopsis thaliana*, is composed of approximately 155 kbp of DNA containing 87 protein coding genes (Sato *et al.*, 1999). Chloroplast genes, like those of bacteria, are generally arranged into operons. This allows for the insertion of multiple foreign genes into the chloroplast that can be expressed from the same polycistronic transcript. The chloroplast is highly polyploid – each chloroplast may contain multiple copies of the genome – and photosynthetic cells may contain many thousands of chloroplasts. Consequently, the expression of genes inserted into the chloroplast can greatly exceed that from a gene in the nucleus. Importantly, however, the chloroplast genome is not found in pollen. Thus, the chances of spreading a foreign gene inserted into the chloroplast genome during crosses are remote.

Unlike all of the previous methods we have looked at for the transfer of genes into plants, the insertion of DNA into the chloroplast genome occurs *via*

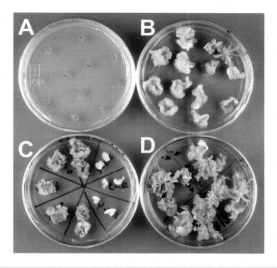

Figure 11.5. Chloroplast transformation. (A) Small leaf sections are biolistically bombarded with DNA fragments containing the *aadA* spectinomycin-resistance gene flanked by regions of homology to the chloroplast genome at both the 5′- and 3′-ends. (B) After four weeks on a medium containing spectinomycin, the leaf discs are bleached due to the inhibition of chloroplast protein synthesis. Successfully transformed chloroplasts initially grow as undifferentiated green calluses, which will, in the presence of appropriate growth hormones, produce shoots. (C) Spectinomycin-positive leaf segments are exposed to spectinomycin and streptomycin. The presence of the *aadA* gene confers broad-spectrum antibiotic resistance. Both antibiotics are therefore used to ensure that spontaneous spectinomycin-resistance mutants have not arisen. The three leaf fragments on the right-hand side of the plate have become bleached in the presence of both antibiotics because they are spontaneous spectinomycin resistance. (D) After several rounds of regeneration and screening, homogenous transgenic plantlets can be produced for the production of plants. This image was kindly provided by Professor Ralf Bock (Universität Münster, Germany), and is reproduced from Bock (2001)

homologous recombination at precise locations. The foreign gene that is being introduced into the chloroplast must be flanked by sequences homologous to the chloroplast genome itself (Staub and Maliga, 1992). To obtain chloroplast transformants at a reasonable frequency, the foreign gene must be flanked by greater than 400 bp of homologous sequence at both its 5′- and 3′-ends. Chloroplast genes are themselves transcribed by chloroplast-specific promoters and use chloroplast-specific termination signals. Therefore, foreign genes to be expressed in the chloroplast must be placed between suitable promoter and terminator sequences. Additionally, a selectable marker is used to identify transformed cells. The chloroplast-specific antibiotic resistance marker *aadA*, conferring resistance to aminoglycoside type antibiotics such as spectinomycin, is commonly used (Goldschmidt-Clermont, 1991). Spectinomycin is a prokaryotic translational inhibitor that inhibits protein production in the chloroplast, but has little effect on plant cells themselves. The kanamycin-resistance gene may also be used, but alternative selectable markers to antibiotic-resistance genes are desirable as there is a risk of transferring antibiotic resistance from the plant to bacteria either in the soil or in the guts of animals that eat the plant. One selection method that does not require antibiotic selection utilizes the betaine aldehyde dehydrogenase (BADH) gene. This gene produces an enzyme that converts betaine aldehyde, a toxic compound, into glycine betaine, a non-toxic derivative, and is an effective selectable marker (Daniell, Muthukumar and Lee, 2001). Alternatively, selectable markers may be selectively removed after the construction of the transgenic plant (Iamtham and Day, 2000). For example, the marker can be flanked by *loxP* sequences and eliminated from the genome using Cre-mediated recombination (Corneille *et al.*, 2001).

The double membrane of the chloroplast contained within plant cells would appear to be a serious physical barrier to the introduction of foreign DNA. However, biolistic methods (see above) are capable of delivering DNA into them (Svab, Hajdukiewicz and Maliga, 1990). Chloroplast transformation in this way results the transformation of only one, or a few, genome copies within a single plant cell. This produces genetically unstable cells containing a mixture of transformed and wild-type chloroplast genomes. Strong selective pressure (high antibiotic levels) must be applied for two to four cycles of plant regeneration to ensure that cells contain uniform recombinant chloroplasts (Figure 11.6). This is, however, a time consuming process and imposes limits on the speed at which transformants may be generated.

A range of proteins have been produced as a result of chloroplast transformation. These include proteins resulting in insect resistance herbicide resistance and drought tolerance. In addition, human therapeutic proteins can be produced in this way. For example, human somatotropin (a growth hormone that is used to treat pituitary dwarfism) has been expressed in tobacco chloroplasts (Staub *et al.*, 2000). The protein accumulates at high levels and is both correctly folded and possesses appropriate disulphide bonds. An extremely attractive extension of this work would be to produce therapeutic proteins in edible plants. Even if therapeutic proteins can be produced in this way, mechanisms by which they can be ingested and maintained in an active form still need to be established. It should also be possible to express recombinant proteins in potato and tomato chloroplasts. For example, tomato plants have been produced expressing the *aadA* marker gene, where the recombinant protein accumulates in both fruits and leaves (Ruf *et al.*, 2001).

One unique aspect of chloroplast engineering is the possibility of using operons to express multiple transgenes. In the plant nuclear genome this is not

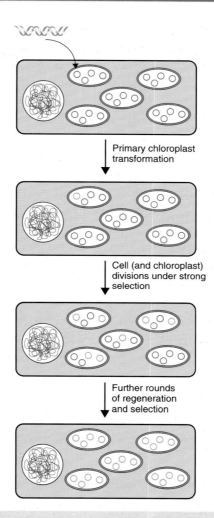

Figure 11.6. Selection of homogenous transgenic chloroplasts. Plant cells can contain between 50 and 100 chloroplasts, and each of these contains 10–20 nucleoids. Each nucleoid contains between five and 10 chloroplast genomes. Therefore, each plant cell may contain >10 000 chloroplast genomes. Transformation is likely to result in the alteration of just one of these genomes. A non-homogenous genome content in a chloroplast appears to be unstable, so multiple rounds of regeneration and selection are required to produce a homogenous population of transgenic chloroplasts (often termed **homoplasmic**) within the plant cell. Adapted from Bock (2001)

possible, because each gene produces a separate mRNA. Therefore, multiple transgene expression relies on either the crossing of plants containing single transgenes, or the concurrent transfer of transgenes (Daniell and Dhingra, 2002). In the chloroplast, most genes are transcribed as polycistronic messages, therefore multiple foreign genes may be expressed within the same transgene.

For example, tobacco chloroplasts have been engineered to express three genes of the *Bacillus thuringiensis cry* operon (De Cosa *et al.*, 2001). We will discuss the use of the *Bacillus thuringiensis cry* system below, but this approach results in very high levels of insecticidal protein accumulation – over 45% total soluble protein – with potent effects on insect pests. The lack of transfer of either the gene or its lethal protein product into pollen eliminates the danger of harming non-target and beneficial insects. Chloroplast engineering therefore appears to be a safe and environmentally friendly alternative to nuclear gene transfer for the plant biotechnology industry (Maliga, 2002).

11.2 Commercial Exploitation of Plant Transgenics

For many thousands of years attempts of been made to improve crop plants. Selective breeding programmes have been used to generate varieties yielding better nutritional qualities, higher yields, or improvements that can aid cultivation and harvesting of the crop. Genetic engineering does, however, provide the opportunity to alter the properties of a plant in a directed fashion. Some examples of commercially released genetically altered plants are listed in Table 11.1.

11.2.1 Delayed Ripening

Ripening is an important part of the commercial production of fruits. Fruits that are either under- or over-ripened are less attractive to consumers. However, the

Table 11.1. Some commercial releases of transgenic plants, from Birch (1997)

Crop and release date	Name	Company	Novel properties
Tomato (1994)	Flavr Savr	Calgene	Vine-ripened flavour, increased shelf life
Tomato (1995)		Zeneca	Consistency of tomato paste
Soybean		Monsanto	Glyphosate herbicide resistance
Canola (rape)	Roundup Ready		
Cotton (1995–96)			
Squash (1995)	Freedom III	Upjohn	Multiple virus resistance
Cotton	Bollgard	Monsanto	*Bacillus thuringiensis* toxin for insect resistance
Potato	NewLeaf		
Maize (1996–97)	YieldGuard		

transportation of many fully ripened soft fruits may result in their damage. This is particularly relevant to the transportation of tomatoes, where any damage can make the fruit unsellable. The protein products of a number of genes control the ripening process. One of these, encoding the enzyme polygalacturonase, is involved in the slow break-down of the polygalacturonic acid component of cell walls in the fruit pericarp. Its effects result in a gradual softening that makes the fruit edible. However, the longer the enzyme is able to act on the cell walls, the softer and more over-ripe fruit will become. Therefore, if the effects of the enzyme can be delayed then the fruit will ripen more slowly and, as a result, tomatoes can be left on the plant for longer to accumulate greater flavour. Tomatoes have been engineered so that they express less of the polygalacturonase enzyme. This was achieved through the insertion of the antisense sequence to a 5′-region of the polygalacturonase gene into the tomato genome. Expression of the antisense sequence was driven from the cauliflower mosaic virus 35S promoter, and the construct was inserted into tomato cells using Agrobacterium (Smith *et al.*, 1988). The resulting transgenic tomatoes expressed reduced levels (6 per cent) of the polygalacturonase gene in comparison to their wild-type counterparts, and the fruit could be stored for prolonged periods before beginning to spoil. Antisense RNA has also been used to reduce the expression of some of the genes required for ethylene synthesis, again resulting in delayed ripening (Fray and Grierson, 1993).

11.2.2 Insecticidal Resistance

Crop plants have been engineered to express the insecticidal toxin gene of *Bacillus thuringiensis* so that insects attempting to eat these plants are killed. *Bacillus thuringiensis* is a Gram-positive spore-forming bacterium that synthesizes a large cytoplasmic crystal containing insecticidal toxins. Different strains of the bacterium produce toxins that are effective against different insect species. The crystals are aggregates of a $\sim$130 kDa protein (Figure 11.7) that is produced as a protoxin that must be activated before it has any effect. The crystal protein is highly insoluble so it is relatively safe to humans, higher animals and most insects. However, the protein is solubilized in reducing conditions at high pH (>9.5) – the conditions commonly found in the mid-gut of lepidopteran larvae (moths and butterflies). Once it has been solubilized in the insect gut, the protoxin is cleaved by a gut protease to produce an active toxin, termed δ-endotoxin, of about 60 kDa. It binds to the midgut epithelial cells, creating pores in the cell membranes and leading to equilibration of ions. As a result, the gut is rapidly immobilized, the epithelial cells lyse, the larva stops feeding, and the gut pH is lowered by equilibration with the blood pH. This lower pH

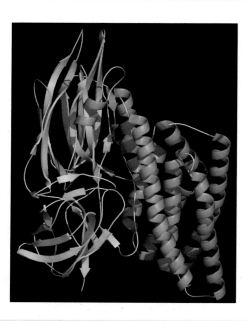

Figure 11.7. The structure of the *Bacillus thuringiensis* δ-endotoxin (Li, Carroll and Ellar, 1991)

enables the bacterial spores to germinate, and the bacterium can then invade the host, causing a lethal septicaemia. Several crops have been engineered to contain a copy of the *Bacillus thuringiensis* cry1Ac gene, encoding the protoxin (Table 11.1). In addition, the gene has been expressed at very high levels in the chloroplasts of tomato plants, resulting plants that are resistant to a range of insect pests (McBride *et al.*, 1995). This approach is highly successful, but has the potential disadvantage that continuous exposure of insects to the toxin will select for the development of toxin resistance.

11.2.3 *Herbicidal Resistance*

Several crops have been engineered for resistance to herbicides such as **glyphosate** (Figure 11.8). Glyphosate is an inhibitor of aromatic amino acid production in both plants and bacteria. Two approaches have been used to engineer resistance so that the herbicide can be used for weed control without damaging the crop. In the first, the target protein of the herbicide (EPSPSase) can be over-produced so that resistance occurs as a consequence of having more enzyme available to the cell (Shah *et al.*, 1986). A second approach results from the expression of a mutant version of EPSPSase that is resistant to the herbicide within cells (Stalker, Hiatt and Comai, *et al.*, 1985). Glyphosate-resistant

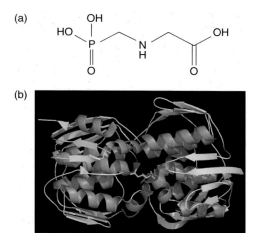

Figure 11.8. The structure of the herbicide glyphosate. (a) Glyphosate, N-phosphono-methyl glycine, is sold as an isopropylamine salt under the trade name Roundup. It is an inhibitor of aromatic amino acid synthesis, specifically inhibiting 5-enolpyruvylshikimate-3-phosphate synthase (EPSPSase). The herbicide is absorbed by foliage, but rapidly moves to apical areas of the plant and inhibits protein synthesis. Plants stop growing and tissues slowly degrade due to lack of proteins. Death ultimately results from dehydration and desiccation. (b) The interaction between glyphosate and EPSPSase. The active site of the enzyme is located in an interdomain cleft in the two-domain enzyme. Glyphosate (blue) binds close to shikimate-3-phosphate (red) and appears to occupy the binding site of phosphoenol pyruvate, the second substrate of EPSPase, thereby mimicking an intermediate state of the ternary enzyme–substrate complex (Schönbrunn *et al.*, 2001)

crops, e.g. the 'Roundup Ready' crop plants marketed by Monsanto, have the advantage of increased yield due to the elimination of weeds. However, the potentially detrimental effects of increased herbicide usage, and the potential for transmission of the herbicide-resistance gene to other plant species, are still relatively unknown (Gressel, 2000). Of course, it is easy to level criticism at agrochemical companies that produce both the herbicide and crops that are resistant to it. The potential for conflicts of inrest in a relatively captive market have led to accusations that farming interests and practices are not being served by the introduction of genetically modified crops.

11.2.4 Viral Resistance

Viral damage to crop plants is responsible for huge commercial losses. The control of viral infections is based on the observations of resistance to superinfection – the infection of a plant with one viral strain protects the plant from infection by a second, related virus. The expression of the tobacco mosaic virus coat protein within plants, which is not sufficient to cause infection, does yield resistance to virus infection (Abel *et al.*, 1986). Additionally, plants expressing RNA sequences complementary to the coat protein coding region and the 3′ untranslated region were protected from infection (Powell *et al.*, 1989). The mechanism of resistance to superinfection is not well understood, although the expression of the coat protein in different tissues of the plant will give rise to different levels of viral resistance (Clark *et al.*, 1990). The expression of viral proteins in plant cells has also raised the possibility that other viruses, that are still able to infect the plant, may be able to 'inherit' the expressed coat protein. To date, there is no clear evidence that this type of problem might actually occur.

11.2.5   Fungal Resistance

Plants have been engineered to have an increased resistance to various fungal infestations. This can be achieved by expressing antifungal proteins within the plant. For example, transgenic tobacco seedlings constitutively expressing a bean chitinase gene under control of the cauliflower mosaic virus 35S promoter showed an increased ability to survive in soil infested with the fungal pathogen *Rhizoctonia solani* and delayed development of disease symptoms (Broglie *et al.*, 1991). Chitinases hydrolyse chitin, an unbranched polymer of $\beta(1{\rightarrow}4)$ linked N-acetyl-D-glucosamine (GlcNAc) that is found in the cell walls of many fungi. An alternative approach is to transfer sets of avirulence genes from the pathogen into the plant (Melchers and Stuiver, 2000).

11.2.6   Terminator Technology

Perhaps the most controversial use of plant manipulation is the ability of producers to sell seeds that will produce plants for a single crop only. That is, seeds can be produced that will grow into normal plants, but the resulting seeds are sterile. This **terminator technology** (Figure 11.9) ensures that growers cannot save some of their own seeds for use in the following year, but must buy fresh seeds from the producer. The ability to grow sterile, but otherwise normal, seeds relies on the insertion of three genes into the plant:

- the gene encoding the tetracycline repressor (TET^R) under the control of a constitutive promoter,

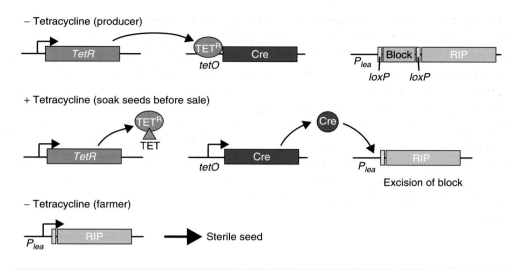

Figure 11.9. Terminator technology for the production of sterile seeds. See the text for details

- the gene encoding the Cre recombinase protein, whose expression can be inhibited by the binding of TET^R to its promoter, and

- a gene that can be expressed late in embryogenesis that will make the seed sterile. As shown in Figure 11.9, the gene for a ribosome inactivating protein (RIP) is controlled from a promoter that is only active during late embryogenesis (*lea*; the promoter from the late embryogenesis abundant gene). RIPs are plant enzymes that are potent inactivators of eukaryotic protein synthesis (Stirpe *et al.*, 1992). RIPs inactivate eukaryotic ribosomes by enzymatically cleaving a specific A residue in the 28S rRNA. This irreversible modification renders the ribosome unable to bind elongation factors and blocks translation.

The transgenic plant is constructed such that the RIP gene cannot be produced due to the presence of a blocking DNA sequence inserted into it. The blocker sequence is flanked by *loxP* recombination sites. The resulting plant is able to grow normally and produce normal, fertile seeds. The producer then soaks the seeds in tetracycline to inhibit the binding of the tetracycline repressor to the Cre gene, which can then be transcribed. The Cre protein catalyses the excision of the blocker DNA from the RIP gene. The seed is still capable of producing apparently normal plants, but the seeds it produces are sterile due to the expression of the RIP gene late in embryogenesis. The sterile seeds retain all the nutritional value of their normal counterparts, but are unable to form new plants.

This immensely elegant way of protecting the investment made in transgenic crops was greeted with a massive public outcry when it was first patented in 1998 and 1999. The potential dangers of the terminator gene spreading from the plant to other organisms, coupled with the, real or perceived, control given to the seed producers, particularly in Third World countries, has meant that the technology has yet to be implemented.

11.3   Ethics of Genetically Engineered Crops

A detailed discussion of the ethics of crop genetic engineering programmes is beyond the scope of this text. The engineering of plant traits has been occurring for thousands of years and the introduction of DNA technology has allowed, and will allow, many novel and important traits to be imparted. The safety of the resulting crops, both in terms of the edible product and potential effects on the environment, need to be rigorously assessed. The only realistic way that this can be achieved is through the careful design and thorough analysis of crop trials in a natural setting. Only then will any potentially harmful side-effects of the engineering process be identified.

12 Engineering animal cells

Key concepts

♦ Higher-eukaryotic cells are relatively difficult to grow in culture – most cells will only divide a few times before dying

♦ Immortalized cells lines can be produced following viral transformation

♦ Foreign DNA may be inserted (transfected) into cultured cells using a variety of chemical, physical or viral methods

Unlike their plant counterparts, most animal cells are either already fully differentiated or are committed to a particular differentiation pathway. Consequently, animal cells into which foreign DNA has been inserted cannot easily be used to regenerate whole animals. An obvious exception to this is the creation of cloned animals from adult cells using nuclear transfer technology. We will discuss nuclear transfer in greater depth in Chapter 13, but here we will concentrate on methods to introduce DNA into **somatic** cells and in particular into cultured animal cell lines.

12.1 Cell Culture

To establish a culture of animal cells, a small amount of tissue (usually as small a sample as 2 mm^3) is removed from a specific organ (skin, liver etc.) and placed in a sterile dish. The tissue is treated with proteases to break down some of the proteins that hold the cells together and then teased apart to separate the individual cells. These cells are then placed in another dish containing a culture medium with serum and allowed to divide. **Primary cells** produced in this way do not easily divide outside the animal, and will usually undergo only a few divisions before undergoing **senescence**. If the cells can be induced to reproduce

Analysis of Genes and Genomes   Richard J. Reece
© 2004 John Wiley & Sons, Ltd   ISBNs: 0-470-84379-9 (HB); 0-470-84380-2 (PB)

for more than a few generations, a **secondary cell** line may be produced. Cell lines contain many cells that are essentially identical. Most cell lines will divide a relatively small number of times (10–20) before entering senescence. Because these cells will die on their own, they are called mortal cells. Some cell lines, however, do not proceed to senescence and are described as immortal. These cells are said to be 'transformed' in that they have undergone a change to make them malignant or immortal. Examples of immortalized cell lines include

- HeLa cells (derived from a 31-year-old black woman named Henrietta Lacks, who died of cervical cancer in 1951) and

- Chinese hamster ovary (CHO) cells.

The changes that occur within these cells to make them immortal may result from a viral infection or other change within the cell that leads to unregulated cell division and growth. Here, the term transformation refers to this process rather than the uptake of foreign DNA as we have used it previously. In higher eukaryotes the process of inserting foreign DNA into cells is called **transfection**.

12.2 Transfection of Animal Cells

The insertion of foreign DNA into animal cells is a two-stage process in which the DNA is firstly taken up by the cells (transfection), and secondly integrated into the genome. It is possible to generate stable animal cell lines that harbour extra-chromosomal vectors. For example, plasmids bearing the Epstein–Barr virus nuclear antigen (EBNA-1) and origin of replication (*oriP*) can be maintained within some primate and canine cell lines but not in rodent cell lines (Yates *et al.*, 1985). However, most stable cell lines that 'permanently' contain the gene of interest depend upon the stable non-homologous integration of foreign DNA into the host chromosome. The transfection of cells is a far more efficient than integration of foreign DNA into the genome. Consequently, large numbers of cells can be transfected with foreign DNA, where it is maintained within the nucleus, but do not integrate it into their genome. The process of DNA uptake into animal cells may thus be described as being either transient or stable, depending how long the DNA is maintained within the cell.

- *Transient transfection.* If the foreign DNA transfected into cells is not integrated into the genome and does not contain an animal origin of replication, the cell will maintain it for a short period before it is either degraded or diluted from the cell during division. During the period in which the foreign DNA is inside the cell, it is treated just as any other DNA sequence – genes contained within it may be expressed provided they are under the control of a suitable promoter, and the encoded protein produced. Transient transfection represents a rapid way to analyse foreign genes and gene products within cells. For example, many types of gene expression assay rely on the transfection of cells with an appropriate reporter gene, and the collection of data some 24–48 h later. Relative to the generation time of most cell lines (~16–24 h), this represents a relatively short time between transfection and cell harvesting. In general, transient transfection protocols need to be optimized for each cell type being analysed due to inherent differences in DNA uptake efficiencies.

- *Stable transfection.* Rather than being maintained in an extra-chromosomal state within the nucleus of a host cell, the foreign DNA becomes integrated into an apparently random location within the genome. Once integrated, the foreign DNA is replicated together with the host chromosome in which it resides and gene expression can occur, providing that the foreign gene is controlled by suitable transcriptional control elements. Stable cell lines are required for the production of large amounts of a recombinant protein over a prolonged period of time. Technical difficulties in producing stable cell lines, arising mainly from the inefficiency of the integration process, mean that a clonal cell line can take several months to produce (Power and Meyer, 2000). This timescale often precludes their production.

The transfection of animal cells with genes from other organisms dates back over 40 years (Szybalska and Szybalski, 1962). In these experiments, the uptake of naked DNA into animal cells was an incredibly rare event. Animal cells are enclosed by a single membrane, called both the cell membrane and the plasma membrane. It is composed of a double layer of lipids containing a variety of proteins. The membrane is relatively impermeable and, under normal circumstances, allows only certain materials to pass into and out of the cell. In tissues, cells are also surrounded by extra-cellular matrix (ECM) – a mixture of proteins and proteoglycans – that are involved in cell–cell signalling, wound repair, cell adhesion and tissue function. Most cultured animal cells do not possess an ECM and are therefore relatively easy to insert DNA into compared with their native counterparts. As we have seen with other forms of transformation, the efficiency of the process can be greatly increased by various treatments, which are discussed below.

12.2.1 Chemical Transfection

The efficiency of naked DNA transfection can be greatly improved if the DNA is precipitated in the presence of the cells. This can be achieved simply by

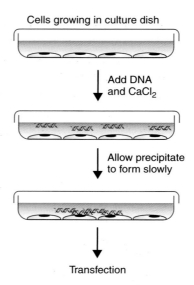

Cells growing in culture dish

Add DNA
and CaCl$_2$

Allow precipitate
to form slowly

Transfection

Figure 12.1. The transfection of animal cells using calcium phosphate precipitation

washing cultured cells in a phosphate buffer, adding the DNA, and then adding calcium chloride to the mixture (Figure 12.1). Under these circumstances, it is thought that the precipitate settles on the surface of cells and is then internalized through **endocytosis** (Orrantia and Chang, 1990). Some of the DNA will be released into the nucleus, where it may be expressed, or become integrated into the genome. Transfection efficiencies using calcium phosphate, as judged by the percentage of cells that take up the DNA, can be quite low, in the range of 1–2 per cent (Graham and van der Eb, 1973), but can be increased if very high-purity DNA is used and the precipitate allowed to form slowly (Chen and Okayama, 1988).

12.2.2 Electroporation

We have previously discussed the transformation of bacterial cells based on electroporation (Chapter 2). The technique also works well for animal cells (Wong and Neumann, 1982). By varying the electric field strength, and the length of time the cells are exposed to the electric field, it is possible to optimize electroporation parameters for almost any cell type.

12.2.3 Liposome-mediated Transfection

Liposomes are spheres of lipid that can be used to transport molecules into cells. Classical liposomes are composed of a sphere of lipid bilayer surrounding the molecule to be transported, and promote transport after fusing with the cell membrane. Cationic lipids (those bearing a positive charge) are used for the transport of nucleic acid. Cationic lipids naturally result in the formation of unilamellar liposome vesicles (Schaefer-Ridder, Wang and Hofschneider, 1982) that, rather than encapsulating the DNA within the liposome, are thought to bind along the surface of the negatively charged DNA to form clusters of aggregated vesicles over its length (Figure 12.2). Cationic liposomes

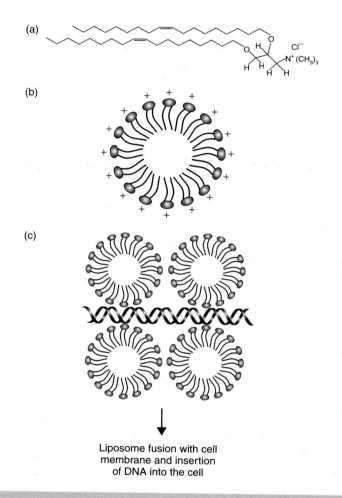

Liposome fusion with cell
membrane and insertion
of DNA into the cell

Figure 12.2. The introduction of DNA into cells using cationic liposomes. (a) The chemical structure of DOTMA (N-[1-(2,3-dioleyloxy)propyl]-N,N,N-trimethylammonium chloride), a cationic lipid used in the formation of liposomes (Felgner *et al.*, 1987). (b) The structure of a cationic liposome. The positively charged 'heads' project outwards from a hydrophobic core formed by the lipid 'tails'. (c) The surrounding of a DNA fragment by cationic liposomes prior to its insertion into the cell after fusion with the cell membrane

can be formed from a variety of cationic lipids, e.g. DOTAP (N-1(-(2,3-dioleoyloxy)propyl)-N,N,N-trimethylammoniumethyl sulphate) and DOTMA (N-(1-(2,3-dioleoyloxy)propyl)-N,N,N-trimethylammonium chloride). These are commercially available lipids that are sold as *in vitro* transfecting agents, with the latter sold as Lipofectin®. Commercially available liposome formulations usually include a neutral lipid such as DOPE (dioleoylphosphatidylethanolamine) in order to facilitate liposome formation and cell membrane fusion. Cationic liposomes are able to interact with the negatively charged cell membrane more readily than uncharged liposomes, with the fusion between cationic liposome and the cell surface resulting in the delivery of the DNA directly across the plasma membrane (Figure 12.2). Additionally, material brought into the cell through cationic liposomes appears to be able to escape endosome mediated degradation that otherwise presents a major barrier to transfection *via* endocytosis. Liposome mediated DNA delivery is technically easy, highly reproducible, and very efficient. It is therefore often the method of choice for routine cell line DNA transfection.

12.2.4 *Peptides*

A number of peptide sequences have been shown to be able to bind to, and condense, DNA to make it more amenable for entry into cells; e.g. the tetrapeptide serine–proline–lysine–lysine (located in the C-terminus of the histone H1 protein) promotes DNA transfection (Schwartz *et al.*, 1999). The positively charged lysine amino acid side chains (see Appendix A) help to counteract the negatively changed phosphate DNA backbone and allow the DNA molecules to pack closely to each other. Rational design of peptide sequences has also been used to develop synthetic DNA binding peptides. For example, the peptide tyrosine–lysine–alanine–(lysine)$_8$–tryptophan–lysine has been shown to be very effective at forming complexes with DNA (Gottschalk *et al.*, 1996). DNA binding peptides that can be coupled to cell-specific ligands can also be synthesized, thereby allowing receptor mediated targeting of the peptide/DNA complexes to specific cell types. One such approach has been to synthesise a cationic peptide based on 16 lysine residues and an arginine–glycine–aspartic acid (RGD) peptide sequence. The RGD sequence facilitates the binding of the peptide/DNA complex to an integrin receptor on Caco-2 cells *in vitro* (Harbottle *et al.*, 1998). Synthetic peptides have also been developed that enhance the release of the peptide/DNA complexes from the endosome following endocytosis.

12.2.5 *Direct DNA Transfer*

DNA may be physically injected directly into the nucleus of cultured cells (Capecchi, 1980). This approach is somewhat technically difficult and labour intensive in that cell nuclei must be injected with the foreign DNA on an individual basis, but is especially useful for large cells. For example, the oocytes of the African clawed frog *Xenopus laevis*, which can be harvested in large numbers from the ovaries of adult female frogs, are approximately 1 mm in diameter and have a correspondingly large nucleus. DNA, or RNA, injected into the nucleus is usually expressed in a mosaic pattern during the development of the frog (Melton, 1987). Integration of injected foreign DNA occurs at a very low frequency, but the expression of the foreign gene can be followed for many cell generations. The efficiency of the integration process varies considerably in different organisms. The direct injection of DNA into certain animal tissues, particularly into muscle and skin, has also been shown to produce high levels of transgene expression (Wolff *et al.*, 1990). For example, patients suffering from the genetic skin condition lamellar ichthyosis (in which the skin is scaled and reddened), caused by a loss of transglutaminase 1 (TGase1) expression, show some skin regeneration following repeat injections of plasmid DNA encoding the TGase1 gene (Choate and Khavari, 1997). The restoration of TGase1 expression occurred in the correct location of the skin. Further analysis, however, revealed that the pattern of expression was non-uniform and failed to correct the underlying histological and functional abnormalities of the disease. Direct DNA injection has also been applied to the gene therapy (see below) of cancer, where the DNA can be injected either directly into the tumour or can be injected into muscle cells in order to express tumour antigens that might function as a cancer vaccine. Although direct injection of plasmid DNA has been shown to lead to gene expression, the overall level of expression is generally lower than that obtained with liposomes. We will return to direct DNA injection later when we look at the production of transgenic animals (Chapter 13).

12.3 Viruses as Vectors

The relatively low efficiency of foreign DNA integration into animal cells, combined with the lack of naturally occurring extra-chromosomal plasmids, led to the manipulation of viruses as potential vectors for gene transfer.

12.3.1 SV40

The first mammalian cell viral vector to be developed was based on the simian virus 40 (SV40) (Hamer and Leder, 1979). SV40 is a primate double-stranded DNA tumour virus whose genome is 5243 bp in size (Figure 12.3). Genes are

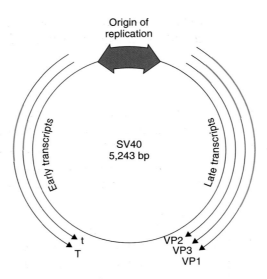

Figure 12.3. The SV40 genome. SV40 has two transcriptional units, whose expression is driven from the early and late promoters. Both transcripts produce multiple protein products through alternative splicing. The large T and small t antigens are required for viral replication, while the VP proteins form the viral capsid. The introns sequences removed during splicing are shown in red

encoded on both strands of the genome such that they overlap each other. Viral infection primarily occurs in monkey kidney cells, but the virus is capable of infecting a variety of mammalian cell types and, depending on the cell type infected, can either undergo a lytic or a lysogenic life-cycle (Das and Niyogi, 1981). SV40 contains two sets of expressed genes.

- *Early genes* are expressed immediately following infection and are required for viral replication. The early genes include the large T antigen and small t antigen.

- *Late genes* are expressed at a later time during the infective cycle and encode the viral capsid proteins required for the formation of new viral particles. These genes are under the control of the SV40 major late promoter (MLP).

Foreign genes may be inserted into the SV40 genome to replace either an early or a late gene. If a foreign gene replaces, for example, a late gene, then the virus cannot replicate properly. Infection can, however, proceed using an SV40 helper virus defective in an early gene to supply 'late' functions in trans. In a similar fashion to λ phages (Chapter 3), SV40 has a limited capacity for the addition of foreign DNA. Even with the replacement of existing genes, the maximum foreign DNA insert size is limited to ~2300 bp (Naim and Roth, 1994).

12.3.2 Adenovirus

Adenoviruses are responsible for approximately 25 per cent of the 'common colds' that we suffer. There are many distinct types that can cause infections in humans, but they are all non-enveloped icosohedral viruses (ranging from 60 to 90 nm long) that contain linear double-stranded DNA genomes ~36 kbp in size (Figure 12.4). The transcription of the adenovirus genome occurs

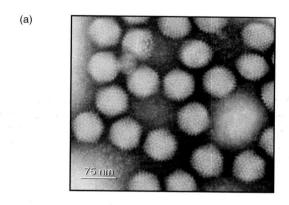

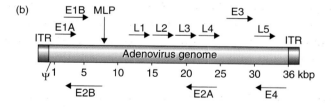

Figure 12.4. Adenovirus. (a) Electron micrograph of a human faecal adenovirus. Image courtesy of Tony Oliver (IBMS London Region Virology Discussion Group). (b) The adenoviral genome. The approximate location of some of the early (E) and late (L) genes are shown, along with the site of the major late promoter (MLP), and an approximately 300 bp packaging site (Ψ), which is essential for the packaging of viral genome into the assembled capsid. Both ends of the genome are composed of 100 bp inverted terminal repeats (ITRs). ITRs serve as origins of DNA replication by priming the replication process and acting as assembly points for the proteins of the replication complex. The adenovirus genome also encodes several viral associated RNAs (VA RNAs) that are transcribed constitutively at a high level using RNA polymerase III. VA RNAs are ~160 nucleotides in length and have a high GC content and a high degree of secondary structure. They may serve to regulate mRNA splicing or to control of the rate of translation of late genes (Mathews, 1995)

in two phases – early and late – occurring either before or after virus DNA replication, respectively. Transcription is accompanied by a complex series of splicing events, with four early regions of gene transcription (E1–E4, which are required for viral replication), and an MLP producing the late genes (L1–L5, which produce the viral capsid) (Logan and Shenk, 1984). The virus will replicate in many different mammalian cell lines when the naked genomic DNA is transfected into them. Additionally, the virus produces large numbers of progeny (up to 10^5 virions per infected cell), which means that viral particles, recombinant or otherwise, can be purified in large amounts with ease.

Most vectors derived from the adenoviral genome are replication deficient. They lack the essential E1 genes and often the non-essential E3 gene. Since they are replication defective, they must be propagated in cell lines that have also been transfected with DNA containing the E1 genes – e.g. the human embryonic kidney cell line 293 – to supply E1 function in *trans* (Graham *et al.*, 1977). This type of vector has a maximum capacity for foreign DNA of about 7 kbp. Other adenoviral vectors lack either the E2 or E4 genes in addition to E1 and E3 to increase the capacity to carry foreign DNA, these functions again, being supplied in *trans* from the cell line (Gorziglia *et al.*, 1996).

Adenoviruses bearing foreign DNA can be used to produce the foreign protein in many different cell types, but gene expression is usually transient because the viral DNA does not integrate into the host genome. The lack of integration may, however, be advantageous if adenoviral vectors are used in gene therapy trails (see below). Additionally, tissue specific gene expression is possible with adenoviral vectors if the foreign gene is placed under the control of cell-specific promoter and enhancer elements, e.g. the myosin light chain 1 promoter (Shi, Wang and Worton, 1997) or the smooth muscle cell SM22a promoter (Kim *et al.*, 1997), or by direct delivery to a local area *in vivo* (Rome *et al.*, 1994).

Adenoviral vectors are useful because they are highly efficient at getting DNA into cells. They are capable of containing DNA inserts up to about 8 kbp in size and they can infect both replicating and differentiated cells. Additionally, since they do not integrate into the host genome, they cannot bring about mutagenic effects caused by random integration events. The disadvantages of adenoviral vectors include the following.

- Expression is transient since the viral DNA does not integrate into the host.

- Adenoviral vectors are based on an extremely common human pathogen and *in vivo* delivery may be hampered by prior host immune response to one type of virus.

12.3.3 Adeno-associated Virus (AAV)

Adeno-associated virus was first discovered as a contaminant of adenovirus preparations (Atchison *et al.*, 1965). Despite its name, AAV is not related to adenovirus but rather it is a member of the parvovirus family, which has a single-stranded DNA genome approximately 5000 nucleotides in size. The virus is commonly found in human tonsil tissue, although no specific disease of man appears to be associated with it. AAVs are naturally replication deficient, and require the presence of another virus – e.g. adenovirus – in order to complete their own replicative cycle. Some of the early adenovirus genes are required to promote AAV replication. However, the treatment of AAV infected cells with ultraviolet light, cycloheximide or some carcinogens can replace the requirement for helper virus (Bantel-Schaal, 1991). Therefore the requirement for a helper virus appears to be for a modification of the cellular environment rather than a specific viral protein.

In the absence of helper virus, the AAV genome integrates into the host cell, where it remains as a latent provirus. The integration of the viral DNA into the host genome is not, however, a random process; it specifically integrates into the same genetic location on chromosome 19 (Kotin *et al.*, 1990). AAV based vectors exploit two 145-base ITR sequences found within the viral genome, which are the only DNA elements required for replication, transcription, provirus integration and rescue. These repeats form hair-pin loop structures that are essential for viral replication. Cloning into the AAV genome is facilitated by inserting the inverted terminal repeats into a plasmid vector and then cloning foreign DNA between them. Linear DNA fragments containing the foreign gene flanked by the ITRs may be prepared from such plasmids following digestion with restriction enzymes. This DNA is then transfected into cells, along with helper plasmid to supply the AAV genes needed for viral integration. Viral stocks may then be prepared by infecting the cells with adenovirus to stimulate AAV lytic replication and packaging. Finally, the AAV viral particles may purified from the contaminating adenovirus using biochemical methods (Gao *et al.*, 2000). This type of approach suffers from the use of adenovirus to recover the AAV particles. The likelihood of contamination is high and therefore the recombinant AAV produced cannot be used for human gene therapy trials. An alternative approach to the production of recombinant AAV particles is shown in Figure 12.5. Here, the foreign gene flanked by the ITRs is co-transfected into human 293 kidney cells together with two other DNA fragments – one containing the genes required for the production of single-stranded AAV genomic DNA, and the other bearing the adenoviral genes required for lytic replication and packaging (Matsushita *et al.*, 1998). The recombinant AAV

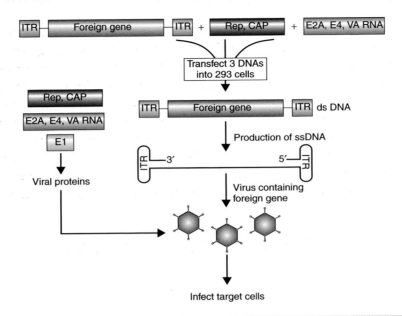

Figure 12.5. The production of recombinant AAV particles without adenoviral infection. A linear DNA fragment containing a foreign gene, under the control of a suitable promoter so that it may be expressed, is transfected into human kidney cell line 293 (which expresses the adenoviral E1 gene) together with DNA containing the AAV Rep and Cap genes, and another DNA fragment bearing the adenoviral E2A, E4 and VA RNA genes. The Rep and Cap proteins are required to produce single-stranded AAV genomes and to encapsulate them in coat proteins. The helper proteins from adenovirus needed for AAV lytic replication (E1, E2A, E4 and VA RNA) are produced within the cells to promote the formation of viral particles containing the foreign gene. These can then be used to infect the target cells

particles harvested from this procedure can then be used to infect target cells, in which the foreign gene will be expressed.

AAV vectors have been used to introduce foreign DNA into a wide variety of cell types, including neurons (Davidson *et al.*, 2000), muscle (Pruchnic *et al.*, 2000) and epithelial cells (Pajusola *et al.*, 2002). The predictable location of DNA integration into the host cell, its ability to transform both dividing and non-dividing cells, and advances in the ability to infect a variety of cell types *in vivo* have made AAV vectors very promising for gene therapy (Sanlioglu *et al.*, 2001).

12.3.4   *Retrovirus*

Retroviruses are the only animal viruses that integrate into the host cell genome during the normal growth cycle. There are two classes of retrovirus that infect humans – the HTLV retroviruses (e.g. human T-cell leukemia virus, HTLV-1) (Yoshida and Seiki, 1987), and the lentiviruses (e.g. the human immunodeficiency virus, HIV-1) (Luciw *et al.*, 1984) – but retroviruses have been found in a wide variety of organisms, including both vertebrates and invertebrates. They are enveloped particles (approximately 100 nm in diameter), where the envelope takes the form of a lipid bilayer extracted from the host cell as the virus buds away from it. The envelope also contains some virally encoded proteins. Infection occurs as a result of the interaction of the envelope proteins with specific receptors on the surface of the host cell (Sommerfelt, 1999). Receptor binding results in the internalization of the viral particle (Figure 12.6).

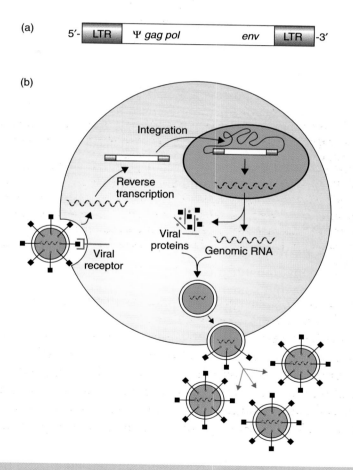

Figure 12.6. Retroviruses. (a) The structure of an integrated form of a typical retroviral genome. The left and right long terminal repeats (LTRs) flank the structural genes (*gag, pol* and *env*). A packaging site (Ψ) is required for the insertion of the viral RNA genome into assembled viral particles. (b) The retroviral life cycle. See the text for details

Partial uncoating of the particle occurs prior to the reverse transcriptase enzyme converting the RNA genome into a linear double-stranded DNA copy. DNA synthesis initiates from a specific tRNA molecule (Mak and Kleiman, 1997) acquired from the host cell in which the virus was produced, which, like the reverse transcriptase, is packaged in the viral particle itself. The newly formed proviral DNA then integrates randomly into the host genome where the viral genes are expressed to make new viral proteins and results in the assembly of new viral particles. The production of new viruses does not directly result in host cell death, but rather the newly formed particles bud from the cell surface. The diseases associated with retroviral infections are usually a consequence of the site of retroviral insertion into the host genome, or as a result of alterations made to the types of cell they infect, e.g. the invasion of the immune system by HIV-1 leading to AIDS.

The genome of a retrovirus is composed of two identical single-stranded RNA molecules (between 8000 and 11 000 nucleotides in length) that, similar to mRNA molecules, each possess a 5' cap and a 3' polyA tail (Ratner *et al.*, 1985). The viral RNA genome cannot be used as a translation template prior to integration into the host. Transcription and subsequent translation only occurs after the DNA version of the genome has been integrated into the host. All retroviruses contain three major structural genes, each of which encodes a number of protein activities (Katz and Skalka, 1994):

- *gag* – encodes a variety of capsid proteins;

- *pol* – encodes the reverse transcriptase required to convert the RNA genome into a DNA copy, the integrase needed for the insertion of the DNA into the host genome and sometimes a protease (*pro*) required for cleavage of the *gag* gene product during maturation;

- *env* – encodes the surface and transmembrane proteins found in the envelope.

Some retroviruses also contain other genes; e.g. *pro* may be encoded by a separate gene. All of the proteins encoded by *gag, pol* and *pro* are expressed from a single full-length genomic RNA transcript. The *env* protein is produced from a spliced mRNA. To compress this number of genes into a small genome, the virus utilizes a number of strategies such as splicing and ribosomal frame-shifting (Farabaugh, 1996). For example, the *pro* gene often overlaps *gag* and/or *pol* but is still produced from the same mRNA molecule. The retroviral structural genes are flanked within the genome by LTRs, which become duplicated as part of the replication process prior to integration into the host genome. Viral RNA synthesis initiates from a single promoter within the left-hand LTR and ends at a polyadenylation signal within the right-hand LTR. Packaging of the transcribed RNA into viral particles requires a packaging signal (Ψ) located between the left-hand LTR and the structural genes (Figure 12.6(a)).

Packaging constraints on the amount of additional nucleic acid that can maintained within the viral genome mean that most vectors based on retroviruses are usually replication defective. The LTRs, the packaging site (Ψ) and the primer binding sites are the only genomic elements required for the replication process. Therefore, as we discussed for adenoviral vectors above, infective viral particles must be produced in specially constructed cell lines that can provide the necessary viral proteins. For example, a foreign gene may be cloned into a plasmid that contains DNA versions of the elements described above such that it is located between the left- and right-hand LTRs. The recombinant viral shuttle vector is then transfected into a cell line that constitutively expresses the viral reverse transcriptase and capsid proteins. Viruses produced from this cell line can then be used to infect the target cells, where the vector will become integrated into the genome and the foreign gene expressed. In addition to the foreign gene, a marker gene must also be used to assess whether an individual cell has taken up the recombinant DNA.

The major advantages to using retroviral based vectors arise from the stability of the integration of the viral genome into the host. The integration of a single copy of the viral DNA at a random location within the host's genome allows for the long-term expression of the integrated foreign gene. Additionally, retroviruses represent a highly efficient mechanism for the transfer of DNA into cells. The disadvantages of such vectors are the random nature of the integration process, which may have deleterious effects on the host cell, and the general requirement that retroviruses have to infect only dividing cells. This may severely limit their use. Only lentiviruses, e.g. HIV1, will also infect and replicate in non-dividing cells. Therefore, much effort has been directed into making suitably safe lentivirus vectors (Zufferey *et al.*, 1998).

12.4 Selectable Markers and Gene Amplification in Animal Cells

Like all of the other DNA transfer processes we have discussed, the incorporation of foreign DNA into animal cells needs to be accompanied by a phenotypic change to distinguish the transfected cells from those that have not taken up the foreign DNA (Figure 12.7). Some of the first experiments to identify transfected animal cells involved the complementation of a nutritional defect in a cell line. For example, human cells defective in the gene encoding the enzyme hypoxanthine guanine phosphoribosyltransferase (HPRT; part of

Figure 12.7. The structure of the bleomycin–bleomycin binding protein complex (Maruyama *et al.*, 2001). Binding sites for the drug (shown in red) are located in deep clefts in the sides of the dimeric protein (green and blue)

the inosinate cycle for the salvage of purine bases and the production of inosine monophosphate) and therefore unable to grow in medium containing hypoxanthine, aminopterin and thymidine (HAT, which blocks *de novo* inosine monophosphate production (Lester *et al.*, 1980)) were transfected with total genomic DNA from a wild-type cell line. Very rarely, cells could be isolated that were able to grow in this medium, indicating that they had acquired the ability to make HPRT from the wild-type cells (Szybalska and Szybalski, 1962). Similarly, mouse cells deficient in the enzyme thymidine kinase (TK, which is part of the nucleotide biosynthesis salvage pathway) are unable to grow on HAT medium, but can be transfected with the herpes simplex virus (HSV) *Tk* gene to allow growth on this medium (Wigler *et al.*, 1977). A number of other such metabolic markers have also been used to monitor the transfection process, but they all suffer from the requirement of a mutant cell line in order to detect transformed cells. This need has been overcome by using dominant selectable markers that confer a drug resistance phenotype to the transfected cells (Table 12.1). Antibiotics such as ampicillin have no effect on eukaryotic cells due to the lack of an animal cell wall. Some other antibiotics, particularly those that are protein synthesis inhibitors, are active against both prokaryotes and eukaryotes. For example, the *E. coli* transposon Tn5 encodes the kanamycin-resistance gene (Yamamoto and Yokota, 1980). Kanamycin is an aminoglycoside antibiotic that interferes with translation and induces bacterial cell death through site-specific targeting of ribosomal 16S RNA. At higher concentrations, aminoglycosides also inhibit protein synthesis in mammalian cells probably through non-specific binding to eukaryotic ribosomes and/or nucleic acids (Mingeot-Leclercq, Glupczynski and Tulkens, 1999). To achieve resistance as a method of selection, another gene encoded within the Tn5 transposon (producing neomycin phosphotransferase) is placed under the

Table 12.1. Markers for the selection of DNA fragments added to higher eukaryotic cells

Marker	Gene product	Selection method	Reference
neo	Aminoglycoside phosphotransferase; *neo* gene from the bacterial transposon Tn5	Select cells in G418 (0.1–1.0 μg/mL), an aminoglycoside that blocks protein synthesis and is similar to kanamycin	(Colbère-Garapin *et al.*, 1981)
hyg	Hygromycin-B-transferase; *hyg* gene from *E. coli*	Select cells in Hygromycin B (10–300 μg/mL), an aminocyclitol that inhibits protein synthesis	(Blochlinger and Diggelmann, 1984)
pac	Puromycin-N-acetyl transferase; *pac* gene from *Streptomyces alboniger*	Select cells in puromycin (0.5–5 μg/mL), an antibiotic that inhibits protein synthesis	(Vara *et al.*, 1986)
ble	Bleomycin binding protein; a *ble* gene is located on the bacterial transposon Tn5, or the *Sa ble* gene from *Staphylococcus aureus*	Select cells in bleomycin or Zeocin (50–500 μg/mL), an antibiotic that binds DNA and blocks RNA synthesis	(Genilloud, Garrido and Moreno, 1984)
gpt	Xanthine–guanine phosphoribosyltransferase; *gpt* gene isolated from *E. coli*	Select cells in guanine-deficient media that contains inhibitors of *de novo* GMP synthesis and xanthine; this selects for *gpt*$^+$ cells that can synthesize guanine from xanthine	(Mulligan and Berg, 1981)

control of a mammalian promoter, e.g. the HSV *Tk* promoter or the SV40 early promoter (Berg, 1981). We will discuss promoters used to drive expression of genes in animal cells in more detail below.

The exposure of animal cells to high levels of toxic drugs for prolonged periods for the purposes of recombinant selection can give rise, at a very low frequency, to the formation of cells that have become spontaneously highly resistant to the drug. For example, mouse cells exposed to methotrexate (a folic acid analogue that is an inhibitor of the enzyme dihydrofolate reductase, DHFR) have been shown to undergo three types of mutation to generate resistance (Schimke *et al.*, 1978):

- mutations within the DHFR enzyme to generate inhibitor resistance,

- mutations that prevent cellular uptake of the drug and

- amplification of the *Dhfr* locus to increase the copy number of the *Dhfr* gene to produce sufficient quantities of the enzyme to overcome the effects of the drug. The amplification process appears to be quite random, with large regions of flanking DNA surrounding the *Dhfr* locus also becoming amplified.

Mutations in this last class are particularly important for the high-level expression of foreign genes. The foreign DNA is cloned into a plasmid vector that also bears the Dhfr gene. This is then transfected into methotrexate-resistant cells and recombinants selected for in the presence of high levels of the drug. Cells that amplify the *Dhfr* locus should also contain large numbers of copies of the foreign DNA (Wigler *et al.*, 1980).

12.5   Expressing Genes in Animal Cells

We have previously looked at the expression of foreign gene in baculovirus infected cells (Chapter 8), but recombinant proteins can also be produced in mammalian cells. The insertion of a foreign gene into an animal cell is usually insufficient to direct its efficient expression and the production of the encoded protein. The foreign gene to be expressed must be associated with transcriptional and translational control elements appropriate for the cell type in which the protein will be produced. Most promoters used to drive the expression of foreign genes in animal cells are constitutive. We have previously discussed the Tet expression system for producing proteins in mammalian cells (Chapter 8). Many of the constitutive promoters used to drive gene expression in transfected cells are transcriptionally active in a wide range of cell types and tissues, but most exhibit some degree of tissue specificity. For example, the widely used cytomegalovirus (CMV) promoter exhibits low transcriptional activity in hepatocytes (Najjar and Lewis, 1999). Strong constitutive promoters which drive expression in many cell types include the adenovirus MLP, the human cytomegalovirus immediate early promoter, the SV40 and Rous sarcoma virus promoters, and the murine 3-phosphoglycerate kinase promoter (Makrides, 1999).

In addition to a suitable promoter, genes to be expressed in animal cells also require a polyadenylation site, a transcriptional termination signal and a variety of translational control elements. In general, it has been noted that genes containing introns are expressed at a higher level than the equivalent cDNA copy of the gene (Buchman and Berg, 1988). This may be due to the coupling of transcription, splicing and mRNA processing in higher-eukaryotic cells (Maniatis and Reed, 2002).

13 Engineering animals

Key concepts

- To create a modified animals, new or altered genes may be integrated into the genome

  - Pronuclear injection – the injection of DNA fragments into the nuclei of newly fertilized eggs

- An increased understanding of the events that take place during early embryogenesis has allowed mechanisms to be developed by which whole animals can be produced from the DNA contained within a single cell

  - Embryonic stem cells isolated from the blastocyst embryo can be maintained in culture indefinitely, extensively manipulated *in vitro* and then returned to a blastocyst, where the modified cells will form parts of the animal

- The transfer of the nucleus of an apparently fully differentiated adult cell into an enucleated egg can result in the reprogramming of the adult cell DNA to produce a cloned animal

- The correction of human genetic disorders with gene therapy has great potential and some recent successes, but still requires an enormous amount of development before it can be applied to many diseases

The engineering of specific traits in whole animals has huge potential benefits in understanding complex biological phenomenon such as development and disease progression. To understand the basis of creating whole animals that contain altered genes, we must first look at some early embryology (Burki, 1986) (Figure 13.1). Immediately after the sperm enters the egg, the fertilized

Analysis of Genes and Genomes   Richard J. Reece
© 2004 John Wiley & Sons, Ltd   ISBNs: 0-470-84379-9 (HB); 0-470-84380-2 (PB)

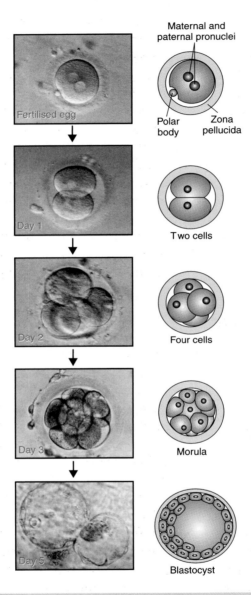

cell, now called a zygote, contains two nuclei – called **pronuclei**. The maternal and paternal pronuclei then fuse with each other to form a single fertilized nucleus. The zygote then begins to divide – first into two cells, then four, then eight and so on, forming a ball of cells called a **morula** – from the Latin for mulberry. The morula continues to divide and a cavity forms within it that fills with fluid from the uterus. At this stage, the zygote is called a **blastocyst** and the cavity is called the **blastocoele**. The cavity divides the cells of the blastocyst into an inner cell mass (which will become the embryo) and an outer **trophoblast** (which will form the placenta). Before implanting into the wall of the uterus, the blastocyst floats in the uterine cavity for 2 days and sheds the zona pellucida, allowing its adherence to the uterine wall. The implanted embryo continues to divide and specialize until birth and beyond. Not all of the newly divided cells will go on to form parts of the animal; some are programmed to die as part of the normal developmental process (Sulston and Horvitz, 1977).

Three main methods have been developed to introduce foreign DNA into animals. The mouse has long been the organism of choice for this type of manipulation as a laboratory mammal that has relatively well understood and amenable genetics. The production of altered mouse embryos for the creation of transgenic mice is certainly well advanced but other animals, particularly farm animals, have also been modified using similar techniques.

13.1 Pronuclear Injection

As with the methods we have previously discussed for the direct injection of DNA fragments into *Xenopus* oocytes (Chapter 12), DNA can be injected directly into the pronuclei of freshly fertilized mouse eggs (Palmiter and Brinster, 1986). Immediately following fertilization, the large male and small female pronuclei are visible under the microscope as discrete entities. DNA injections are usually made into the larger male pronucleus while the egg is being held in position using a suction pipette in a micromanipulation device (Figure 13.2). The injected DNA may integrate into the pronuclear DNA and, upon fusion with the female pronucleus, will be incorporated into the zygote. The injected embryos are cultured *in vitro* until the morula stage and then implanted into a pseudo-pregnant female mouse that has been previously mated with a vasectomized male. The stimulus of mating elicits the appropriate hormonal changes needed to make her uterus receptive. The implanted embryo is then allowed to develop into a mouse pup. If the foreign DNA has been successfully transferred to the mouse, then the pup will be heterozygous for the new DNA. A small piece of the newly born pup's tail is usually taken for DNA analysis (Southern blotting, PCR etc.) to check for the presence of the foreign DNA. Mating two of the heterozygotes can produce homozygous mice, with one in four of their offspring being homozygous for the transgene.

Pronuclear injection has been used to introduce a variety of foreign DNA fragments into mice. For example, a linear DNA fragment containing the promoter of the mouse metallothionein-I gene fused to the structural gene of

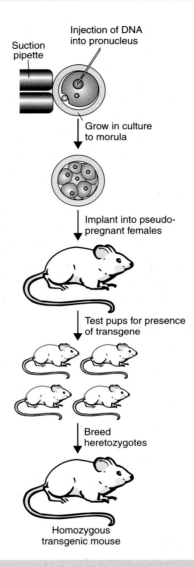

Figure 13.2. The production of transgenic mice by pronuclear injection. DNA is injected into the larger male pronucleus and grown in culture until several divisions have occurred. The embryos are then implanted into a pseudo-pregnant female. Assuming that the transgene integrated before the first cell division, the pups should be heterozygous for the transgene. Inbreeding of the heterozygotes will generate homozygous individuals

rat growth hormone was microinjected into the pronuclei of fertilized mouse eggs (Palmiter *et al.*, 1982). Of 21 mice that developed from the injected eggs, seven carried the fusion gene and six of these grew significantly larger than their littermates. Several of these transgenic mice were found to have extraordinarily high levels of growth hormone mRNA in their liver and growth hormone in their serum. At 74 days of age, the transgenic mice weighed up to 44 g, while their non-transgenic littermates weighed approximately 29 g. The technique has also been used to attempt to produce therapeutic proteins within transgenic animals. For example, human α_1-antitrypsin (AAT) has been produced in mice for the treatment of cystic fibrosis lung disease and other conditions in which connective tissue is broken down irreversibly. AAT is a plasma protein that inhibits elastase, a key player in the inflammatory response that, unchecked, will lead to excessive tissue destruction. A DNA fragment containing the genomic form of the human AAT gene, whose natural promoter had been replaced by the sheep β-lactoglobulin milk promoter, was injected into the pronucleus of mice embryos (Archibald *et al.*, 1990). Mice that expressed the transgene in the mammary gland secreted the human form of the AAT protein into their milk at high levels (up to 7 mg of protein per mL milk). Subsequently, transgenic sheep expressing AAT in their milk have been produced in the same way (Wright *et al.*, 1991). In this case, sheep expressing up to 60 mg of AAT per mL milk were reported.

One of the major advantages of pronuclear injection is that the foreign DNA to be inserted does not necessarily need to be contained within a vector. Linear DNA fragments may be injected into the pronucleus, where they often integrate as multiple (varying from a few to several hundred) head-to-tail copies at an apparently random location within the mouse genome. The potential disadvantages of pronuclear injection include the following.

- The nature of the DNA integration event means that pronuclear injection can only be used to add genes to the animal. It cannot be used to delete genes (knock-out), or to alter existing genes within the genome.

- The randomness of the insertion can have dramatic effects on the expression of the foreign gene depending on the precise site of the insertion within individual animals. Therefore, the expression of the transgene cannot readily be controlled.

- The expression of the transgene is not strictly inherited. That is, the offspring of highly expressing parent animals may show considerably different levels of expression. In some cases, this may be due to altered genomic methylation patterns at the site of the transgene (Palmiter, Chen and Brinster, 1982).

- The production of transgenic mice by pronuclear injection can occasionally result in a mosaic animal, where the transgene is only present in a limited set of tissues and organs of the animal. This happens when integration of the transgene is delayed until after the first cell division. There can also be multiple insertion events at different genomic loci and at different times. Thus, a single founder can be mosaic for one insertion site but not the other.

13.2   Embryonic Stem Cells

Embryonic stem (ES) cells are undifferentiated cells isolated from the inner cell mass of a blastocyst (Evans and Kaufman, 1981) (see Figure 13.1). They can be cultured *in vitro* by growing them in a dish coated with mouse embryonic skin cells that have been treated so they will not divide. This coating layer of cells (called a feeder layer) provides a surface to which the ES cells can attach and, in addition, releases nutrients into the culture medium. Unlike most other animal cells, they can be maintained in culture, through successive cell divisions, for long periods. ES cells in culture remain undifferentiated provided that they are grown well separated from each other. If they are allowed to clump together, they begin to differentiate spontaneously. ES cells have the potential to form all of the cell types, of the mature animal (muscle, nerve, skin etc.) including the gametes (Nagy *et al.*, 1993). In addition, systems for the specific differentiation of cultured ES cells have been developed (Keller, 1995). For example, ES cells cultured in the presence of stromal cells and various cytokines resulted in the generation of primitive erythrocytes and other haematopoietic precursor cells (Nakano, Kodama and Honjo, 1994; Kennedy *et al.*, 1997).

The ability of ES cells to be maintained in culture for extended periods, combined with their ability to differentiate into a variety of different cell types, makes them an attractive target for genetic manipulation. The basic method for ES cell based animal production is shown in Figure 13.3. Foreign DNA can be introduced into the cultured ES cells, using the methods discussed previously (Chapter 12), and transfected cells selected. The recombinant ES cells are then introduced into a fresh blastocyst, where they mix with the cells of the inner cell mass. The blastocyst is then implanted into the uterus of a pseudo-pregnant female and pups produced. Since the implanted blastocyst contains two different types of ES cell (normal and recombinant), the resulting offspring will be chimeric – some cells will contain the transgene, while other will not. The chimeric pups are then crossed with wild-type animals to generate true heterozygotes, which can then subsequently be inbred to create a homozygote. Thus ES cell animal production requires two rounds of breeding to generate a homozygote.

One of the major advantages of ES cells is that they are relatively efficient at homologous recombination in comparison to other animal cells. This means that targeted transgenes can be produced in which specific genes of the genome are either deleted or altered (Thomas and Capecchi, 1987). Recombination

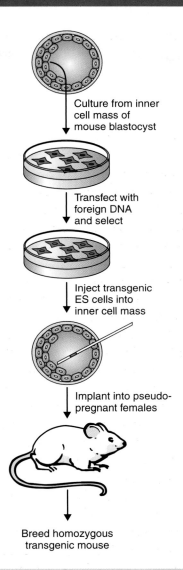

Culture from inner
cell mass of
mouse blastocyst

Transfect with
foreign DNA
and select

Inject transgenic
ES cells into
inner cell mass

Implant into pseudo-
pregnant females

Breed homozygous
transgenic mouse

Figure 13.3. Embryonic stem cells. ES cells are harvested from the inner cell mass of a blastocyst and cultured *in vitro*. Here they can be genetically modified before being returned to a fresh blastocyst

between homologous sequences in the vector DNA and the genome is used to target the insertion of the foreign DNA fragment to a specific sequence within the genome. Although ES cells are able to perform homologous recombination, a significant level of non-homologous recombination still occurs. Therefore, it is important to be able to separate the two types of event. A mechanism to

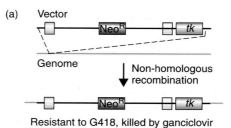

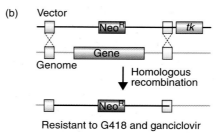

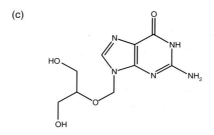

Figure 13.4. Selection of gene knockouts in ES cell cultures. (a) Non-homologous recombination results in the transfer of both the neomycin resistance and thymidine kinase (*tk*) genes to the host cell. (b) Homologous recombination results in the transfer of only the neomycin resistance gene to the host cell. (c) The structure of ganciclovir. Cells containing the *tk* gene may be killed by treatment with ganciclovir, which is phosphorylated by thymidine kinase, and then undergoes further phosphorylation by cellular kinases. In its triphosphorylated form, the drug inhibits DNA polymerase by acting as a terminator of DNA synthesis

delete a gene by homologous recombination is shown in Figure 13.4. A vector is constructed in which DNA sequences corresponding to the regions immediately flanking the 5′- and 3′-ends of the gene that is to be deleted from the genome are cloned either side of a selectable marker gene (e.g. the neomycin resistance gene, whose expression allows the cells to grow in the presence of G418). The vector also contains the HSV thymidine kinase (*tk*) gene. A linear DNA fragment bearing these sequences is transfected into cultured ES cells and selection is made in a medium containing G418. Only ES cells that have taken up the DNA fragment will be able to grow. To distinguish between cells that have that have integrated the DNA fragment in an homologous fashion and those that have done so non-homologously, selection is then made on **ganciclovir**. Ganciclovir is a synthetic analogue of 2′-deoxyguanosine (Figure 13.4(c)) that is phosphorylated by thymidine kinase to form a dGTP analogue that inhibits DNA polymerase activity. If the DNA inserted randomly, then the *tk* gene will still be associated with the transgene, and cells will die due to the drug treatment. If, however, homologous integration has occurred, then the *tk* gene will be lost and cells will survive ganciclovir treatment (Mansour, Thomas and Capecchi, 1988). In addition to supplying a mechanism to delete genes (knock-out), specific genes may also be replaced with mutated versions of themselves. The mutant version of the gene is simply cloned into the vector next to the neomycin resistance gene and then transfected into ES cells. The regions of homology at the ends of the linear DNA fragment determine the genomic location (or individual gene) into which the transgene is inserted.

The ability to specifically knock out genes can provide an immensely powerful approach to assigning gene function in whole animals, especially the mouse (Osada and Maeda, 1998). Perhaps more importantly, knockouts can provide excellent model systems for the analysis of human disease. We have previously discussed the potential difficulties with this type of analysis in other organisms (Chapter 10), and many of the same problems can also be encountered with animal knock-outs. Three main classes of knock-out may be generated.

- *Lethal.* The deletion of the molecular chaperone hsp47 is lethal to mouse embryos, predominately as a function of defective collagen biosynthesis (Nagai *et al.*, 2000).

- *Observable phenotype.* The deletion of the tumour suppressor gene p53 results in the formation of mice that develop normally, but are exquisitely sensitive to spontaneous tumours early in their lives (Donehower *et al.*, 1992).

- *No observable phenotype.* The deletion of Matrilin 1, an extracellular matrix protein that is expressed in cartilage, yields transgenic mice with no apparent phenotype in comparison to their wild-type counterparts (Aszodi *et al.*, 1999).

A lethal phenotype generally reflects the earliest non-redundant role of the gene, and precludes an analysis of an analysis of gene function later in development. The diploid nature of higher organisms means that mutants that fall into this class may be analysed in their heterozygous (+/−) state. Additionally,

conditional knock-outs may be produced (see below). Knock-outs that fall into the last category (no observable phenotype) may arise as a result of genes acting in parallel pathways compensating for each others' functions. It is also possible that the techniques are simply too crude to detect any subtle differences between the wild-type and the knock-out animals. The complexity of animal genomes also means that a knock-out may have a profound effect in one strain of mouse, but quite a different effect in another. For example, the deletion of the gene encoding epidermal growth factor in one mouse strain (CF-1) results in embryos that die around the time of implantation into the uterus. If, however, the same knockout is introduced into a different mouse strain (CD-1), then the animals can survive for up to three weeks after birth (Threadgill *et al.*, 1995). Ideally, knockout experiments should be performed in a variety of strain backgrounds, but the length of time required to do that, and the costs involved, often preclude this analysis.

One problem with this type of approach for producing transgenic animals, which we have seem previously when looking at engineering in plants (Chapter 11), is that the selectable maker gene is transferred to the transgenic animal. The high-level expression of an antibiotic-resistance gene within a transgenic animal is generally undesirable. The expression of the marker may induce the abnormal expression of other neighbouring genes, and the potential for transfer of the marker gene to non-transgenic animals should be avoided. The marker gene can effectively be removed after the transgene has been established within the ES cell if its sequences are flanked by *loxP* sites – the recognition sequences for the Cre recombinase (Kilby, Snaith and Murray, 1993). Transfection of the transgenic cell line with a plasmid expressing Cre recombinase catalyses the excision of the DNA between the two *loxP* sites to remove the marker gene and leave a single *loxP* site in its place.

There are many instances where the expression of an inserted transgene is required only in a specific tissue or set of cells. This can readily be achieved by constructing the foreign gene such that it is under the control of a tissue-specific promoter. For example, the promoter of the calcium–calmodulin dependent kinase II (CaMKIIα) gene drives expression only in the neurons of the hippocampus (Mayford *et al.*, 1996). Such an approach works well, provided that a suitable tissue-specific promoter is available (Table 13.1).

Conditional knock-outs can also be produced, again using the *loxP*-Cre site-specific recombination system (Gossen and Bujard, 2002). If, for example, the knock-out of a gene results in an embryonic-lethal phenotype, then it may be necessary to delete the gene from the genome after the animal has been born. A method by which this can be achieved is shown in Figure 13.5 (Kühn *et al.*, 1995). The normal copy of the gene to be deleted is replaced in the

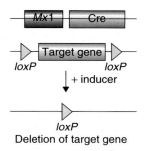

Figure 13.5. Tissue-specific gene knock-outs. See the text for details

Table 13.1. Some tissue-specific promoters in mice. Adapted from Lewandoski (2001)

Promoter	Gene normally controlled	Tissue or cells of expression	Reference
Alb	Albumin	Liver	(Postic *et al.*, 1999)
Camk2α	Ca²⁺/calmodulin-dependent protein kinase II, α	Forebrain	(Mayford *et al.*, 1995)
Cryαa	Crystallin αA	Eye lens	(Lakso *et al.*, 1992)
En2	Engrailed	Mid/hindbrain	(Logan *et al.*, 1993)
Gcg	Glucagon	Pancreatic α-cells	(Herrera, 2000)
Ins2	Insulin II	Pancreatic β-cells	(Rommel *et al.*, 1994)
KRT5	Keratin 5	Epidermis	(Ramirez *et al.*, 1994)
Lck	Lymphocyte-specific tyrosine kinase	T cells	(Chaffin *et al.*, 1990)
Msx2	Msh-like homeobox gene 2	Apical ectodermal ridge of limb bud	(Liu *et al.*, 1994)
Myog	Myogenin	Skeletal muscle	(Yee and Rigby, 1993)
Nes	Nestin	Neuronal cells	(Zimmerman *et al.*, 1994)
Pax6	Paired-box gene 6	Retina	(Gruss and Walther, 1992)
Wnt1	Wingless related MMTV integration site 1	Neural crest	(Echelard, Vassileva and McMahon, 1994)

genome by a version that is flanked by *loxP* sites (often referred to as a **floxed** gene – flanked by *loxP*). In addition, the transgenic animal is also modified to carry a copy of the gene encoding the Cre recombinase under the control of an inducible promoter, e.g. *Mx1*. *Mx1* is part of the mouse viral defence system and is transcriptionally inert in healthy mice (Hug *et al.*, 1988). The promoter can, however, be activated by high levels of interferon or by adding synthetic double-stranded RNA to cells (which induces interferon expression). Transgenic animals produced in this way retain a functional copy of the gene to be deleted until they are injected with double-stranded RNA. The effect of the lost gene may then be investigated.

Rather than constructing a transgenic mouse containing both the tissue-specific promoter expressing the Cre recombinase and the target gene surrounded by *loxP* sites, a series of transgenic mice have been constructed that each contain a different tissue-specific promoter controlling the expressing of Cre. These can then be used as a 'bank' of mice strains to which transgenic mice containing a particular floxed gene can be crossed. Mating these strains will result in the formation of progeny in which the gene in inactivated only in those tissues that express Cre (Gu *et al.*, 1994). This means that a single transgenic floxed gene can be deleted in a variety of tissues without having to resort to further *in vitro* manipulation.

The tetracycline-inducible expression system (see Chapter 8) may be used to drive *Cre* expression to regulate knock-out function. In this system, a transactivator fusion protein composed of the tetracycline repressor (tetR) and the acidic activation domain of the herpes simplex virus 16 (VP16) protein regulate the expression of the *Cre* gene from a promoter containing *tet*-operator (*tetO*) sequences. In the absence of tetracycline, the Cre gene is expressed and will induce site-specific recombination between two *loxP* sites. In the presence of tetracycline, the Cre gene will not be expressed and recombination will not occur (St-Onge, Furth and Gruss, 1996).

13.3 Nuclear Transfer

Although animal cells become increasingly committed as differentiation and development proceeds, the DNA contained within each differentiated cell still retains all the information necessary to form the whole animal. If the nucleus of a differentiated cell is introduced into an enucleated egg then, under appropriate conditions, the nucleus can become 'reprogrammed' such that development of the animal reoccurs. The production of cloned animals – all of which have originated from a single, possibly recombinant, cell line – has several potential uses.

- *Recombinant protein production.* We have discussed previously that the expression level of recombinant protein production is not strictly inherited (Chapter 12). Therefore, the ability to create large number of animals each expressing identical levels of, say, a therapeutic protein can only be achieved using cloned animals.

- *The conservation of endangered species.* Rare animals could be cloned to repopulate dwindling natural levels.

The idea of transferring a nucleus from one cell to another is not new. Over 50 years ago it was discovered that the nuclei of blastocyst frog cells could be implanted into eggs that lacked a nucleus to created a series of cloned frogs that were identical to the donor cells (Briggs and King, 1952). It was found, however, that as the donor cells became more differentiated, it became increasing difficult to reprogramme them to produce new animals. The few embryos cloned from differentiated cells that survived to become tadpoles grew abnormally. This led to the speculation that genetic potential diminished as a cell differentiated and that it was impossible to clone an organism from adult differentiated cells. In 1975, however, John Gurdon developed a method of nuclear transfer using fully differentiated cells and *Xenopus* eggs (Gurdon, Laskey and Reeves, 1975). This is a two-step process.

- *Production of enucleated eggs.* Delicate needles and a powerful microscope were used to suck the nucleus from a frog oocyte to produce an enucleated oocyte. With the genetic material removed the enucleated oocyte would not divide or differentiate even when fertilized.

- *Introduction of a new nucleus.* Using the same equipment, the nuclei of keratinized skin cells of adult *Xenopus* foot-webs were transferred into the enucleated oocytes. Many of these new cells behaved like normal fertilized eggs and were capable of producing tadpoles. Since the tadpoles arose from the cells of the same adult, they all contained the same genetic material and were clones of each other produced from apparently fully differentiated cells. This indicates that DNA is not discarded or permanently inactivated even in highly specialized cells.

A somewhat modified procedure has been used recently to produce cloned mammals (Figure 13.6). This was first achieved by taking cells from the blastocyst stage of a sheep embryo and fusing them with enucleated eggs (Smith and Wilmut, 1989). The reconstituted cells were subjected to a brief electrical pulse to stimulate embryonic development prior to implantation into a surrogate ewe. Live sheep have subsequently been produced from the

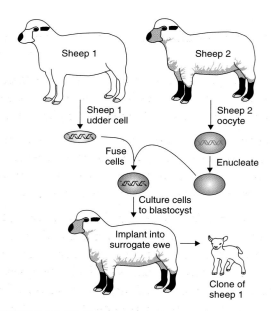

Figure 13.6. Nuclear transfer. The cells of an adult sheep (sheep 1) are fused with the enucleated eggs of a sheep of a different breed (sheep 2). The fusion between the two is grown in culture to the blastocyst stage prior to implantation into a surrogate ewe. The resulting lamb contains the nuclear genome of sheep 1

nuclei of cultured embryonic cells (Campbell *et al.*, 1996), and from cultured adult breast epithelial cells (Wilmut *et al.*, 1997). This last example produced probably the most famous sheep in the world – Dolly (Box 13.1). The success of these experiments appears to be dependent on the synchronization of the cell cycles of the donor and recipient cells that are to be fused. In the case of Dolly, **quiescence** of the donor cell was induced prior to the cell fusion process. Unsynchronized cells appear to be less successful in forming fruitful fusions.

Box 13.1. The life and death of Dolly.

Dolly was the first mammal clone to be produced from an adult cell. She was produced following the procedures described below (Wilmut *et al.*, 1997).

- *Donor cells*. Mammary gland tissue of a 6-year-old Finn Dorset ewe was used to prepare a primary cell culture. This culture contained a mixture of mammary epithelial cells (>90 per cent), myoepithelial cells and fibroblasts. An important step in the success of the cloning process was to induce these donor cells to exit their growth cycle and enter the G_0 phase of the cell cycle before nuclear transfer. This was accomplished by reducing the concentration of serum in which they were grown to starve the cells.

- *Recipient cells.* Oocytes were obtained from Scottish Blackface ewes between 28 and 33 hours after injection of gonadotropin releasing hormone (GnRH) and enucleated using a fine glass pipette.

- *Cell fusion.* Fusion of the donor cell to the enucleated oocyte and activation of the oocyte were induced by the electrical pulses – a single DC pulse to activate the cells and a single AC pulse followed by three DC pulses to promote cell fusion. 277 individual fused cells were produced.

- *Growth and implantation.* The fused cells were cultured in ligated oviducts of sheep. After 6 days of culture, 29 of the 277 reconstructed embryos had developed into a morula or blastocyst. One, two or three embryos were transferred to Scottish Blackface ewes and allowed to develop to term. The 29 morula/blastocysts were transferred to 13 different ewes, and from these only one became pregnant. On July 5 1996 after 148 days pregnancy, the normal duration for her breed, Dolly – a Finn Dorset sheep – was born with a healthy birth weight of 6.6 kg. Dolly, a sheep derived from a mammary gland cell, was named after the singer Dolly Parton.

Box Figure 13.1. Dolly, and her lamb Bonnie. Image courtesy of The Roslin Institute

The precise cell type from which Dolly was derived remains unclear. Further analysis indicated that she was indeed derived from the cells of the mammary gland of the donor sheep, rather than from a contaminating cell (Ashworth *et al.*, 1998). She is not, however, an exact clone of the sheep whose cells were used to create her. The DNA of her mitochondria are derived exclusively from recipient enucleated oocytes (Evans *et al.*, 1999). Therefore she is a chimera, containing somatic cell derived nuclear DNA but oocyte derived mitochondrial DNA. It also is interesting to note that the scientific paper in which Dolly was introduced to the world (Wilmut *et al.*, 1997) does not include the words 'clone' or 'cloning' anywhere within its text. Perhaps the authors realized the potential impact of their findings and chose less inflammatory language to describe their results. Dolly subsequently grew into an adult sheep have bore her own offspring (Box Figure 13.1). Finn Dorset sheep have an average life expectancy of about 12 years, but in January 2002 Dolly was reported to be suffering from arthritis, which is highly unusual for a sheep of her age. On 14 February 2003, aged only six, Dolly was put to sleep following a diagnosis that she was suffering from a progressive lung disease.

The method of nuclear transfer to produce viable offspring from differentiated adult cells is not without its problems (Wilmut *et al.*, 2002). It is likely that not all of the difficulties described below are due to the nuclear transfer process itself, as some similar abnormalities have been reported after embryo culture.

- The process is extremely inefficient. In the case of Dolly, only one of the 277 cell fusions produced was capable of developing into a lamb. Similar efficiency levels have also been reported for other whole animal cloning experiments.

- Many of the embryos produced by nuclear transfer suffer gross abnormalities. In addition to embryonic loss, nuclear transfer is also associated with very high rates of foetal, perinatal and neonatal loss, and production of abnormal offspring.

- Although Dolly was born following a normal gestation period and was a normal weight, many offspring produced by nuclear transfer suffer from large offspring syndrome (LOS) in which gestation period and birth-weight are greatly increased (Lazzari *et al.*, 2002). The frequency and severity of the symptoms of LOS appear to vary widely even under similar experimental conditions. Early deviations from the normal developmental pattern, particularly with regard to embryonic gene expression, may be involved in this phenomenon.

- Since Dolly was created from a cell that was potentially 6 years old, what genetic age was she when she was born? This has been addressed by looking at the length of the telomeres at the ends of chromosomes. Telomeres generally shorten as aging progresses, although the precise effects of this phenomenon are not well understood. Dolly has been found to have short telomeres when compared with other sheep of the same age (Shiels *et al.*, 1999). It was recently reported that Dolly developed arthritis, which is highly unusual in a sheep of her age (Williams, 2002). It remains to be seen whether this and other potential age-related effects, including Dolly's death, are a result of the nuclear transfer process.

- Widespread disruptions in the DNA methylation patterns have been described in cloned embryos of a number of cloned animals (Fairburn, Young and Hendrich, 2002). The effects of these changes remain unclear.

- The technique of nuclear transfer is still in its infancy. This means that the effects of aging and genetic inheritance have not been fully assessed. In two independent studies, animals cloned from one cell type became obese in adult life (Tamashiro *et al.*, 2002) whereas those from another cell type died at an unusually early age (Ogonuki *et al.*, 2002). Further work in this area is required.

The technique of nuclear transfer by which Dolly was produced has been replicated or modified to produced clones from adult cells using a variety of other farm animals, e.g. cows, goats and pigs (Cibelli *et al.*, 1998; Baguisi *et al.*, 1999; Polejaeva *et al.*, 2000), and in more experimentally amenable laboratory animals such as mice (Wakayama *et al.*, 1998). In addition, cloned domestic pets such as cats (Shin *et al.*, 2002) and rabbits (Chesné *et al.*, 2002) have also been reported. In early 2003, news reports suggested that the first cloned human child had been born. Although such claims have not been scrutinized scientifically, it seems inevitable that a cloned human will be produced at some stage. The difficulties encountered with cloned animals described above should serve as a warning to anyone considering the procedure. The temptation to replace a dead or dying child with an 'exact copy' may be more than some parents can bear, but the potentially disastrous consequences should not be underestimated.

Aside from the very negative impact of nuclear transfer technology described above, the process has proved useful for the creation of animals with specific traits. The ability to recreate a whole animal from cells that have been extensively manipulated *in vitro* could have a profound positive impact on medicine. For example, there is great potential for the replacement of damaged human organs (e.g. liver, heart) with their equivalents from animals. This process, termed **xenotransplantation,** is often unsuccessful because some of the cell surface carbohydrates are different between humans and animals. With the exception of catarrhines (Old World monkeys, apes and humans), all animals possess the enzyme $\alpha(1,3)$-galactosyl transferase, which catalyses the formation of the disaccharide galactose-$\alpha(1,3)$-galactose that is found on the cell surface. The presence of the disaccharide causes hyperactue rejection of the organ in humans. This problem can only be partially overcome by temporarily removing antibodies to galactose-$\alpha(1,3)$-galactose from the recipient through affinity adsorption. However, returning antibodies can damage the transplanted organ and severely limit its survival even in the presence of high levels of immunosuppressive drugs. Sheep have been produced that lack the *GGTA1* gene encoding the $\alpha(1,3)$-galactosyl transferase enzyme (Denning *et al.*, 2001). *GGTA1* was replaced in tissue culture cells by a copy of the neomycin-resistance gene, and nuclear transfer was used to generate sheep embryos. Unfortunately, the foetuses died before birth, so it remains to be seen whether organs from animals produced in this way may be suitable for human transplantation. More recently, pigs knocked out for either one (Lai *et al.*, 2002; Dai *et al.*, 2002) or both (Phelps *et al.*, 2002) alleles of *GGTA1* have been produced. Some of the knock-out pigs are apparently healthy and further work will assess the suitability of their organs for human transplantation.

13.4 Gene Therapy

Gene therapy is an approach to treat, cure or ultimately prevent disease by changing the expression of genes within an individual. The idea seems simple – a healthy copy of a mutated gene is introduced into an affected individual such that the normal protein can be made, and the disease symptoms thereby alleviated (Morgan and Anderson, 1993). Although the idea of gene therapy has been around for some time, actual treatments are still in their infancy. Most human clinical trials are only in the research stages. Gene therapy is most applicable to the correction of single gene disorders, especially recessive diseases where a functional copy of the defective gene will restore the activity of the mis-functional protein (Table 13.2). The insertion of the transgene to bring about the desired change can be targeted to either germ (egg and sperm) or somatic (body) cells.

- *Germ-line gene therapy.* The egg or sperm cells are changed with the goal of passing on the changes to their offspring. Human germ-line gene therapy is

Table 13.2. Examples of some single-gene human genetic disorders

Disorder	Symptoms
Autosomal recessive:	
Cystic fibrosis	Recurrent lung infection, increased mucus production
α_1-antitrypsin deficiency	Liver failure, emphysema
Phenylketonuria	Mental retardation
Tay-Sachs disease	Neurological degeneration, blindness, paralysis
Sickle cell anaemia	Anaemia
Thalassemia	Anaemia
Autosomal dominant:	
Neurofibromatosis type 1	Tumours of peripheral nerves
Huntington's disease	Involuntary dance-like movements, dementia
Mytonic dystrophy	Heart defects and cataracts
Familial retinoblastoma	Tumours of the eye
X-linked:	
Haemophilia	Deficient blood clotting
Duschenne muscular dystrophy	Progressive muscle wasting
Fragile-X syndrome	Mental retardation

prohibited in most countries since the consequences of producing a human with artificially altered genetic traits are far from clear.

- *Somatic gene therapy*. The genome of the recipient is altered, but this change is not passed to the next generation. Somatic gene therapy can be classed as being performed either *in vivo* or *ex vivo* (Figure 13.7). *In vivo* therapy involves the addition of a gene directly to a patient. *Ex vivo* therapy involves the removal of cells from the patient and their culturing and genetic manipulation *in vitro* before the return of the modified cells to the patient.

The type of therapy used depends on the sorts of cell that need to be modified. If the cells in which the gene defect is apparent can be easily cultured, then the *ex vivo* route offers tremendous advantages. For example, all blood cells are derived from **multipotent stem cells** in the bone marrow. These differ from ES cells that we have previously discussed in that they can only differentiate into a limited number of different cell types. Multipotent stem cells can, however, be

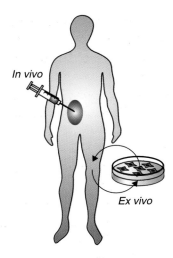

Figure 13.7. *In vivo* and *ex vivo* human gene therapy. See the text for details

cultured *in vitro* for extended periods. Therefore, disorders of the blood system may be treated *ex vivo* through the isolation and culturing of bone marrow stem cells. These cells can be modified *in vitro* and the resulting transgenic cells can then be returned to the patient. The modified stem cells will then produce the various modified differentiated cells that may cure the disease. *In vitro* manipulation of the cells permits the use of a wide variety of methods to insert the transgene – e.g. viral infection, injection and other methods (see Chapter 12). Other cells and tissues are less amenable to *ex vivo* treatment. For example, lung epithelial cells – whose function is severely impaired in cystic fibrosis patients – grow very poorly in culture. Even if they could be cultured, it would not be possible to repopulate an affected lung with transgenic lung epithelial cells. Therefore, diseases such as cystic fibrosis must be treated *in vivo* so that the cells of the defective lung can be modified. This limits the type of transgene insertion that can take place. Viruses, e.g. adenovirus that naturally infect epithelial cells, are usually used to transport the transgene into the affected cells.

13.5   Examples and Potential of Gene Therapy

The history of human gene therapy trails is not a particularly happy one. With one notable exception (see below), the effect of introducing a gene into cells rarely promotes more than a transient relief from the symptoms of the disease being treated. Worse still, there have been highly publicized cases where gene therapy trial patients have suffered as a consequence of the treatment itself. For example, in 1999 an 18-year-old gene therapy trial volunteer from Philadelphia, Jesse Gelsinger, died following a gene therapy trial (Teichler Zallen, 2000). Gelsinger had an ornithine transcarbamylase (OTC) disorder, a rare genetic defect of the liver that renders the body unable to clear ammonia from the bloodstream. He was treated with an adenoviral vector as a mechanism to insert a healthy copy of the gene into his liver, but the virus itself elicited a massive immune reaction that resulted in his death. Cases such as this graphically illustrate the need for the development of vectors characterized by maximum transfection efficiency and minimal toxicity.

Some gene therapy successes have, however, been noted. Children born with severe combined immune deficiency, X-SCID, have a poor prognosis using traditional medicines. The disease is caused by a mutation on the X chromosome in the gene encoding the gamma chain (γc) of the interleukin-2 receptor. Mutations in this gene prevent two types of white blood cell, the T-cells and natural killer cells, from developing normally (Sugamura *et al.*, 1996). With little or no defence against infection, sufferers usually die within the first year of life unless a bone marrow donor can be found. Stem cells were collected from the bone marrow of an affected infant and treated with a retrovirus carrying a wild-type copy of the γc gene (Cavazzana-Calvo *et al.*, 2000). When the transgenic stem cells were returned to the infant they were capable of generating all of the cells required for a fully functional immune system for at least 10 months (Fischer, Hacein-Bey and Cavazzana-Calvo, 2002). Removing the bone marrow cells from the body prior to infection with the retrovirus eliminates the danger of acute reaction to the virus itself, and also ensures that the virus only infects the correct cells. Repopulating the immune system with a relatively small number of transgenic bone marrow cells may also cause problems. The treatment specifically selects for proliferating cells and may therefore increase the risk of bone marrow related cancers. It has been noted that some of patients treated in this way develop leukaemia (one out of 10 patients successfully treated), attributed to a result of the integration of the foreign DNA fragments into the genome at random locations. In this case, the retrovirus inserted the therapeutic gene into the regulatory region of a gene called *Lmo2* on chromosome 11 (Gänsbacher *et al.*, 2003). The activation of the therapeutic gene appeared to cause the expression of *Lmo2* which is an oncogene (Davenport, Neale and Goorha, 2000). Even with these problems, these experiments represent the only example to date where a patient is apparently completely cured using gene therapy.

Some of the problems associated with random integration of the transgene during gene therapy may be addressed by utilizing site-specific recombination systems. For example, DNA fragments have been constructed such that they contain a therapeutic gene adjacent to the recognition sequence of a site-specific recombinase enzyme. If these DNA fragments are injected into the tail veins of mice together with a DNA fragment encoding the integrase, then site-specific genomic integration of the transgene occurs (Olivares *et al.*, 2002). This could be developed from the mouse model into a human therapy.

Other gene therapy trails are currently ongoing for both genetic and non-heritable diseases.

- *Haemophilia B.* Sufferers lack the gene for factor IX, a critical agent in the blood clotting process. Parvoviruses have been used to insert the missing gene into skeletal muscle cells (High, 2001). The cells then generate the missing factor, thereby removing the need for daily injections of the protein itself.

- *Cancer.* Some cancer treatments may be amenable to gene therapy (Wadhwa *et al.*, 2002). Modified viral vectors can be used to prime the immune system to attack cancer cells, while other approaches employ viruses to carry suicide genes into the cancer cells.

- *HIV.* Specifically engineered HIV may eventually be recruited to help control HIV-1 infection (Statham and Morgan, 1999).

Currently, the promise of gene therapy remains just that. Even single gene defect diseases can manifest themselves as deficiencies in a wide variety of different cell types. Being able to correct the defect in one cell type may not be sufficient to cure the disease fully. However, the development and refinement of transgene delivery systems, combined with advances in our understanding of stem cells may generate many more opportunities in the future where gene therapy may be clinically important.

Glossary

Adenine – a purine base found in DNA and RNA. Adenine base pairs with thymine in DNA and uracil in RNA

Alanine scanning mutagenesis – the conversion of amino acids within a protein to alanine to determine the role of specific amino acid side chains

Alkaline lysis – a method for breaking open bacterial cells for the isolation of extra-chromosomal DNA

Allele – one of several alternative versions of a gene located at the same locus of a chromosome

α-complementation – in mutants of *E. coli* which express an inactive version of *β*-galactosidase, subunit assembly (and enzyme activity) may be restored by the presence of a small amino-terminal fragment of the *lacZ* product (the a-polypeptide) usually produced from a cloning vector

Antibiotic – a substance able to inhibit or kill microorganisms

Antibody – a protein produced by B lymphocytes that recognizes an antigen and triggers an immune response

Anticodon – a triplet of nucleotide bases in tRNA that identifies the amino acid carried and binds to a complementary codon in mRNA during protein synthesis at a ribosome

Antigen – a protein or substance capable of stimulating an immune response

BAC – bacterial artificial chromosomes

Bacteriophage – a bacterial virus

Base pair – bp – the pairing of A with T and G with C in duplex DNA

Bermuda principle – the rapid, public release of genome DNA sequence data, without restrictions on use

Blastocyst – an early embryo typically having the form of a hollow fluid-filled cavity bounded by a single layer of cells

Catabolite repression – the decreased expression of genes when organisms are grown in glucose

cDNA – a single strand of DNA that is synthesized from, and is therefore complementary to, an RNA molecule

cDNA library – a collection of double-stranded cDNA molecules contained within a vector

Cell cycle – the period from one cell division to the next

Cell-cycle checkpoints – systems for interrupting the cell cycle if something has gone wrong

Analysis of Genes and Genomes   Richard J. Reece
© 2004 John Wiley & Sons, Ltd   ISBNs: 0-470-84379-9 (HB); 0-470-84380-2 (PB)

Centromere – the point or region on a chromosome to which the spindle attaches during mitosis and meiosis

Chromatid – one of the usually paired and parallel strands of a duplicated chromosome joined by a single centromere

Chromatin – a complex of DNA and proteins in the nucleus of a cell

Chromatin immunoprecipitation (ChIP) – a method for identifying proteins bound to particular sequences of DNA

Chromosome – a discrete unit of the genome that is visible as a morphological entity during cell division. Each chromosome is a single DNA molecule

Chromosome walking – the sequential isolation of clones carrying overlapping DNA sequences that allows the sequencing of large regions of the chromosome from a single starting point

Clone – an organism, cell or molecule produced from a single ancestor

Cloning vector – a plasmid or phage that is used to carry inserted foreign DNA

Codon – the triplet of nucleotides that result in the insertion of an amino acid or a termination signal into a polypeptide

Codon usage – the frequency at which amino acid codons are used for the production of proteins

Complementary – the sequences on one strand of a nucleic acid molecule can bind to their complementary partners on another strand. A = T, G = C

Conjugation – the transfer of all or part of a chromosome that occurs during bacterial mating

Conservative replication – a disproved model for DNA synthesis in which the newly synthesized DNA strands bind to each other

Contig – a continuous sequence of DNA produced from a number of smaller, overlapping fragments

Cosmid – a plasmid onto which phage lambda *cos* sites have been inserted. Consequently, the plasmid DNA can be packaged in vitro into the lambda phage coat

Cytological map – a type of chromosome map where genes are located on the basis of the effect that chromosome mutations have on staining patterns

Cytosine – a pyrimidine base found in DNA and RNA. Cytosine bases pairs with guanine

Denatured – in DNA, the conversion of the double-stranded form to a single-stranded form. In proteins, the conversion from an active to an inactive form

Differential display – a technique to visualize difference in the expression of genes from different sources

Dinucleotide – the joining of two nucleotides through the formation of a phosphodiester linkage

Dispersive replication – a disproved model of DNA synthesis in which a random interspersion of parental and new segments are found in daughter DNA molecules

DNA – deoxyribonucleic acid

DNA ligase – the enzyme that catalyses the formation of a phosphodiester bond between two DNA chains

DNA polymerase – the enzyme that synthesizes new DNA strands from a DNA template

DNA topoisomerase – an enzyme that changes the linking number of DNA molecules

Electrophoresis – the application of an electric current to separate molecules (as proteins and nucleic acids) through a gel

Electroporation – a physical way to introducing DNA into cells using an electric current

Epitope – the molecular region on the surface of an antigen capable of eliciting an immune response

Euchromatic DNA – the gene-rich areas (including both exons and introns) of a genome

Exon – a segment of a gene that is represented in the mature mRNA

Expressed sequence tag (EST) – small pieces of cDNA sequence generated by sequencing either one or both ends of an expressed gene

Expression vector – a cloning vector designed so that the genes cloned into it may be transcribed and translated

F episome – a large extrachromosomal circular double-stranded DNA molecule that carries bacterial fertility genes

FISH – fluorescent *in situ* hybridization

Frame-shift mutation – occurs when the coding sequence of a gene contains a deletion or insertion of bases that are not in multiples of three. This changes the reading frame in which translation occurs

Functional complementation – the identification of genes from one organism by their ability to counteract the defect caused by the lack of a gene in another organism

Gene – a discrete unit of genetic information that is required for the production of a polypeptide. It includes the coding sequence, the promoter and terminator, and introns

Gene knockout – the removal of a gene from the genome

Gene knock-down – the use of silencing techniques to reduce or eliminate the expression of a particular gene

Genetic engineering – the deliberate modification of the characters of an organism by the manipulation of DNA and the transformation of certain genes

Genetic marker – any DNA sequence that can be used to identify a gene or phenotypic trait associated with it

Genetic switch – the control of transcription in response to particular signals

Genome – the genetic make-up of an organism

Genomic library – a collection of DNA fragments, derived from the genome of an organism, cloned into a vector

Guanine – a purine base found in DNA and RNA. Guanine base pairs with cytosine

HAC – human artificial chromosome

Helper phage – provides certain functions, that are absent from a defective phage, to allow complete phage replication during a mixed infection

Hemimethylated DNA – DNA that contains methylated bases on one strand only

Heterochromatic DNA – regions of the chromosome that are highly condensed and not transcribed

Heterozygous – an individual with different alleles at a particular genomic locus

Histone – a DNA binding protein that forms part of the nucleosome

Holliday junction – a central intermediate formed during recombination

Homologous recombination – DNA crossovers that occur between two homologous DNA molecules

Homology – a similarity often attributable to common origin

Homozygous – an individual with the same allele at a particular genomic locus

Hybridization – the pairing of complementary nucleic acids (DNA–DNA or DNA–RNA)

Hyperchromic effect – the increase in optical density that occurs when DNA become single stranded

Insertional inactivation – the destruction of the function of a gene by cloning a DNA sequence into it

Intron – a segment of DNA that is transcribed, but is removed from the transcript during splicing

Isopycnic centrifugation – the separation of the components of a mixture on the basis of differences in density

Isoschizomer – different restriction enzymes that recognize the same target DNA sequence

Karyotype – the chromosome content of an organism

Kozak sequence – the consensus sequence for the initiation of translation. A binding site for the small subunit of the ribosome

lac **operon** – the genes required in bacteria for the metabolism of lactose

Lagging strand – short DNA fragments are replicated discontinuously in a $5'-3'$ direction and later covalently joined

Lariat – an intermediate of RNA splicing

Leading strand – the strand of DNA that is synthesized continuously during replication in a $5'-3'$ direction

Melting temperature $(T_{\mathbf{m}})$ – the temperature at the mid-point of DNA denaturation

Methylated DNA – the modification of DNA bases by the addition of methyl (CH_3) groups

Microarray – sets of miniaturized reaction areas that may also be used to test the binding of DNA fragments

Microsatellites – highly polymorphic DNA markers comprised of a variable number of tandem repeats

Mini-prep – the purification of extra-chromosomal DNA (usually a plasmid) from a small culture (e.g. 1.5 mL) of bacterial cells

Monocistronic – an RNA molecule that codes for one protein

Monoclonal antibody – a highly specific protein that can bind to a single epitope within an antigen

Morula – a globular solid mass of cells formed by the cleavage of a zygote that precedes the blastula

mRNA – messenger RNA

Northern blot – a technique for the separation of RNA molecules through agarose gels followed by detection of specific RNAs through hybridization with single-stranded DNA

Nucleoside – a purine or pyrimidine base combined with either deoxyribose or ribose found in DNA or RNA

Nucleosome – the basic structural unit of chromatin

Nucleotide – a purine or pyrimidine base combined with either deoxyribose or ribose and a phosphate group found in DNA or RNA

Oligonucleotide – a short molecule of single-stranded DNA, usually synthesized chemically

Oocyte – an egg before maturation

Opine – a product produced by the condensation product of an amino acid with either a keto-acid or a sugar

Origin of replication – the nucleotide sequence at which DNA synthesis (replication) is initiated

PAC – P1 derived artificial chromosome. A vector, based on the bacteriophage P1 genome, used to clone large DNA fragments in *E. coli*

Palindromic – a sequence of DNA that read on one strand in a 5′–3′ direction is the same as than on the other strand in a 5′–3′ direction

PCR – polymerase chain reaction – cycles of DNA denaturation, primer annealing and extension with DNA polymerase lead to a amplification of the target DNA sequence

PFGE – Pulsed field gel electrophoresis

Phage – bacteriophage – a bacterial virus

Phage display – a technique that fuses peptides to capsid proteins on the surface of phages. Libraries of phage displayed peptides may be screened for binding to specific ligands

Phenotype – the functional and structural characteristics of an organism as determined by the interaction of the genotype with the environment

Pilus – a filament-like projection from the surface of a bacterial cell

Phosphodiester bond – a covalent bond between the 5′ phosphate group on one nucleotide and the 3′ hydroxyl group on an adjacent nucleotide

Plasmid – an autonomous self-replicating extra-chromosomal closed-circular DNA

Polycistronic – an mRNA molecule that contains more than one coding region

Polyclonal antibody – a mixture of many antibodies, each raised against the same antigen

Polynucleotide – long chains of nucleotides linked together by phosphodiester bonds

Polypeptide – a chain of amino acids connected by peptide linkages

Polysome – multiple ribosomes actively translating a single mRNA molecule into polypeptides

Primer – a short DNA or RNA sequence that is paired with one strand of DNA and provides a free 3′ hydroxyl group at which DNA replication can initiate

Promoter – a DNA sequence which serves as the binding site for transcription factors and RNA polymerase during the initiation of transcription

Pronucleus – the haploid nucleus of an egg or sperm cell prior to fertilization, and immediately after fertilization before the sperm and egg nuclei have fused into a single diploid nucleus

Propeller twist – the rotation of individual DNA base pairs within the double helix

Purine – a nitrogen containing, double-ring basic compound that occurs in nucleic acids. The purines in DNA and RNA are adenine and guanine

Pyrimidine – a nitrogen containing, single-ring basic compound that occurs in nucleic acids. The pyrimidines in DNA are cytosine and thymine, and those in RNA are cytosine and uracil

Quiescence – a dormant phase of cell growth. All but the most basic functions of a cell or group of cells have stopped, usually in response to an unfavourable environment. The cell remains dormant until its surroundings are more favourable

Replicon – part of the genome in which replication is initiated. Contains origins of replication

Restriction enzyme – (or restriction endonuclease) recognizes short, often palindromic, DNA sequences (recognition sites) and cleaves the DNA

Restriction map – a linear map of the restriction enzyme recognition sites with a DNA molecule

Reverse transcriptase – the enzyme that synthesizes DNA from RNA templates

Reverse transcription – the synthesis of DNA from RNA

RFLP – restriction fragment length polymorphism – changes in DNA sequence that result in altered lengths of DNA when cleaved with a restriction enzyme

Ribosome – RNA-rich cytoplasmic granules that are sites of protein synthesis

RNA – ribonucleic acid

RNA interference (RNAi) – the process in which the introduction of double-stranded RNA into a cell inhibits the expression of genes

RNA polymerase – the enzyme that synthesizes RNA from a DNA template

RT-PCR – a method for the amplification of a specific mRNA. Reverse transcriptase is used to form a cDNA which is then amplified using PCR

Semi–conservative replication – the separation of the strands of DNA during replication with each acting as a template for the synthesis of a new complementary DNA strand

Shine-Dalgarno sequence – an mRNA sequence that precedes the translation initiation codon and is complementary to a ribosomal RNA

Signal sequence – a short amino acid sequence that determines the localization of a protein within the cell

Silent mutation – a mutation with DNA that does not cause an alternation to the encoded amino acid sequence

Single-nucleotide polymorphisms (SNPs) – single-base-pair variations scattered within the genetic code of the individuals within a population

siRNA – small interfering RNAs. Synthetic short double-stranded RNA molecules that inhibit gene expression

Site-specific recombination – occurs between two specific, but not necessarily homologous, DNA sequences

Southern blot – a technique for the separation of DNA molecules through agarose gels followed by detection of specific DNAs after denaturation through hybridization with single-stranded DNA

Spliceosome – a complex consisting of RNA and small nuclear ribonucleoproteins (snRNPs). The spliceosome splices RNA transcripts by excising introns and ligating the ends of exons

Splicing – the removal of introns and joining of exons to form a mature mRNA

Stem cell – a cell from the embryo, foetus or adult that has capability to reproduce itself. It can give rise to specialized cells that make up the tissues and organs of the body

Supercoiling – the way in which closed-circular DNA crosses over its own axis in three-dimensional space

T-DNA – the portion of the *Agrobacterium tumefaciens* Ti plasmid that is inserted into the genome of the host plant cell

TA cloning – a method for cloning PCR products that relies on the addition of template independent A residues to PCR products by Taq DNA polymerase

Telomere – the repetitive DNA sequences at the end of a chromosome

Terminal transferase – an enzyme that catalyses the addition of nucleotides to the 3′-terminus of DNA

Terminator – the sequence of DNA that causes RNA polymerase to stop transcription

Terminator technology – plants are engineered so when crops are harvested, all new seeds produced are sterile

Thymine – a pyrimidine base found in DNA. Thymine base pairs with adenine

Transcription – the synthesis of RNA from a DNA template

Transfection – the addition of DNA to eukaryotic cells

Transformation – the addition of DNA to bacteria. In eukaryotes, this term also refers to the state of cells that undergo tumour-like growth

Transforming principle – the factor identified by Griffith as being able to convert one bacterial type to another

Transgenic animals – created by inserting DNA sequences into the germ line through addition to the egg

Transition mutation – a mutation in DNA in which one purine–pyrimidine base pair is changed to a different purine–pyrimidine base pair

Translation – the synthesis of protein using an mRNA template

Transversion mutation – a mutation in DNA in which a purine–pyrimidine base pair is changed to a pyrimidine–purine base pair

Trinucleotide – the joining of three nucleotides through phosphodiester linkages

Two-hybrid screen – a technique to identify proteins that are able to interact with each other

Uracil – a pyrimidine base usually only found in RNA. Uracil bases pairs with adenine

Variable number tandem repeats (VNTRs) – different numbers of tandemly repeated DNA sequences at a given locus.

Vector – a DNA molecule that possesses the ability to self-replicate. Used to introduce foreign DNA into host cells, where it is replicated autonomously in large quantities

Western blot – the detection of specific proteins following electrophoresis using antibodies

Xenobiotic – a compound with a chemical structure that is foreign to an organism

Xenotransplantation – the surgical transplantation of tissue or organs from an individual of one species into an individual of another species

YAC – yeast artificial chromosome

Zygote – a cell formed by the union of two gametes (egg and sperm)

Proteins

A1.1

APPENDICES

The structures of the 20 amino acids that are encoded by the genetic code.

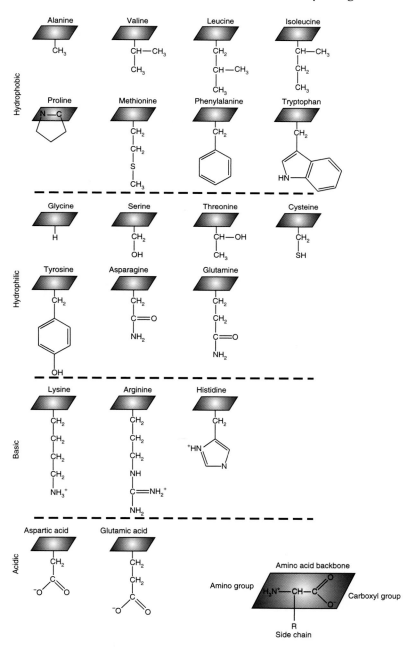

Analysis of Genes and Genomes   Richard J. Reece
© 2004 John Wiley & Sons, Ltd   ISBNs: 0-470-84379-9 (HB); 0-470-84380-2 (PB)

A1.2

The α-helix and the β-sheet are common secondary structure protein elements. An α-helix from the *Saccharomyces cerevisiae* transcriptional activator protein Gcn4p. A β-sheet from concanavalin A. In each case, the views depicted show, (A) all atoms of both the polypeptide backbone and the amino acid side chains, (B) the residues of the polypeptide backbone only, and (C) a trace of the peptide backbone itself.

The α-helix

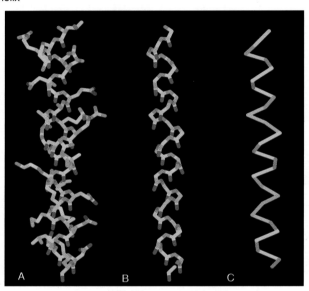

The β-sheet

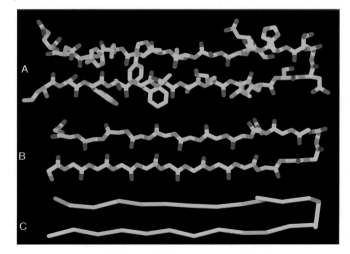

A1.3

The genetic code. The translational initiating methionine codon is highlighted in green and the termination codons are highlighted in red.

First position (5′ end)	Second position				Third position (3′ end)
	U	C	A	G	
U	UUU ⎤ Phe UUC ⎦ UUA ⎤ Leu UUG ⎦	UCU ⎤ UCC UCA Ser UCG ⎦	UAU ⎤ Tyr UAC ⎦ UAA Stop UAG Stop	UGU ⎤ Cys UGC ⎦ UGA Stop UGG Trp	U C A G
C	CUU ⎤ CUC CUA Leu CUG ⎦	CCU ⎤ CCC CCA Pro CCG ⎦	CAU ⎤ His CAC ⎦ CAA ⎤ Gln CAG ⎦	CGU ⎤ CGC CGA Arg CGG ⎦	U C A G
A	AUU ⎤ AUC Ile AUA ⎦ AUG Met	ACU ⎤ ACC ACA Thr ACG ⎦	AAU ⎤ Asn AAC ⎦ AAA ⎤ Lys AAG ⎦	AGU ⎤ Ser AGC ⎦ AGA ⎤ Arg AGG ⎦	U C A G
G	GUU ⎤ GUC GUA Val GUG ⎦	GCU ⎤ GCC GCA Ala GCG ⎦	GAU ⎤ Asp GAC ⎦ GAA ⎤ Glu GAG ⎦	GGU ⎤ GGC GGA Gly GGG ⎦	U C A G

Nobel prize winners

The following is a list of Nobel Prize winners whose work has had influence on, or is related to, the topics we have discussed in this book.

2002
Sydney Brenner, H. Robert Horvitz and John E. Sulston for their discoveries concerning genetic regulation of organ development and programmed cell death (Physiology or Medicine)

2001
Leland H. Hartwell, R. Tim Hunt and Paul M. Nurse for their discoveries of key regulators of the cell cycle (Physiology or Medicine)

1999
Günter Blobel for the discovery that proteins have intrinsic signals that govern their transport and localization in the cell (Physiology or Medicine)

1995
Edward B. Lewis, Christiane Nüsslein-Volhard and Eric F. Wieschaus for their discoveries concerning the genetic control of early embryonic development (Physiology or Medicine)

1993
Richard J. Roberts and Phillip A. Sharp for their independent discoveries of split genes (Physiology or Medicine)

Kary B. Mullis for his invention of the polymerase chain reaction (PCR) method.

Michael Smith for his fundamental contributions to the establishment of oligonucleotide based, site directed mutagenesis and its development for protein studies (Chemistry)

Analysis of Genes and Genomes   Richard J. Reece
© 2004 John Wiley & Sons, Ltd   ISBNs: 0-470-84379-9 (HB); 0-470-84380-2 (PB)

1989

Sidney Altman and Thomas R. Cech for their discovery of catalytic properties of RNA (Chemistry)

J. Michael Bishop and Harold E. Varmus for their discovery of the cellular origin of retroviral oncogenes (Physiology or Medicine)

1987

Susumu Tonegawa for his discovery of the genetic principle for generation of antibody diversity (Physiology or Medicine)

1985

Michael S. Brown and Joseph L. Goldstein for their discoveries concerning the regulation of cholesterol metabolism (Physiology or Medicine)

1984

Niels K. Jerne, Georges J. F. Köhler and César Milstein for theories concerning the specificity in development and control of the immune system and the discovery of the principle for production of monoclonal antibodies (Physiology or Medicine)

1983

Barbara McClintock for her discovery of mobile genetic elements (Physiology or Medicine)

1982

Aaron Klug for his development of crystallographic electron microscopy and his structural elucidation of biologically important nuclei acid–protein complexes (Chemistry)

1980

Paul Berg for his fundamental studies of the biochemistry of nucleic acids, with particular regard to recombinant DNA.

Walter Gilbert and Frederick Sanger for their contributions concerning the determination of base sequences in nucleic acids (Chemistry)

1978

Werner Arber, Daniel Nathans and Hamilton O. Smith for the discovery of restriction enzymes and their application to problems of molecular genetics (Physiology or Medicine)

1975

David Baltimore, Renato Dulbecco and Howard M. Temin for their discoveries concerning the interaction between tumour viruses and the genetic material of the cell (Physiology or Medicine)

1972

Christian B. Anfinsen for his work on ribonuclease, especially concerning the connection between the amino acid sequence and the biologically active confirmation.

Stanford Moore and William H. Stein for their contribution to the understanding of the connection between chemical structure and catalytic activity of the active centre of the ribonuclease molecule (Chemistry)

1969

Max Delbrück, Alfred D. Hershey and Salvador E. Luria for their discoveries concerning the replication mechanism and the genetic structure of viruses (Physiology or Medicine)

1968

Robert W. Holley, Har Gobind Khorana and Marshall W. Nirenberg for their interpretation of the genetic code and its function in protein synthesis. (Physiology or Medicine)

1966

Peyton Rous for his discovery of tumour inducing viruses (Physiology or Medicine)

1965

François Jacob, André L. Woff and Jacoues Monod for their discoveries concerning genetic control of enzyme and virus synthesis (Physiology or Medicine)

1964

Dorothy C. Hodgkin for her determinations by X-ray techniques of the structures of important biochemical substances (Chemistry)

1962

Francis H. C. Crick, James D. Watson and Maurice H. F. Wilkins for their discoveries concerning the molecular structure of nuclear acids and its significance for information transfer in living material (Physiology or Medicine)

Max F. Perutz and John C. Kendrew for their studies of the structures of globular proteins (Chemistry)

1959

Severo Ochoa and Arthur Kornberg for their discovery of the mechanisms in the biological synthesis of ribonucleic acid and deoxyribonucleic acid (Physiology or Medicine)

1958

George W. Beadle and Edward L. Tatum for their discovery that genes act by regulating definite chemical events.

Joshua Lederberg for his discoveries concerning genetic recombination and the organization of the genetic material of bacteria (Physiology or Medicine)

Frederick Sanger for his work on the structure of proteins, especially that of insulin (Chemistry)

1957

Alexander R. Todd for his work on nucleotides and nucleotide co-enzymes (Chemistry)

1954

Linus C. Pauling for his research into the nature of the chemical bond and its application to the elucidation of the structure of complex substances (Chemistry)

1946

James B. Sumner for his discovery that enzymes can be crystallized.

John H. Northrop and Wendell M. Stanley for their preparation of enzymes and virus proteins in a pure form (Chemistry)

Hermann J. Muller for the discovery of the production of mutations by means of X-ray irradiation (Physiology or Medicine)

1933

Thomas H. Morgan for his discoveries concerning the role played by the chromosome in heredity (Physiology or Medicine)

1902

H. Emil Fischer for his work on sugar and purine syntheses (Chemistry)

References

Abel, P.P., Nelson, R.S., De, B., Hoffmann, N., Rogers, S.G., Fraley, R.T. and Beachy, R.N. (1986) Delay of disease development in transgenic plants that express the tobacco mosaic virus coat protein gene. *Science*, **232**, 738–743.

Abremski, K. and Hoess, R. (1984) Bacteriophage P1 site-specific recombination: purification and properties of the Cre recombinase protein. *J. Biol. Chem.*, **259**, 1509–1514.

Abremski, K., Hoess, R. and Sternberg, N. (1983) Studies on the properties of P1 site-specific recombination: evidence for topologically unlinked products following recombination. *Cell*, **32**, 1301–1311.

Adams, M.D. *et al.* (2000) The genome sequence of *Drosophila melanogaster*. *Science*, **287**, 2185–2195.

Adesnik, M., Salditt, M., Thomas, W. and Darnell, J.E. (1972) Evidence that all messenger RNA molecules (except histone messenger RNA) contain Poly(A) sequences and that the Poly(A) has a nuclear function. *J. Mol. Biol.*, **71**, 21–30.

Ahlquist, P. and Janda, M. (1984) cDNA cloning and *in vitro* transcription of the complete brome mosaic virus genome. *Mol. Cell. Biol.*, **4**, 2876–2882.

Allfrey, V., Falkner, R.M. and Mirsky, A.E. (1964) Acetylation and methylation of histones and their possible role in the regulation of RNA synthesis. *Proc. Natl. Acad. Sci. USA*, **51**, 786–794.

Alt, F.W. and Baltimore, D. (1982) Joining of immunoglobulin heavy chain gene segments: implications from a chromosome with evidence of three D-JH fusions. *Proc. Natl. Acad. Sci. USA*, **79**, 4118–4122.

Amann, E., Brosius, J. and Ptashne, M. (1983) Vectors bearing a hybrid *trp–lac* promoter useful for regulated expression of cloned genes in *Escherichia coli*. *Gene*, **25**, 167–178.

An, G., Costa, M.A. and Ha, S.B. (1990) Nopaline synthase promoter is wound inducible and auxin inducible. *Plant Cell*, **2**, 225–233.

Anand, R., Villasante, A. and Tyler-Smith, C. (1989) Construction of yeast artificial chromosome libraries with large inserts using fractionation by pulsed-field gel electrophoresis. *Nucleic Acids Res.*, **17**, 3425–3433.

Arber, W. (1965) Host-controlled modification of bacteriophage. *Annu. Rev. Microbiol.*, **19**, 365–378.

Archibald, A.L., McClenaghan, M., Hornsey, V., Simons, J.P. and Clark, A.J. (1990) High-level expression of biologically active human α_1-antitrypsin in the milk of transgenic mice. *Proc. Natl. Acad. Sci. USA*, **87**, 5178–5182.

Aronheim, A., Zandi, E., Hennemann, H., Elledge, S.J. and Karin, M. (1997) Isolation of an AP-1 repressor by a novel method for detecting protein–protein interactions. *Mol. Cell. Biol.*, **17**, 3094–3102.

Ashworth, D., Bishop, M., Campbell, K., Colman, A., Kind, A., Schnieke, A., Blott, S., Griffin, H., Haley, C., McWhir, J. and Wilmut, I. (1998) DNA microsatellite analysis of Dolly. *Nature*, **394**, 329.

Aszodi, A., Bateman, J.F., Hirsch, E., Baranyi, M., Hunziker, E.B., Hauser, N., Bosze, Z. and Fassler, R. (1999) Normal skeletal development of mice lacking matrilin 1: redundant function of matrilins in cartilage? *Mol. Cell. Biol.*, **19**, 7841–7845.

Atchison, R.W., Castro, B.C. and Hammon, W.M. (1965) Adenovirus-associated defective virus particles. *Science*, **149**, 754–756.

Avery, O.T., MacLeod, C.M. and McCarty, M. (1944) Studies on the chemical nature of the substance inducing transformation of pneumococcal types. *J. Exp. Med.*, **79**, 137–158.

Ayres, M.D., Howard, S.C., Kuzio, J., Lopez-Ferber, M. and Possee, R.D. (1994) The complete DNA sequence of *Autographa californica* nuclear polyhedrosis virus. *Virology*, **202**, 586–605.

Bachmair, A., Finley, D. and Varshavsky, A. (1986) *In vivo* half-life of a protein is a function of its amino-terminal residue. *Science*, **234**, 179–186.

Baguisi, A. *et al.* (1999) Production of goats by somatic cell nuclear transfer. *Nat. Biotechnol.*, **17**, 456–461.

Baltimore, D. (1970) RNA-dependent DNA polymerase in virions of RNA tumour viruses. *Nature*, **226**, 1209–1211.

Banfi, S., Guffanti, A. and Borsani, G. (1998) How to get the best of dbEST. *Trends Genet.*, **14**, 80–81.

Bantel-Schaal, U. (1991) Infection with adeno-associated parvovirus leads to increased sensitivity of mammalian cells to stress. *Virology*, **182**, 260–268.

Barnes, W.M. (1992) The fidelity of Taq polymerase catalyzing PCR is improved by an N-terminal deletion. *Gene*, **112**, 29–35.

Barnes, W.M. (1994) PCR amplification of up to 35-kb DNA with high fidelity and high yield from λ bacteriophage templates. *Proc. Natl. Acad. Sci. USA*, **91**, 2216–2220.

Barr, F.G. and Emanuel, B.S. (1990) Application of a subtraction hybridization technique involving photoactivatable biotin and organic extraction to solution hybridization analysis of genomic DNA. *Anal. Biochem.*, **186**, 369–373.

Bass, S.H., Mulkerrin, M.G. and Wells, J.A. (1991) A systematic mutational analysis of hormone-binding determinants in the human growth hormone receptor. *Proc. Natl. Acad. Sci. USA*, **88**, 4498–4502.

Bates, A.D. and Maxwell, A. (1993) *DNA Topology*. IRL Press, Oxford.

Baugh, C.M., Malone, J.H. and Butterworth, C.E., Jr. (1968) Human biotin deficiency. A case history of biotin deficiency induced by raw egg consumption in a cirrhotic patient. *Am. J. Clin. Nutr.*, **21**, 173–182.

Beard, D.A. and Schlick, T. (2001) Computational modeling predicts the structure and dynamics of chromatin fiber. *Structure*, **9**, 105–114.

Beck, Z.Q., Hervio, L., Dawson, P.E., Elder, J.H. and Madison, E.L. (2000) Identification of efficiently cleaved substrates for HIV-1 protease using a phage display library and use in inhibitor development. *Virology*, **274**, 391–401.

Becker, D.M., Fikes, J.D. and Guarente, L. (1991) A cDNA encoding a human CCAAT-binding protein cloned by functional complementation in yeast. *Proc. Natl. Acad. Sci. USA*, **88**, 1968–1972.

Bentley, S.D. *et al.* (2002) Complete genome sequence of the model actinomycete *Streptomyces coelicolor* A3(2). *Nature*, **417**, 141–147.

Benton, W.D. and Davis, R.W. (1977) Screening λgt recombinant clones by hybridization to single plaques *in situ*. *Science*, **196**, 180–182.

Benvenisti, L., Rogel, A., Kuznetzova, L., Bujanover, S., Becker, Y. and Stram, Y. (2001) Gene gun-mediate DNA vaccination against foot-and-mouth disease virus. *Vaccine*, **19**, 3885–3895.

Berg, P. (1981) Dissections and reconstructions of genes and chromosomes. *Science*, **213**, 296–303.

Berget, S.M., Moore, C. and Sharp, P.A. (1977) Spliced segments at the 5′ terminus of adenovirus 2 late mRNA. *Proc. Natl. Acad. Sci. U.S.A.*, **74**, 3171–3175.

Berman, B.P., Nibu, Y., Pfeiffer, B.D., Tomancak, P., Celniker, S.E., Levine, M., Rubin, G.M. and Eisen, M.B. (2002) Exploiting transcription factor binding site clustering to identify cis-regulatory modules involved in pattern formation in the *Drosophila* genome. *Proc. Natl. Acad. Sci. USA*, **99**, 757–762.

Bernard, H.U., Remaut, E., Hershfield, M.V., Das, H.K., Helinski, D.R., Yanofsky, C. and Franklin, N. (1979) Construction of plasmid cloning vehicles that promote gene expression from the bacteriophage lambda pL promoter. *Gene*, **5**, 59–76.

Bernardi, A. and Bernardi, F. (1984) Complete sequence of pSC101. *Nucleic Acids Res.*, **12**, 9415–9426.

Bernstein, E., Caudy, A.A., Hammond, S.M. and Hannon, G.J. (2001) Role for a bidentate ribonuclease in the initiation step of RNA interference. *Nature*, **409**, 363–366.

Bevan, M. (1984) Binary *Agrobacterium* vectors for plant transformation. *Nucleic Acids Res.*, **12**, 8711–8721.

Bhaumik, S.R. and Green, M.R. (2001) SAGA is an essential *in vivo* target of the yeast acidic activator Gal4p. *Genes Dev.*, **15**, 1935–1945.

Bilban, M., Buehler, L.K., Head, S., Desoye, G. and Quaranta, V. (2002) Normalizing DNA microarray data. *Curr. Issues Mol. Biol.*, **4**, 57–64.

Birch, R.G. (1997) Plant transformation: problems and strategies for practical application. *Annu. Rev. Plant Physiol. Plant Mol. Biol.*, **48**, 297–326.

Bird, A.P. (1980) DNA methylation and the frequency of CpG in animal DNA. *Nucleic Acids Res.*, **8**, 1499–1504.

Birnboim, H.C. and Doly, J. (1979) A rapid alkaline extraction procedure for screening recombinant plasmid DNA. *Nucleic Acids Res.*, **7**, 1513–1523.

Bittner, M. *et al.* (2000) Molecular classification of cutaneous malignant melanoma by gene expression profiling. *Nature*, **406**, 536–540.

Blattner, F.R. *et al.* (1997) The complete genome sequence of *Escherichia coli* K-12. *Science*, **277**, 1453–1474.

Blau, H.M., Brazelton, T.R. and Weimann, J.M. (2001) The evolving concept of a stem cell: entity or function? *Cell*, **105**, 829–841.

Blochlinger, K. and Diggelmann, H. (1984) Hygromycin B phosphotransferase as a selectable marker for DNA transfer experiments with higher eucaryotic cells. *Mol. Cell. Biol.*, **4**, 2929–2931.

Blumenthal, R.M. and Cheng, X. (2001) A Taq attack displaces bases. *Nat. Struct. Biol.*, **8**, 101–103.

Bock, R. (2001) Transgenic plastids in basic research and plant biotechnology. *J. Mol. Biol.*, **312**, 425–438.

Bolivar, F., Rodriguez, R.L., Greene, P.J., Betlach, M.C., Heyneker, H.L., Boyer, H.W., Crosa, J.H. and Falkow, S. (1977) Construction and characterization of new cloning vehicles. II. A multipurpose cloning system. *Gene*, **2**, 95–113.

Boos, W. and Shuman, H. (1998) Maltose/maltodextrin system of *Escherichia coli*: transport, metabolism, and regulation. *Microbiol. Mol. Biol. Rev.*, **62**, 204–229.

Boshart, M., Weber, F., Jahn, G., Dorsch-Hasler, K., Fleckenstein, B. and Schaffner, W. (1985) A very strong enhancer is located upstream of an immediate early gene of human cytomegalovirus. *Cell*, **41**, 521–530.

Botstein, D. and Fink, G.R. (1988) Yeast: an experimental organism for modern biology. *Science*, **240**, 1439–1443.

Bradshaw, R.A., Brickey, W.W. and Walker, K.W. (1998) N-terminal processing: the methionine aminopeptidase and N^{α}-acetyl transferase families. *Trends Biochem. Sci.*, **23**, 263–267.

Brady, G., Billia, F., Knox, J., Hoang, T., Kirsch, I.R., Voura, E.B., Hawley, R.G., Cumming, R., Buchwald, M., Siminovitch, K., Miyamoto, N., Boehmelt, G. and Iscove, N. (1995) Analysis of gene expression in a complex differentiation hierarchy by global amplification of cDNA from single cells. *Curr. Biol.*, **5**, 909–922.

Brady, G., Funk, A., Mattern, J., Schutz, G. and Brown, R. (1985) Use of gene transfer and a novel cosmid rescue strategy to isolate transforming sequences. *EMBO J.*, **4**, 2583–2588.

Brandhorst, B.P. and McConkey, E.H. (1974) Stability of nuclear RNA in mammalian cells. *J. Mol. Biol.*, **85**, 451–563.

Brent, R. and Finley, R.L., Jr. (1997) Understanding gene and allele function with two-hybrid methods. *Annu. Rev. Genet.*, **31**, 663–704.

Brent, R. and Ptashne, M. (1985) A eukaryotic transcriptional activator bearing the DNA specificity of a prokaryotic repressor. *Cell*, **43**, 729–736.

Briggs, R. and King, T.J. (1952) Transplantation of living nuclei from blastula cells into enucleated frogs' eggs. *Proc. Natl. Acad. Sci. USA*, **38**, 455–463.

Brinkmann, U., Mattes, R.E. and Buckel, P. (1989) High-level expression of recombinant genes in *Escherichia coli* is dependent on the availability of the *dnaY* gene product. *Gene*, **85**, 109–114.

Brock, T.D. and Freeze, H. (1969) *Thermus aquaticus* gen. n. and sp. n., a nonsporulating extreme thermophile. *J. Bacteriol.*, **98**, 289–297.

Broglie, K., Chet, I., Holliday, M., Cressman, R., Biddle, P., Knowlton, S., Mauvais, C.J. and Broglie, R. (1991) Transgenic plants with enhanced resistance to the fungal pathogen *Rhizoctonia solani*. *Science*, **254**, 1194–1197.

Broome, S. and Gilbert, W. (1978) Immunological screening method to detect specific translation products. *Proc. Natl. Acad. Sci. USA*, **75**, 2746–2749.

Brosius, J., Ullrich, A., Raker, M.A., Gray, A., Dull, T.J., Gutell, R.R. and Noller, H.F. (1981) Construction and fine mapping of recombinant plasmids containing the *rrnB* ribosomal RNA operon of *E. coli*. *Plasmid*, **6**, 112–118.

Buchman, A.R. and Berg, P. (1988) Comparison of intron-dependent and intron-independent gene expression. *Mol. Cell. Biol.*, **8**, 4395–4405.

Bult, C.J. *et al.* (1996) Complete genome sequence of the methanogenic archaeon, *Methanococcus jannaschii*. *Science*, **273**, 1058–1073.

Burke, D.T., Carle, G.F. and Olson, M.V. (1987) Cloning of large segments of exogenous DNA into yeast by means of artificial chromosome vectors. *Science*, **236**, 806–812.

Burki, K. (1986) Experimental embryology of the mouse. *Monogr. Dev. Biol.*, **19**, 1–77.

Burnette, W.N. (1981) 'Western blotting': electrophoretic transfer of proteins from sodium dodecyl sulfate–polyacrylamide gels to unmodified nitrocellulose and radiographic detection with antibody and radioiodinated protein A. *Anal. Biochem.*, **112**, 195–203.

Campbell, J.L. and Kleckner, N. (1988) The rate of *Dam*-mediated DNA adenine methylation in *Escherichia coli*. *Gene*, **74**, 189–190.

Campbell, K.H., McWhir, J., Ritchie, W.A. and Wilmut, I. (1996) Sheep cloned by nuclear transfer from a cultured cell line. *Nature*, **380**, 64–66.

Canatella, P.J., Karr, J.F., Petros, J.A. and Prausnitz, M.R. (2001) Quantitative study of electroporation-mediated molecular uptake and cell viability. *Biophys. J.*, **80**, 755–764.

Capecchi, M.R. (1980) High efficiency transformation by direct microinjection of DNA into cultured mammalian cells. *Cell*, **22**, 479–488.

Causton, H.C., Ren, B., Koh, S.S., Harbison, C.T., Kanin, E., Jennings, E.G., Lee, T.I., True, H.L., Lander, E.S. and Young, R.A. (2001) Remodeling of yeast genome expression in response to environmental changes. *Mol. Biol. Cell.*, **12**, 323–337.

Cavazzana-Calvo, M., Hacein-Bey, S., de Saint Basile, G., Gross, F., Yvon, E., Nusbaum, P., Selz, F., Hue, C., Certain, S., Casanova, J.L., Bousso, P., Deist, F.L. and Fischer, A. (2000) Gene therapy of human severe combined immunodeficiency (SCID)-X1 disease. *Science*, **288**, 669–672.

Celis, R.T., Leadlay, P.F., Roy, I. and Hansen, A. (1998) Phosphorylation of the periplasmic binding protein in two transport systems for arginine incorporation in *Escherichia coli* K-12 is unrelated to the function of the transport system. *J. Bacteriol.*, **180**, 4828–4833.

Cereghino, G.P. and Cregg, J.M. (1999) Applications of yeast in biotechnology: protein production and genetic analysis. *Curr. Opin. Biotechnol.*, **10**, 422–427.

Cesareni, G., Helmer-Citterich, M. and Castagnoli, L. (1991) Control of ColE1 plasmid replication by antisense RNA. *Trends Genet.*, **7**, 230–235.

Chabot, B. (1996) Directing alternative splicing: cast and scenarios. *Trends Genet.*, **12**, 472–478.

Chaffin, K.E., Beals, C.R., Wilkie, T.M., Forbush, K.A., Simon, M.I. and Perlmutter, R.M. (1990) Dissection of thymocyte signaling pathways by *in vivo* expression of pertussis toxin ADP-ribosyltransferase. *EMBO J.*, **9**, 3821–3829.

Chalfie, M., Tu, Y., Euskirchen, G., Ward, W.W. and Prasher, D.C. (1994) Green fluorescent protein as a marker for gene expression. *Science*, **263**, 802–805.

Chang, J.Y. (1985) Thrombin specificity. Requirement for apolar amino acids adjacent to the thrombin cleavage site of polypeptide substrate. *Eur. J. Biochem.*, **151**, 217–224.

Chang, L.M. and Bollum, F.J. (1986) Molecular biology of terminal transferase. *CRC Crit. Rev. Biochem.*, **21**, 27–52.

Chargaff, E., Lipshitz, R. and Green, C. (1952) Composition of the desoxypentose nucleic acids of four genera of sea-urchin. *J. Biol. Chem.*, **195**, 155–160.

Chargaff, E., Lipshitz, R., Green, C. and Hodes, M.E. (1951) The composition of the desoxyribonucleic acid of salmon sperm. *J. Biol. Chem.*, **192**, 223–230.

Chen, C.A. and Okayama, H. (1988) Calcium phosphate-mediated gene transfer: a highly efficient transfection system for stably transforming cells with plasmid DNA. *Biotechniques*, **6**, 632–638.

Cheng, M., Fry, J.E., Pang, S., Zhou, H., Hironaka, C.M., Duncan, D.R., Conner, T.W. and Wan, Y. (1997) Genetic transformation of wheat mediated by *Agrobacterium tumefaciens*. *Plant Physiol.*, **115**, 971–980.

Cheng, X. and Roberts, R.J. (2001) AdoMet-dependent methylation, DNA methyltransferases and base flipping. *Nucleic Acids Res.*, **29**, 3784–3795.

Chesné, P., Adenot, P.G., Viglietta, C., Baratte, M., Boulanger, L. and Renard, J.P. (2002) Cloned rabbits produced by nuclear transfer from adult somatic cells. *Nat. Biotechnol.*, **20**, 366–369.

Chien, A., Edgar, D.B. and Trela, J.M. (1976) Deoxyribonucleic acid polymerase from the extreme thermophile *Thermus aquaticus*. *J. Bacteriol.*, **127**, 1550–1557.

Chien, C.T., Bartel, P.L., Sternglanz, R. and Fields, S. (1991) The two-hybrid system: a method to identify and clone genes for proteins that interact with a protein of interest. *Proc. Natl. Acad. Sci. USA.*, **88**, 9578–9582.

Choate, K.A. and Khavari, P.A. (1997) Direct cutaneous gene delivery in a human genetic skin disease. *Hum. Gene Ther.*, **8**, 1659–1665.

Chong, S., Mersha, F.B., Comb, D.G., Scott, M.E., Landry, D., Vence, L.M., Perler, F.B., Benner, J., Kucera, R.B., Hirvonen, C.A., Pelletier, J.J., Paulus, H. and Xu, M.Q. (1997) Single-column purification of free recombinant proteins using a self-cleavable affinity tag derived from a protein splicing element. *Gene*, **192**, 271–281.

Chong, S. and Xu, M.Q. (1997) Protein splicing of the *Saccharomyces cerevisiae* VMA intein without the endonuclease motifs. *J. Biol. Chem.*, **272**, 15 587–15 590.

Chow, L.T., Gelinas, R.E., Broker, T.R. and Roberts, R.J. (1977) An amazing sequence arrangement at the 5′ ends of adenovirus 2 messenger RNA. *Cell*, **12**, 1–8.

Chu, S., DeRisi, J., Eisen, M., Mulholland, J., Botstein, D., Brown, P.O. and Herskowitz, I. (1998) The transcriptional program of sporulation in budding yeast. *Science*, **282**, 699–705.

Chung, C.H. (1993) Proteases in *Escherichia coli*. *Science*, **262**, 372–374.

Cibelli, J.B., Stice, S.L., Golueke, P.J., Kane, J.J., Jerry, J., Blackwell, C., Ponce de Leon, F.A. and Robl, J.M. (1998) Cloned transgenic calves produced from nonquiescent fetal fibroblasts. *Science*, **280**, 1256–1258.

Clark, W.G., Register, J.C., III, Nejidat, A., Eichholtz, D.A., Sanders, P.R., Fraley, R.T. and Beachy, R.N. (1990) Tissue-specific expression of the TMV coat protein in transgenic tobacco plants affects the level of coat protein-mediated virus protection. *Virology*, **179**, 640–647.

Clarke, L. (1990) Centromeres of budding and fission yeasts. *Trends Genet.*, **6**, 150–154.

Clarke, L. and Carbon, J. (1976) A colony bank containing synthetic Col El hybrid plasmids representative of the entire *E. coli* genome. *Cell*, **9**, 91–99.

Clewell, D.B. (1972) Nature of ColE1 plasmid replication in *Escherichia coli* in the presence of chloramphenicol. *J. Bacteriol.*, **110**, 667–676.

Cline, J., Braman, J.C. and Hogrefe, H.H. (1996) PCR fidelity of *Pfu* DNA polymerase and other thermostable DNA polymerases. *Nucleic Acids Res.*, **24**, 3546–3551.

Cohen, J.S. (1991) Oligonucleotides as therapeutic agents. *Pharmacol. Ther.*, **52**, 211–225.

Colas, P. and Brent, R. (1998) The impact of two-hybrid and related methods on biotechnology. *Trends Biotechnol.*, **16**, 355–363.

Colbère-Garapin, F., Horodniceanu, F., Kourilsky, P. and Garapin, A.C. (1981) A new dominant hybrid selective marker for higher eukaryotic cells. *J. Mol. Biol.*, **150**, 1–14.

Collins, F.S., Patrinos, A., Jordan, E., Chakravarti, A., Gesteland, R. and Walters, L. (1998) New goals for the U.S. Human Genome Project: 1998–2003. *Science*, **282**, 682–689.

Collins, J. and Brüning, H.J. (1978) Plasmids useable as gene-cloning vectors in an *in vitro* packaging by coliphage lambda: 'cosmids'. *Gene*, **4**, 85–107.

Cooper, A.A. and Stevens, T.H. (1995) Protein splicing: self-splicing of genetically mobile elements at the protein level. *Trends Biochem. Sci.*, **20**, 351–356.

Corneille, S., Lutz, K., Svab, Z. and Maliga, P. (2001) Efficient elimination of selectable marker genes from the plastid genome by the CRE-*lox* site-specific recombination system. *Plant J.*, **27**, 171–178.

Cramer, P., Bushnell, D.A., Fu, J., Gnatt, A.L., Maier-Davis, B., Thompson, N.E., Burgess, R.R., Edwards, A.M., David, P.R. and Kornberg, R.D. (2000) Architecture of RNA polymerase II and implications for the transcription mechanism. *Science*, **288**, 640–649.

Cramer, P., Bushnell, D.A. and Kornberg, R.D. (2001) Structural basis of transcription: RNA polymerase II at 2.8 Å resolution. *Science*, **292**, 1863–1876.

Crawford, D.R., Kochheiser, J.C., Schools, G.P., Salmon, S.L. and Davies, K.J. (2002) Differential display: a critical analysis. *Gene Expr.*, **10**, 101–107.

Cregg, J.M., Vedvick, T.S. and Raschke, W.C. (1993) Recent advances in the expression of foreign genes in *Pichia pastoris*. *Bio/technology*, **11**, 905–910.

Cross, S.H. and Bird, A.P. (1995) CpG islands and genes. *Curr. Opin. Genet. Dev.*, **5**, 309–314.

Crowlesmith, I. and Gamon, K. (1982) Rate of translation and kinetics of processing of newly synthesized molecules of two major outer-membrane proteins, the *Omp*A and *Omp*F proteins, of *Escherichia coli* K12. *Eur. J. Biochem.*, **124**, 577–583.

Cunningham, B.C. and Wells, J.A. (1989) High-resolution epitope mapping of hGH-receptor interactions by alanine-scanning mutagenesis. *Science*, **244**, 1081–1085.

Dai, Y. *et al.* (2002) Targeted disruption of the α1,3-galactosyltransferase gene in cloned pigs. *Nat. Biotechnol.*, **20**, 251–255.

Daniell, H. and Dhingra, A. (2002) Multigene engineering: dawn of an exciting new era in biotechnology. *Curr. Opin. Biotechnol.*, **13**, 136–141.

Daniell, H., Muthukumar, B. and Lee, S.B. (2001) Marker free transgenic plants: engineering the chloroplast genome without the use of antibiotic selection. *Curr. Genet.*, **39**, 109–116.

Das, G.C. and Niyogi, S.K. (1981) Structure, replication, and transcription of the SV40 genome. *Prog. Nucleic Acid Res. Mol. Biol.*, **25**, 187–241.

Daubert, S., Shepherd, R.J. and Gardner, R.C. (1983) Insertional mutagenesis of the cauliflower mosaic virus genome. *Gene*, **25**, 201–208.

Daugherty, P.S., Olsen, M.J., Iverson, B.L. and Georgiou, G. (1999) Development of an optimized expression system for the screening of antibody libraries displayed on the *Escherichia coli* surface. *Protein Eng.*, **12**, 613–621.

Davenport, J., Neale, G.A. and Goorha, R. (2000) Identification of genes potentially involved in *LMO2*-induced leukemogenesis. *Leukemia*, **14**, 1986–1996.

Davidson, B.L., Stein, C.S., Heth, J.A., Martins, I., Kotin, R.M., Derksen, T.A., Zabner, J., Ghodsi, A. and Chiorini, J.A. (2000) Recombinant adeno-associated virus type 2, 4, and 5 vectors: transduction of variant cell types and regions in the mammalian central nervous system. *Proc. Natl. Acad. Sci. USA*, **97**, 3428–3432.

Davies, D.R., Padlan, E.A. and Sheriff, S. (1990) Antibody–antigen complexes. *Annu. Rev. Biochem.*, **59**, 439–473.

Davis, G.D., Elisee, C., Newham, D.M. and Harrison, R.G. (1999) New fusion protein systems designed to give soluble expression in *Escherichia coli*. *Biotechnol. Bioeng.*, **65**, 382–388.

de Boer, H.A., Comstock, L.J. and Vasser, M. (1983) The *tac* promoter: a functional hybrid derived from the *trp* and *lac* promoters. *Proc. Natl. Acad. Sci. USA*, **80**, 21–25.

De Cosa, B., Moar, W., Lee, S.B., Miller, M. and Daniell, H. (2001) Overexpression of the *Bt cry2Aa2* operon in chloroplasts leads to formation of insecticidal crystals. *Nat. Biotechnol.*, **19**, 71–74.

de Wet, J.R., Fukushima, H., Dewji, N.N., Wilcox, E., O'Brien, J.S. and Helinski, D.R. (1984) Chromogenic immunodetection of human serum albumin and α-L-fucosidase clones in a human hepatoma cDNA expression library. *DNA*, **3**, 437–447.

de Wet, J.R., Wood, K.V., DeLuca, M., Helinski, D.R. and Subramani, S. (1987) Firefly luciferase gene: structure and expression in mammalian cells. *Mol. Cell. Biol.*, **7**, 725–737.

De Zoeten, G.A., Penswick, J.R., Horisberger, M.A., Ahl, P., Schultze, M. and Hohn, T. (1989) The expression, localization, and effect of a human interferon in plants. *Virology*, **172**, 213–222.

Denning, C., Burl, S., Ainslie, A., Bracken, J., Dinnyes, A., Fletcher, J., King, T., Ritchie, M., Ritchie, W.A., Rollo, M., de Sousa, P., Travers, A., Wilmut, I. and Clark, A.J. (2001) Deletion of the $\alpha(1,3)$galactosyl transferase (*GGTA1*) gene and the prion protein (*PrP*) gene in sheep. *Nat. Biotechnol.*, **19**, 559–562.

Dente, L., Cesareni, G. and Cortese, R. (1983) pEMBL: a new family of single stranded plasmids. *Nucleic Acids Res.*, **11**, 1645–1655.

DeRisi, J.L., Iyer, V.R. and Brown, P.O. (1997) Exploring the metabolic and genetic control of gene expression on a genomic scale. *Science*, **278**, 680–686.

di Guan, C., Li, P., Riggs, P.D. and Inouye, H. (1988) Vectors that facilitate the expression and purification of foreign peptides in *Escherichia coli* by fusion to maltose-binding protein. *Gene*, **67**, 21–30.

Dib, C., Faure, S., Fizames, C., Samson, D., Drouot, N., Vignal, A., Millasseau, P., Marc, S., Hazan, J., Seboun, E., Lathrop, M., Gyapay, G., Morissette, J. and Weissenbach, J. (1996) A comprehensive genetic map of the human genome based on 5,264 microsatellites. *Nature*, **380**, 152–154.

Dickerson, R.E. (1983) Base sequence and helix structure variation in *B* and *A* DNA. *J. Mol. Biol.*, **166**, 419–441.

Doherty, A.J., Ashford, S.R., Subramanya, H.S. and Wigley, D.B. (1996) Bacteriophage T7 DNA ligase. Overexpression, purification, crystallization, and characterization. *J. Biol. Chem.*, **271**, 11 083–11 089.

Donahue, W.F., Turczyk, B.M. and Jarrell, K.A. (2002) Rapid gene cloning using terminator primers and modular vectors. *Nucleic Acids Res.*, **30**, e95.

Donehower, L.A., Harvey, M., Slagle, B.L., McArthur, M.J., Montgomery, C.A., Jr., Butel, J.S. and Bradley, A. (1992) Mice deficient for p53 are developmentally normal but susceptible to spontaneous tumours. *Nature*, **356**, 215–221.

Dotto, G.P., Enea, V. and Zinder, N.D. (1981) Functional analysis of bacteriophage f1 intergenic region. *Virology*, **114**, 463–473.

Dotto, G.P. and Horiuchi, K. (1981) Replication of a plasmid containing two origins of bacteriophage. *J. Mol. Biol.*, **153**, 169–176.

Dougherty, W.G., Parks, T.D., Cary, S.M., Bazan, J.F. and Fletterick, R.J. (1989) Characterization of the catalytic residues of the tobacco etch virus 49-kDa proteinase. *Virology*, **172**, 302–310.

Dryden, D.T., Murray, N.E. and Rao, D.N. (2001) Nucleoside triphosphate-dependent restriction enzymes. *Nucleic Acids Res.*, **29**, 3728–3741.

Dubendorff, J.W. and Studier, F.W. (1991) Controlling basal expression in an inducible T7 expression system by blocking the target T7 promoter with *lac* repressor. *J. Mol. Biol.*, **219**, 45–59.

Dunham, I. *et al.* (1999) The DNA sequence of human chromosome 22. *Nature*, **402**, 489–495.

Echelard, Y., Vassileva, G. and McMahon, A.P. (1994) Cis-acting regulatory sequences governing *Wnt-1* expression in the developing mouse CNS. *Development*, **120**, 2213–2224.

Efstratiadis, A., Kafatos, F.C., Maxam, A.M. and Maniatis, T. (1976) Enzymatic *in vitro* synthesis of globin genes. *Cell*, **7**, 279–288.

Eide, D.J. (1998) The molecular biology of metal ion transport in *Saccharomyces cerevisiae*. *Annu. Rev. Nutr.*, **18**, 441–469.

Einhauer, A. and Jungbauer, A. (2001) The FLAG peptide, a versatile fusion tag for the purification of recombinant proteins. *J. Biochem. Biophys. Methods*, **49**, 455–465.

Elbashir, S.M., Harborth, J., Lendeckel, W., Yalcin, A., Weber, K. and Tuschl, T. (2001) Duplexes of 21-nucleotide RNAs mediate RNA interference in cultured mammalian cells. *Nature*, **411**, 494–498.

Endow, S.A. and Roberts, R.J. (1977) Two restriction-like enzymes from *Xanthomonas malvacearum*. *J. Mol. Biol.*, **112**, 521–529.

Erkman, L., McEvilly, R.J., Luo, L., Ryan, A.K., Hooshmand, F., O'Connell, S.M., Keithley, E.M., Rapaport, D.H., Ryan, A.F. and Rosenfeld, M.G. (1996) Role of transcription factors Brn-3.1 and Brn-3.2 in auditory and visual system development. *Nature*, **381**, 603–606.

Evan, G.I., Lewis, G.K., Ramsay, G. and Bishop, J.M. (1985) Isolation of monoclonal antibodies specific for human *c-myc* proto-oncogene product. *Mol. Cell. Biol.*, **5**, 3610–3616.

Evans, M.J., Gurer, C., Loike, J.D., Wilmut, I., Schnieke, A.E. and Schon, E.A. (1999) Mitochondrial DNA genotypes in nuclear transfer-derived cloned sheep. *Nat. Genet.*, **23**, 90–93.

Evans, M.J. and Kaufman, M.H. (1981) Establishment in culture of pluripotential cells from mouse embryos. *Nature*, **292**, 154–156.

Fairburn, H.R., Young, L.E. and Hendrich, B.D. (2002) Epigenetic reprogramming: how now, cloned cow? *Curr. Biol.*, **12**, R68–70.

Fan, F., Lorenzen, J.A. and Plapp, B.V. (1991) An aspartate residue in yeast alcohol dehydrogenase I determines the specificity for coenzyme. *Biochemistry*, **30**, 6397–6401.

Farabaugh, P.J. (1996) Programmed translational frameshifting. *Annu. Rev. Genet.*, **30**, 507–528.

Felgner, P.L., Gadek, T.R., Holm, M., Roman, R., Chan, H.W., Wenz, M., Northrop, J.P., Ringold, G.M. and Danielsen, M. (1987) Lipofection: a highly efficient, lipid-mediated DNA-transfection procedure. *Proc. Natl. Acad. Sci. USA*, **84**, 7413–7417.

Ferretti, L. and Sgaramella, V. (1981) Specific and reversible inhibition of the blunt end joining activity of the T4 DNA ligase. *Nucleic Acids Res.*, **9**, 3695–3705.

Fickett, J.W. (1996) Finding genes by computer: the state of the art. *Trends Genet.*, **12**, 316–320.

Fields, S. and Johnston, M. (2002) Genomics. A crisis in postgenomic nomenclature. *Science*, **296**, 671–672.

Fields, S. and Song, O. (1989) A novel genetic system to detect protein-protein interaction. *Nature*, **340**, 245–246.

Figge, J., Wright, C., Collins, C.J., Roberts, T.M. and Livingston, D.M. (1988) Stringent regulation of stably integrated chloramphenicol acetyl transferase genes by *E. coli lac* repressor in monkey cells. *Cell*, **52**, 713–722.

Finer, J.J., Finer, K.R. and Ponappa, T. (1999) Particle bombardment mediated transformation. *Curr. Top. Microbiol. Immunol.*, **240**, 59–80.

Finley, R.L., Jr. and Brent, R. (1994) Interaction mating reveals binary and ternary connections between *Drosophila* cell cycle regulators. *Proc. Natl. Acad. Sci. USA*, **91**, 12 980–12 984.

Fire, A., Xu, S., Montgomery, M.K., Kostas, S.A., Driver, S.E. and Mello, C.C. (1998) Potent and specific genetic interference by double-stranded RNA in *Caenorhabditis elegans*. *Nature*, **391**, 806–811.

Fischer, A., Hacein-Bey, S. and Cavazzana-Calvo, M. (2002) Gene therapy of severe combined immunodeficiencies. *Nat. Rev. Immunol.*, **2**, 615–621.

Fleischmann, R.D. *et al.* (1995) Whole-genome random sequencing and assembly of *Haemophilus influenzae* Rd. *Science*, **269**, 496–512.

Foord, O.S. and Rose, E.A. (1994) Long-distance PCR. *PCR Methods Appl.*, **3**, S149–S161.

Foote, S., Vollrath, D., Hilton, A. and Page, D.C. (1992) The human Y chromosome: overlapping DNA clones spanning the euchromatic region. *Science*, **258**, 60–66.

Foster, T.J., Davis, M.A., Roberts, D.E., Takeshita, K. and Kleckner, N. (1981) Genetic organization of transposon Tn10. *Cell*, **23**, 201–213.

Franklin, R.E. and Gosling, R.G. (1953) Molecular configuration in sodium thymonucleate. *Nature*, **171**, 740–741.

Fraser, A.G., Kamath, R.S., Zipperlen, P., Martinez-Campos, M., Sohrmann, M. and Ahringer, J. (2000) Functional genomic analysis of *C. elegans* chromosome I by systematic RNA interference. *Nature*, **408**, 325–330.

Fraser, M.J. (1992) The baculovirus-infected insect cell as a eukaryotic gene expression system. *Curr. Top. Microbiol. Immunol.*, **158**, 131–172.

Fray, R.G. and Grierson, D. (1993) Molecular genetics of tomato fruit ripening. *Trends Genet.*, **9**, 438–443.

Frohman, M.A., Dush, M.K. and Martin, G.R. (1988) Rapid production of full-length cDNAs from rare transcripts: amplification using a single gene-specific oligonucleotide primer. *Proc. Natl. Acad. Sci. USA*, **85**, 8998–9002.

Fung, E.T., Thulasiraman, V., Weinberger, S.R. and Dalmasso, E.A. (2001) Protein biochips for differential profiling. *Curr. Opin. Biotechnol.*, **12**, 65–69.

Gänsbacher, B., Danos, O., Dickson, G., Thielemans, K., Cosset, F.L., Deglon, N., Dilber, M.S., Galun, E., Klatzmann, D., Mavilio, F. and Taylor, N. (2003) French gene therapy group reports on the adverse event in a clinical trial of gene therapy for X-linked severe combined immune deficiency (X-SCID). *J. Gene Med.*, **5**, 82–84.

Gao, G., Qu, G., Burnham, M.S., Huang, J., Chirmule, N., Joshi, B., Yu, Q.C., Marsh, J.A., Conceicao, C.M. and Wilson, J.M. (2000) Purification of recombinant adeno-associated virus vectors by column chromatography and its performance *in vivo*. *Hum. Gene Ther.*, **11**, 2079–2091.

Garcia-Sáez, I., Párraga, A., Phillips, M.F., Mantle, T.J. and Coll, M. (1994) Molecular structure at 1.8 Å of mouse liver class pi glutathione S-transferase complexed with S-(p-nitrobenzyl)glutathione and other inhibitors. *J. Mol. Biol.*, **237**, 298–314.

Garrick, D., Fiering, S., Martin, D.I. and Whitelaw, E. (1998) Repeat-induced gene silencing in mammals. *Nat. Genet.*, **18**, 56–59.

Gatignol, A., Durand, H. and Tiraby, G. (1988) Bleomycin resistance conferred by a drug-binding protein. *FEBS Lett.*, **230**, 171–175.

Gavin, A.C. *et al.* (2002) Functional organization of the yeast proteome by systematic analysis of protein complexes. *Nature*, **415**, 141–147.

Genilloud, O., Garrido, M.C. and Moreno, F. (1984) The transposon Tn5 carries a bleomycin-resistance determinant. *Gene*, **32**, 225–233.

Georgiadis, M.M., Jessen, S.M., Ogata, C.M., Telesnitsky, A., Goff, S.P. and Hendrickson, W.A. (1995) Mechanistic implications from the structure of a catalytic fragment of Moloney murine leukemia virus reverse transcriptase. *Structure*, **3**, 879–892.

Giaever, G. *et al.* (2002) Functional profiling of the *Saccharomyces cerevisiae* genome. *Nature*, **418**, 387–391.

Giga-Hama, Y. and Kumagai, H. (1999) Expression system for foreign genes using the fission yeast *Schizosaccharomyces pombe*. *Biotechnol. Appl. Biochem.*, **30**, 235–244.

Giniger, E., Varnum, S.M. and Ptashne, M. (1985) Specific DNA binding of GAL4, a positive regulatory protein of yeast. *Cell*, **40**, 767–774.

Glazer, A.N. and Mathies, R.A. (1997) Energy-transfer fluorescent reagents for DNA analyses. *Curr. Opin. Biotechnol.*, **8**, 94–102.

Gnatt, A.L., Cramer, P., Fu, J., Bushnell, D.A. and Kornberg, R.D. (2001) Structural basis of transcription: an RNA polymerase II elongation complex at 3.3 Å resolution. *Science*, **292**, 1876–1882.

Goff, S.A. *et al.* (2002) A draft sequence of the rice genome (*Oryza sativa* L. ssp. *japonica*). *Science*, **296**, 92–100.

Goffeau, A. *et al.* (1997) The yeast genome directory. *Nature*, **387 supp**, 5–6.

Goldfarb, D.S., Gariepy, J., Schoolnik, G. and Kornberg, R.D. (1986) Synthetic peptides as nuclear localization signals. *Nature*, **322**, 641–644.

Goldschmidt-Clermont, M. (1991) Transgenic expression of aminoglycoside adenine transferase in the chloroplast: a selectable marker of site-directed transformation of *Chlamydomonas*. *Nucleic Acids Res.*, **19**, 4083–4089.

Gönczy, P. *et al.* (2000) Functional genomic analysis of cell division in *C. elegans* using RNAi of genes on chromosome III. *Nature*, **408**, 331–336.

Gorman, C.M., Merlino, G.T., Willingham, M.C., Pastan, I. and Howard, B.H. (1982) The Rous sarcoma virus long terminal repeat is a strong promoter when introduced into a variety of eukaryotic cells by DNA-mediated transfection. *Proc. Natl. Acad. Sci. USA*, **79**, 6777–6781.

Gorman, C.M., Moffat, L.F. and Howard, B.H. (1982) Recombinant genomes which express chloramphenicol acetyltransferase in mammalian cells. *Mol. Cell. Biol.*, **2**, 1044–1051.

Gorziglia, M.I., Kadan, M.J., Yei, S., Lim, J., Lee, G.M., Luthra, R. and Trapnell, B.C. (1996) Elimination of both E1 and E2 from adenovirus vectors further improves prospects for *in vivo* human gene therapy. *J. Virol.*, **70**, 4173–4178.

Gossen, M. and Bujard, H. (1992) Tight control of gene expression in mammalian cells by tetracycline-responsive promoters. *Proc. Natl. Acad. Sci. USA*, **89**, 5547–5551.

Gossen, M. and Bujard, H. (2002) Studying gene function in eukaryotes by conditional gene inactivation. *Annu. Rev. Genet.*, **36**, 153–173.

Gossen, M., Freundlieb, S., Bender, G., Muller, G., Hillen, W. and Bujard, H. (1995) Transcriptional activation by tetracyclines in mammalian cells. *Science*, **268**, 1766–1769.

Gottesman, S. (1996) Proteases and their targets in *Escherichia coli*. *Annu. Rev. Genet.*, **30**, 465–506.

Gottschalk, A., Bartels, C., Neubauer, G., Lührmann, R. and Fabrizio, P. (2001) A novel yeast U2 snRNP protein, Snu17p, is required for the first catalytic step of splicing and for progression of spliceosome assembly. *Mol. Cell. Biol.*, **21**, 3037–3046.

Gottschalk, S., Sparrow, J.T., Hauer, J., Mims, M.P., Leland, F.E., Woo, S.L. and Smith, L.C. (1996) A novel DNA-peptide complex for efficient gene transfer and expression in mammalian cells. *Gene Ther.*, **3**, 48–57.

Graham, F.L., Smiley, J., Russell, W.C. and Nairn, R. (1977) Characteristics of a human cell line transformed by DNA from human adenovirus type 5. *J. Gen. Virol.*, **36**, 59–74.

Graham, F.L. and van der Eb, A.J. (1973) A new technique for the assay of infectivity of human adenovirus 5 DNA. *Virology*, **52**, 456–467.

Green, E.D., Riethman, H.C., Dutchik, J.E. and Olson, M.V. (1991) Detection and characterization of chimeric yeast artificial-chromosome clones. *Genomics*, **11**, 658–669.

Green, N.M. (1963) Avidin. 1. The use of [^{14}C] biotin for kinetic studies and for assay. *Biochem. J.*, **89**, 585–591.

Green, P.J., Pines, O. and Inouye, M. (1986) The role of antisense RNA in gene regulation. *Annu. Rev. Biochem.*, **55**, 569–597.

Gressel, J. (2000) Molecular biology of weed control. *Transgenic Res.*, **9**, 355–382.

Griffin, H.G. and Griffin, A.M. (1993) Dideoxy sequencing reactions using Sequenase version 2.0. *Methods Mol. Biol.*, **23**, 103–108.

Griffith, F. (1928) The significance of pneumococcal types. *J. Hyg.*, **27**, 113–159.

Griggs, D. and Johnston, M. (1991) Regulated expression of the GAL4 activator gene in yeast provides a sensitive genetic switch for glucose repression. *Proc. Natl. Acad. Sci. USA*, **88**, 8597–8601.

Gronenborn, B. (1976) Overproduction of phage lambda repressor under control of the *lac* promotor of *Escherichia coli*. *Mol. Gen. Genet.*, **148**, 243–250.

Gross, C., Kelleher, M., Iyer, V.R., Brown, P.O. and Winge, D.R. (2000) Identification of the copper regulon in *Saccharomyces cerevisiae* by DNA microarrays. *J. Biol. Chem.*, **275**, 32 310–32 316.

Grunstein, M. and Hogness, D.S. (1975) Colony hybridization: a method for the isolation of cloned DNAs that contain a specific gene. *Proc. Natl. Acad. Sci. USA*, **72**, 3961–3965.

Grunstein, M. and Wallis, J. (1979) Colony hybridization. *Methods Enzymol.*, **68**, 379–389.

Gruss, P. and Walther, C. (1992) Pax in development. *Cell*, **69**, 719–722.

Gu, H., Marth, J.D., Orban, P.C., Mossmann, H. and Rajewsky, K. (1994) Deletion of a DNA polymerase β-gene segment in T cells using cell type-specific gene targeting. *Science*, **265**, 103–106.

Gubler, U. and Hoffman, B.J. (1983) A simple and very efficient method for generating cDNA libraries. *Gene*, **25**, 263–269.

Gurdon, J.B., Laskey, R.A. and Reeves, O.R. (1975) The developmental capacity of nuclei transplanted from keratinized skin cells of adult frogs. *J. Embryol. Exp. Morphol.*, **34**, 93–112.

Hamer, D.H. and Leder, P. (1979) Expression of the chromosomal mouse β-maj-globin gene cloned in SV40. *Nature*, **281**, 35–40.

Hammond, S.M., Bernstein, E., Beach, D. and Hannon, G.J. (2000) An RNA-directed nuclease mediates post-transcriptional gene silencing in *Drosophila* cells. *Nature*, **404**, 293–296.

Hampsey, M. (1998) Molecular genetics of the RNA polymerase II general transcriptional machinery. *Microbiol. Mol. Biol. Rev.*, **62**, 465–503.

Hanahan, D. (1983) Studies on transformation of *Escherichia coli* with plasmids. *J. Mol. Biol.*, **166**, 557–580.

Hanahan, D. and Meselson, M. (1980) Plasmid screening at high colony density. *Gene*, **10**, 63–67.

Handler, M. and Iozzo, R.V. (2001) Constitutive and inducible antisense expression. *Methods Mol. Biol.*, **171**, 251–259.

Hansen, G. and Wright, M.S. (1999) Recent advances in the transformation of plants. *Trends Plant. Sci.*, **4**, 226–231.

Harbottle, R.P., Cooper, R.G., Hart, S.L., Ladhoff, A., McKay, T., Knight, A.M., Wagner, E., Miller, A.D. and Coutelle, C. (1998) An RGD-oligolysine peptide: a prototype construct for integrin-mediated gene delivery. *Hum. Gene Ther.*, **9**, 1037–1047.

Hardy, E., Martinez, E., Diago, D., Diaz, R., Gonzalez, D. and Herrera, L. (2000) Large-scale production of recombinant hepatitis B surface antigen from *Pichia pastoris*. *J. Biotechnol.*, **77**, 157–167.

Hargreaves, D., Rice, D.W., Sedelnikova, S.E., Artymiuk, P.J., Lloyd, R.G. and Rafferty, J.B. (1998) Crystal structure of *E. coli* RuvA with bound DNA Holliday junction at 6Å resolution. *Nat. Struct. Biol.*, **5**, 441–446.

Harlow, E. and Lane, D. (1999) *Using Antibodies: a Laboratory Manual*. Cold Spring Harbor Laboratory Press, Cold Spring Harbor, NY.

Harrington, J.J., Van Bokkelen, G., Mays, R.W., Gustashaw, K. and Willard, H.F. (1997) Formation of *de novo* centromeres and construction of first-generation human artificial microchromosomes. *Nat. Genet.*, **15**, 345–355.

Harris, L.J., Larson, S.B., Hasel, K.W., Day, J., Greenwood, A. and McPherson, A. (1992) The three-dimensional structure of an intact monoclonal antibody for canine lymphoma. *Nature*, **360**, 369–372.

Haughn, G.W., Smith, J., Mazur, B. and Somerville, C. (1988) Transformation with a mutant *Arabidopsis* acetolactate synthase gene renders tobacco resistant to sulfonylurea herbicides. *Mol. Gen. Genet.*, **211**, 266–271.

Hayashi, M.N. and Hayashi, M. (1985) Cloned DNA sequences that determine mRNA stability of bacteriophage ϕX174 *in vivo* are functional. *Nucleic Acids Res.*, **13**, 5937–5948.

Hecht, A. and Grunstein, M. (1999) Mapping DNA interaction sites of chromosomal proteins using immunoprecipitation and polymerase chain reaction. *Methods Enzymol.*, **304**, 399–414.

Hedgpeth, J., Goodman, H.M. and Boyer, H.W. (1972) DNA nucleotide sequence restricted by the RI endonuclease. *Proc. Natl. Acad. Sci. USA*, **69**, 3448–3452.

Helmer-Citterich, M., Anceschi, M.M., Banner, D.W. and Cesareni, G. (1988) Control of ColE1 replication: low affinity specific binding of Rop (Rom) to RNAI and RNAII. *EMBO J.*, **7**, 557–566.

Hendrix, R.W. (1983) *Lambda II*. Cold Spring Harbor Laboratory, Cold Spring Harbor, NY.

Henning, K.A., Novotny, E.A., Compton, S.T., Guan, X.Y., Liu, P.P. and Ashlock, M.A. (1999) Human artificial chromosomes generated by modification of a yeast artificial chromosome containing both human alpha satellite and single-copy DNA sequences. *Proc. Natl. Acad. Sci. USA*, **96**, 592–597.

Herr, W., Sturm, R.A., Clerc, R.G., Corcoran, L.M., Baltimore, D., Sharp, P.A., Ingraham, H.A., Rosenfeld, M.G., Finney, M., Ruvkun, G. and Horvitz, H.R. (1988) The POU domain: a large conserved region in the mammalian pit-1, oct-1, oct-2, and Caenorhabditis elegans unc-86 gene products. *Genes Dev.*, **2**, 1513–1516.

Herrera, P.L. (2000) Adult insulin- and glucagon-producing cells differentiate from two independent cell lineages. *Development*, **127**, 2317–2322.

Hershey, A.D. (1947) Spontaneous mutations in bacterial viruses. *Cold Spring Harbor Symp. Quant. Biol.*, **11**, 67–77.

Hershey, A.D. and Chase, M. (1952) Independent functions of viral protein and nucleic acid in growth of bacteriophage. *J. Gen. Physiol.*, **36**, 39–56.

Hexham, J.M. (1998) Production of human Fab antibody fragments from phage display libraries. *Methods Mol. Biol.*, **80**, 461–474.

Hiei, Y., Ohta, S., Komari, T. and Kumashiro, T. (1994) Efficient transformation of rice (*Oryza sativa L.*) mediated by *Agrobacterium* and sequence analysis of the boundaries of the T-DNA. *Plant J.*, **6**, 271–282.

High, K.A. (2001) AAV-mediated gene transfer for hemophilia. *Ann. NY Acad. Sci.*, **953**, 64–74.

Higuchi, R., Krummel, B. and Saiki, R. (1988) A general method of *in vitro* preparation and specific mutagenesis of DNA fragments: study of protein and DNA interactions. *Nucleic Acids Res.*, **16**, 7351–7367.

Hinnebusch, A.G. (1997) Translational regulation of yeast GCN4. A window on factors that control initiator–tRNA binding to the ribosome. *J. Biol. Chem.*, **272**, 21 661–21 664.

Hirose, Y. and Manley, J.L. (1998) RNA polymerase II is an essential mRNA polyadenylation factor. *Nature*, **395**, 93–96.

Hirt, R.P., Poulain-Godefroy, O., Billotte, J., Kraehenbuhl, J.P. and Fasel, N. (1992) Highly inducible synthesis of heterologous proteins in epithelial cells carrying a glucocorticoid-responsive vector. *Gene*, **111**, 199–206.

Ho, S.Y. and Mittal, G.S. (1996) Electroporation of cell membranes: a review. *Crit. Rev. Biotechnol.*, **16**, 349–362.

Hoffman, A. and Roeder, R.G. (1991) Purification of his-tagged proteins in nondenaturing conditions suggests a convenient method for protein interaction studies. *Nucleic Acids Res.*, **19**, 6337–6338.

Holland, P.M., Abramson, R.D., Watson, R. and Gelfand, D.H. (1991) Detection of specific polymerase chain reaction product by utilizing the 5′–3′ exonuclease activity of *Thermus aquaticus* DNA polymerase. *Proc. Natl. Acad. Sci. USA*, **88**, 7276–7280.

Holley, R.W. (1966) The nucleotide sequence of a nucleic acid. *Sci. Am.*, **214**, 30–39.

Holliday, R. (1964) A mechanism for gene conversion in fungi. *Genet. Res.*, **5**, 282–304.

Holt, R.A. *et al.* (2002) The genome sequence of the malaria mosquito *Anopheles gambiae*. *Science*, **298**, 129–149.

Hooykaas, P.J. (1989) Transformation of plant cells *via Agrobacterium*. *Plant Mol. Biol.*, **13**, 327–336.

Hope, I.A. (2001) RNAi surges on: application to cultured mammalian cells. *Trends Genet.*, **17**, 440.

Horsch, R.B., Fry, J.E., Hoffmann, N.L., Eichholtz, D., Rogers, S.G. and Fraley, R.T. (1985) A simple and general method for transferring genes into plants. *Science*, **227**, 1229–1231.

Hosoda, F., Arai, Y., Kitamura, E., Inazawa, J., Fukushima, M., Tokino, T., Nakamura, Y., Jones, C., Kakazu, N., Abe, T. and Ohki, M. (1997) A complete *Not*I restriction map covering the entire long arm of human chromosome 11. *Genes Cells*, **2**, 345–357.

Hu, J.C., Kornacker, M.G. and Hochschild, A. (2000) *Escherichia coli* one- and two-hybrid systems for the analysis and identification of protein–protein interactions. *Methods*, **20**, 80–94.

Huang, J. and Schreiber, S.L. (1997) A yeast genetic system for selecting small molecule inhibitors of protein–protein interactions in nanodroplets. *Proc. Natl. Acad. Sci. USA*, **94**, 13 396–13 401.

Hudson, J.R., Jr., Dawson, E.P., Rushing, K.L., Jackson, C.H., Lockshon, D., Conover, D., Lanciault, C., Harris, J.R., Simmons, S.J., Rothstein, R. and Fields, S. (1997) The complete set of predicted genes from *Saccharomyces cerevisiae* in a readily usable form. *Genome Res.*, **7**, 1169–1173.

Hudson, T.J. *et al.* (1995) An STS-based map of the human genome. *Science*, **270**, 1945–1954.

Hug, H., Costas, M., Staeheli, P., Aebi, M. and Weissmann, C. (1988) Organization of the murine *Mx* gene and characterization of its interferon- and virus-inducible promoter. *Mol. Cell. Biol.*, **8**, 3065–3079.

Hull, R., Covey, S.N. and Maule, A.J. (1987) Structure and replication of caulimovirus genomes. *J. Cell. Sci. Suppl.*, **7**, 213–229.

Hull, R. and Shepherd, R.J. (1977) The structure of cauliflower mosaic virus genome. *Virology*, **79**, 216–230.

Hutvágner, G., McLachlan, J., Pasquinelli, A.E., Bálint, E., Tuschl, T. and Zamore, P.D. (2001) A cellular function for the RNA-interference enzyme Dicer in the maturation of the *let*-7 small temporal RNA. *Science*, **293**, 834–838.

Hutvágner, G. and Zamore, P.D. (2002) RNAi: nature abhors a double-strand. *Curr. Opin. Genet. Dev.*, **12**, 225–232.

Iamtham, S. and Day, A. (2000) Removal of antibiotic resistance genes from transgenic tobacco plastids. *Nat. Biotechnol.*, **18**, 1172–1176.

Ichikawa, H., Hosoda, F., Arai, Y., Shimizu, K., Ohira, M. and Ohki, M. (1993) A *Not*I restriction map of the entire long arm of human chromosome 21. *Nat. Genet.*, **4**, 361–366.

Innis, M.A., Gelfand, D.H. and Sninsky, J.J. (eds.). (1999) *PCR Applications: Protocols for Functional Genomics.* Academic, San Diego, CA.

Ioannou, P.A., Amemiya, C.T., Garnes, J., Kroisel, P.M., Shizuya, H., Chen, C., Batzer, M.A. and de Jong, P.J. (1994) A new bacteriophage P1-derived vector for the propagation of large human DNA fragments. *Nat. Genet.*, **6**, 84–89.

Ish-Horowicz, D. and Burke, J.F. (1981) Rapid and efficient cosmid cloning. *Nucleic Acids Res.*, **9**, 2989–2998.

Isono, K., McIninch, J.D. and Borodovsky, M. (1994) Characteristic features of the nucleotide sequences of yeast mitochondrial ribosomal protein genes as analyzed by computer program GeneMark. *DNA Res.*, **1**, 263–269.

Ivanova, N.B., Dimos, J.T., Schaniel, C., Hackney, J.A., Moore, K.A. and Lemischka, I.R. (2002) A stem cell molecular signature. *Science*, **298**, 601–604.

Jackson, D.A., Symons, R.H. and Berg, P. (1972) Biochemical method for inserting new genetic information into DNA of Simian Virus 40: circular SV40 DNA molecules containing lambda phage genes and the galactose operon of *Escherichia coli. Proc. Natl. Acad. Sci. USA*, **69**, 2904–2909.

Jacob, F. and Monod, J. (1961) Genetic regulatory mechanisms in the synthesis of proteins. *J. Mol. Biol.*, **3**, 318–356.

James, P., Halladay, J. and Craig, E.A. (1996) Genomic libraries and a host strain designed for highly efficient two-hybrid selection in yeast. *Genetics*, **144**, 1425–1436.

Janknecht, R., de Martynoff, G., Lou, J., Hipskind, R.A., Nordheim, A. and Stunnenberg, H.G. (1991) Rapid and efficient purification of native histidine-tagged protein expressed by recombinant vaccinia virus. *Proc. Natl. Acad. Sci. USA*, **88**, 8972–8976.

Jefferson, R.A., Burgess, S.M. and Hirsh, D. (1986) β-Glucuronidase from *Escherichia coli* as a gene-fusion marker. *Proc. Natl. Acad. Sci. USA*, **83**, 8447–8451.

Jen-Jacobson,L., Engler, L.E., Lesser, D.R., Kurpiewski, M.R., Yee, C. and McVerry, B. (1996) Structural adaptations in the interaction of *Eco*RI endonuclease with methylated GAATTC sites. *EMBO J.*, **15**, 2870–2882.

Johnsson, N. and Varshavsky, A. (1994) Split ubiquitin as a sensor of protein interactions *in vivo*. *Proc. Natl. Acad. Sci. USA*, **91**, 10 340–10 344.

Johnston, S.A., Salmeron, J.M. and Dincher, S.S. (1987) Interaction of positive and negative regulatory proteins in the galactose regulon of yeast. *Cell*, **50**, 143–146.

Johnston, S.A. and Tang, D.C. (1994) Gene gun transfection of animal cells and genetic immunization. *Methods Cell. Biol.*, **43 Pt A**, 353–365.

Joshi, L., Davis, T.R., Mattu, T.S., Rudd, P.M., Dwek, R.A., Shuler, M.L. and Wood, H.A. (2000) Influence of baculovirus-host cell interactions on complex N-linked glycosylation of a recombinant human protein. *Biotechnol. Prog.*, **16**, 650–656.

Juers, D.H., Jacobson, R.H., Wigley, D., Zhang, X.J., Huber, R.E., Tronrud, D.E. and Matthews, B.W. (2000) High resolution refinement of β-galactosidase in a new crystal form reveals multiple metal-binding sites and provides a structural basis for α-complementation. *Protein Sci.*, **9**, 1685–1699.

Kamada, K., Horiuchi, T., Ohsumi, K., Shimamoto, N. and Morikawa, K. (1996) Structure of a replication-terminator protein complexed with DNA. *Nature*, **383**, 598–603.

Kamath, R.S., Fraser, A.G., Dong, Y., Poulin, G., Durbin, R., Gotta, M., Kanapin, A., Le Bot, N., Moreno, S., Sohrmann, M., Welchman, D.P., Zipperlen, P. and Ahringer, J. (2003) Systematic functional analysis of the *Caenorhabditis elegan*s genome using RNAi. *Nature*, **421**, 231–237.

Karger, A.E., Harris, J.M. and Gesteland, R.F. (1991) Multiwavelength fluorescence detection for DNA sequencing using capillary electrophoresis. *Nucleic Acids Res.*, **19**, 4955–4962.

Karimi, M., Inze, D. and Depicker, A. (2002) GATEWAY™ vectors for *Agrobacterium*-mediated plant transformation. *Trends Plant. Sci.*, **7**, 193–195.

Katz, R.A. and Skalka, A.M. (1994) The retroviral enzymes. *Annu. Rev. Biochem.*, **63**, 133–173.

Keller, G.M. (1995) *In vitro* differentiation of embryonic stem cells. *Curr. Opin. Cell. Biol.*, **7**, 862–869.

Kellermann, O.K. and Ferenci, T. (1982) Maltose-binding protein from *Escherichia coli*. *Methods Enzymol.*, **90**, 459–463.

Kellogg, D.E., Rybalkin, I., Chen, S., Mukhamedova, N., Vlasik, T., Siebert, P.D. and Chenchik, A. (1994) TaqStart Antibody: 'hot start' PCR facilitated by a neutralizing monoclonal antibody directed against Taq DNA polymerase. *Biotechniques*, **16**, 1134–1137.

Kennedy, M., Firpo, M., Choi, K., Wall, C., Robertson, S., Kabrun, N. and Keller, G. (1997) A common precursor for primitive erythropoiesis and definitive haematopoiesis. *Nature*, **386**, 488–493.

Keohavong, P., Ling, L., Dias, C. and Thilly, W.G. (1993) Predominant mutations induced by the *Thermococcus litoralis*, vent DNA polymerase during DNA amplification *in vitro*. *PCR Methods Appl.*, **2**, 288–292.

Keohavong, P. and Thilly, W.G. (1989) Fidelity of DNA polymerases in DNA amplification. *Proc. Natl. Acad. Sci. USA*, **86**, 9253–9257.

Kilby, N.J., Snaith, M.R. and Murray, J.A. (1993) Site-specific recombinases: tools for genome engineering. *Trends Genet.*, **9**, 413–421.

Kim, K.K., Chamberlin, H.M., Morgan, D.O. and Kim, S.H. (1996a) Three-dimensional structure of human cyclin H, a positive regulator of the CDK-activating kinase. *Nat. Struct. Biol.*, **3**, 849–855.

Kim, S., Lin, H., Barr, E., Chu, L., Leiden, J.M. and Parmacek, M.S. (1997) Transcriptional targeting of replication-defective adenovirus transgene expression to smooth muscle cells *in vivo. J. Clin. Invest.*, **100**, 1006–1014.

Kim, U.J., Birren, B.W., Slepak, T., Mancino, V., Boysen, C., Kang, H.L., Simon, M.I. and Shizuya, H. (1996b) Construction and characterization of a human bacterial artificial chromosome library. *Genomics*, **34**, 213–218.

Knapp, J.E., Carroll, D., Lawson, J.E., Ernst, S.R., Reed, L.J. and Hackert, M.L. (2000) Expression, purification, and structural analysis of the trimeric form of the catalytic domain of the *Escherichia coli* dihydrolipoamide succinyltransferase. *Protein Sci.*, **9**, 37–48.

Knittel, T. and Picard, D. (1993) PCR with degenerate primers containing deoxyinosine fails with Pfu DNA polymerase. *PCR Methods Appl.*, **2**, 346–347.

Kodadek, T. (2002) Development of protein-detecting microarrays and related devices. *Trends Biochem. Sci.*, **27**, 295–300.

Koleske, A. and Young, R.A. (1994) An RNA polymerase II holoenzyme responsive to activators. *Nature*, **368**, 466–469.

Konisky, J. and Tokuda, H. (1979) Mode of action of colicins Ia, E1 and K. *Zentralbl. Bakteriol. [Orig. A]*, **244**, 105–120.

Kornberg, A. and Baker, T.A. (1992) *DNA Replication*. Freeman, New York.

Kotin, R.M., Siniscalco, M., Samulski, R.J., Zhu, X.D., Hunter, L., Laughlin, C.A., McLaughlin, S., Muzyczka, N., Rocchi, M. and Berns, K.I. (1990) Site-specific integration by adeno-associated virus. *Proc. Natl. Acad. Sci. USA*, **87**, 2211–2215.

Koutz, P., Davis, G.R., Stillman, C., Barringer, K., Cregg, J. and Thill, G. (1989) Structural comparison of the *Pichia pastoris* alcohol oxidase genes. *Yeast*, **5**, 167–177.

Kovall, R.A. and Matthews, B.W. (1999) Type II restriction endonucleases: structural, functional and evolutionary relationships. *Curr. Opin. Chem. Biol.*, **3**, 578–583.

Kozak, M. (1986) Point mutations define a sequence flanking the AUG initiator codon that modulates translation by eukaryotic ribosomes. *Cell*, **44**, 283–292.

Kramer, B., Kramer, W. and Fritz, H.J. (1984) Different base/base mismatches are corrected with different efficiencies by the methyl-directed DNA mismatch-repair system of *E. coli. Cell*, **38**, 879–887.

Kühn, R., Schwenk, F., Aguet, M. and Rajewsky, K. (1995) Inducible gene targeting in mice. *Science*, **269**, 1427–1429.

Kumar, S. and Fladung, M. (2001) Controlling transgene integration in plants. *Trends Plant Sci.*, **6**, 155–159.

Kunkel, T.A. (1985) Rapid and efficient site-specific mutagenesis without phenotypic selection. *Proc. Natl. Acad. Sci. USA*, **82**, 488–492.

Kunst, F. *et al.* (1997) The complete genome sequence of the gram-positive bacterium *Bacillus subtilis. Nature*, **390**, 249–256.

Lai, L., Kolber-Simonds, D., Park, K.W., Cheong, H.T., Greenstein, J.L., Im, G.S., Samuel, M., Bonk, A., Rieke, A., Day, B.N., Murphy, C.N., Carter, D.B., Hawley, R.J. and Prather, R.S. (2002) Production of α-1,3-galactosyltransferase knockout pigs by nuclear transfer cloning. *Science*, **295**, 1089–1092.

Lakso, M., Sauer, B., Mosinger, B., Jr., Lee, E.J., Manning, R.W., Yu, S.H., Mulder, K.L. and Westphal, H. (1992) Targeted oncogene activation by site-specific recombination in transgenic mice. *Proc. Natl. Acad. Sci. USA*, **89**, 6232–6236.

Lama, J. and Carrasco, L. (1992) Inducible expression of a toxic poliovirus membrane protein in *Escherichia coli*: comparative studies using different expression systems based on T7 promoters. *Biochem. Biophys. Res. Commun.*, **188**, 972–981.

Lander, E.S. *et al.* (2001) Initial sequencing and analysis of the human genome. *Nature*, **409**, 860–921.

Larschan, E. and Winston, F. (2001) The *S. cerevisiae* SAGA complex functions *in vivo* as a coactivator for transcriptional activation by Gal4. *Genes Dev.*, **15**, 1946–1956.

LaVallie, E.R., DiBlasio, E.A., Kovacic, S., Grant, K.L., Schendel, P.F. and McCoy, J.M. (1993a) A thioredoxin gene fusion expression system that circumvents inclusion body formation in the *E. coli* cytoplasm. *Bio/technology*, **11**, 187–193.

LaVallie, E.R., Rehemtulla, A., Racie, L.A., DiBlasio, E.A., Ferenz, C., Grant, K.L., Light, A. and McCoy, J.M. (1993b) Cloning and functional expression of a cDNA encoding the catalytic subunit of bovine enterokinase. *J. Biol. Chem.*, **268**, 23 311–23 317.

Lawlor, D.A., Dickel, C.D., Hauswirth, W.W. and Parham, P. (1991) Ancient HLA genes from 7,500-year-old archaeological remains. *Nature*, **349**, 785–788.

Lawyer, F.C., Stoffel, S., Saiki, R.K., Myambo, K., Drummond, R. and Gelfand, D.H. (1989) Isolation, characterization, and expression in *Escherichia coli* of the DNA polymerase gene from *Thermus aquaticus*. *J. Biol. Chem.*, **264**, 6427–6437.

Lazzari, G., Wrenzycki, C., Herrmann, D., Duchi, R., Kruip, T., Niemann, H. and Galli, C. (2002) Cellular and molecular deviations in bovine *in vitro*-produced embryos are related to the large offspring syndrome. *Biol. Reprod.*, **67**, 767–775.

Leanna, C.A. and Hannink, M. (1996) The reverse two-hybrid system: a genetic scheme for selection against specific protein/protein interactions. *Nucleic Acids Res.*, **24**, 3341–3347.

Lederberg, J. and Tatum, E.L. (1946) Gene recombination in *Escherichia coli*. *Nature*, **158**, 558.

Lee, J. and Goldfarb, A. (1991) *lac* repressor acts by modifying the initial transcribing complex so that it cannot leave the promoter. *Cell*, **66**, 793–798.

Lee, P., Morley, G., Huang, Q., Fischer, A., Seiler, S., Horner, J.W., Factor, S., Vaidya, D., Jalife, J. and Fishman, G.I. (1998) Conditional lineage ablation to model human diseases. *Proc. Natl. Acad. Sci. USA*, **95**, 11 371–11 376.

Lee, T.I. *et al.* (2002) Transcriptional regulatory networks in *Saccharomyces cerevisiae*. *Science*, **298**, 799–804.

Lee, T.I. and Young, R.A. (2000) Transcription of eukaryotic protein-coding genes. *Annu. Rev. Genet.*, **34**, 77–137.

Legrain, P. and Selig, L. (2000) Genome-wide protein interaction maps using two-hybrid systems. *FEBS Lett.*, **480**, 32–36.

Lesley, S.A. *et al.* (2002) Structural genomics of the *Thermotoga maritima* proteome implemented in a high-throughput structure determination pipeline. *Proc. Natl. Acad. Sci. USA*, **99**, 11 664–11 669.

Lester, S.C., LeVan, S.K., Steglich, C. and DeMars, R. (1980) Expression of human genes for adenine phosphoribosyltransferase and hypoxanthine–guanine phosphoribosyltransferase after genetic transformation of mouse cells with purified human DNA. *Somatic Cell Genet.*, **6**, 241–259.

Levene, P.A. and Simms, H.S. (1926) Nucleic acid structure as determined by electrometric titration data. *J. Biol. Chem.*, **70**, 327–341.

Lewandoski, M. (2001) Conditional control of gene expression in the mouse. *Nat. Rev. Genet.*, **2**, 743–755.

Li, H.H., Gyllensten, U.B., Cui, X.F., Saiki, R.K., Erlich, H.A. and Arnheim, N. (1988) Amplification and analysis of DNA sequences in single human sperm and diploid cells. *Nature*, **335**, 414–417.

Li, J.D., Carroll, J. and Ellar, D.J. (1991) Crystal structure of insecticidal δ-endotoxin from *Bacillus thuringiensis* at 2.5 Å resolution. *Nature*, **353**, 815–821.

Li, J.J. and Herskowitz, I. (1993) Isolation of ORC6, a component of the yeast origin recognition complex by a one-hybrid system. *Science*, **262**, 1870–1874.

Li, Q.B. and Guy, C.L. (1996) Prolonged final extension time increases cloning efficiency of PCR products. *Biotechniques*, **21**, 192–196.

Li, S., Crenshaw, E.B., III, Rawson, E.J., Simmons, D.M., Swanson, L.W. and Rosenfeld, M.G. (1990) Dwarf locus mutants lacking three pituitary cell types result from mutations in the POU-domain gene pit-1. *Nature*, **347**, 528–533.

Liang, P., Averboukh, L., Keyomarsi, K., Sager, R. and Pardee, A.B. (1992) Differential display and cloning of messenger RNAs from human breast cancer versus mammary epithelial cells. *Cancer Res.*, **52**, 6966–6968.

Liang, P. and Pardee, A.B. (1992) Differential display of eukaryotic messenger RNA by means of the polymerase chain reaction. *Science*, **257**, 967–971.

Licitra, E.J. and Liu, J.O. (1996) A three-hybrid system for detecting small ligand–protein receptor interactions. *Proc. Natl. Acad. Sci. USA*, **93**, 12 817–12 821.

Lillycrop, K.A., Budrahan, V.S., Lakin, N.D., Terrenghi, G., Wood, J.N., Polak, J.M. and Latchman, D.S. (1992) A novel POU family transcription factor is closely related to Brn-3 but has a distinct expression pattern in neuronal cells. *Nucleic Acids Res.*, **20**, 5093–5096.

Lin, H.J. and Chargaff, E. (1966) On the denaturation of deoxyribonucleic acid. *Biochim. Biophys. Acta*, **123**, 65–75.

Lin-Goerke, J.L., Robbins, D.J. and Burczak, J.D. (1997) PCR-based random mutagenesis using manganese and reduced dNTP concentration. *Biotechniques*, **23**, 409–412.

Lipardi, C., Wei, Q. and Paterson, B.M. (2001) RNAi as random degradative PCR: siRNA primers convert mRNA into dsRNAs that are degraded to generate new siRNAs. *Cell*, **107**, 297–307.

Lisser, S. and Margalit, H. (1993) Compilation of *E. coli* mRNA promoter sequences. *Nucleic Acids Res.*, **21**, 1507–1516.

Liu, H.Y. and Rashidbaigi, A. (1990) Comparison of various competent cell preparation methods for high efficiency DNA transformation. *Biotechniques*, **8**, 21–25.

Liu, M., Cao, D., Russell, R., Handschumacher, R.E. and Pizzorno, G. (1998) Expression, characterization, and detection of human uridine phosphorylase and identification of variant uridine phosphorolytic activity in selected human tumors. *Cancer Res.*, **58**, 5418–5424.

Liu, Y.H., Ma, L., Wu, L.Y., Luo, W., Kundu, R., Sangiorgi, F., Snead, M.L. and Maxson, R. (1994) Regulation of the *Msx2* homeobox gene during mouse embryogenesis: a transgene with 439 bp of 5' flanking sequence is expressed exclusively in the apical ectodermal ridge of the developing limb. *Mech. Dev.*, **48**, 187–197.

Lockhart, D.J. and Winzeler, E.A. (2000) Genomics, gene expression and DNA arrays. *Nature*, **405**, 827–836.

Logan, C., Khoo, W.K., Cado, D. and Joyner, A.L. (1993) Two enhancer regions in the mouse *En-2* locus direct expression to the mid/hindbrain region and mandibular myoblasts. *Development*, **117**, 905–916.

Logan, J. and Shenk, T. (1984) Adenovirus tripartite leader sequence enhances translation of mRNAs late after infection. *Proc. Natl. Acad. Sci. USA*, **81**, 3655–3659.

Lopez-Ferber, M., Sisk, W.P. and Possee, R.D. (1995) Baculovirus transfer vectors. *Methods Mol. Biol.*, **39**, 25–63.

Luciw, P.A., Potter, S.J., Steimer, K., Dina, D. and Levy, J.A. (1984) Molecular cloning of AIDS-associated retrovirus. *Nature*, **312**, 760–763.

Luckow, V.A., Lee, S.C., Barry, G.F. and Olins, P.O. (1993) Efficient generation of infectious recombinant baculoviruses by site-specific transposon-mediated insertion of foreign genes into a baculovirus genome propagated in *Escherichia coli*. *J. Virol.*, **67**, 4566–4579.

Lue, N.F., Chasman, D.I., Buchman, A.R. and Kornberg, R.D. (1987) Interaction of *GAL4* and *GAL80* gene regulatory proteins *in vitro*. *Mol. Cell. Biol.*, **7**, 3446–3451.

Luger, K., Mader, A.W., Richmond, R.K., Sargent, D.F. and Richmond, T.J. (1997) Crystal structure of the nucleosome core particle at 2.8 Å resolution. *Nature*, **389**, 251–260.

Luo, Y., Batalao, A., Zhou, H. and Zhu, L. (1997) Mammalian two-hybrid system: a complementary approach to the yeast two-hybrid system. *Biotechniques*, **22**, 350–352.

Luqmani, Y.A. and Lymboura, M. (1994) Subtraction hybridization cloning of RNA amplified from different cell populations microdissected from cryostat tissue sections. *Anal. Biochem.*, **222**, 102–109.

Ma, J. and Ptashne, M. (1987a) The carboxy-terminal 30 amino acids of GAL4 are recognized by GAL80. *Cell*, **50**, 137–142.

Ma, J. and Ptashne, M. (1987b) Deletion analysis of GAL4 defines two transcriptional activating segments. *Cell*, **48**, 847–853.

Ma, J. and Ptashne, M. (1987c) A new class of yeast transcriptional activators. *Cell*, **51**, 113–119.

Madhani, J., Movsowitz, H. and Kotler, M.N. (1993) Tissue plasminogen activator (t-PA). *Ther. Drug Monit.*, **15**, 546–551.

Mak, J. and Kleiman, L. (1997) Primer tRNAs for reverse transcription. *J. Virol.*, **71**, 8087–8095.

Makrides, S.C. (1996) Strategies for achieving high-level expression of genes in *Escherichia coli. Microbiol. Rev.*, **60**, 512–538.

Makrides, S.C. (1999) Components of vectors for gene transfer and expression in mammalian cells. *Protein Expr. Purif.*, **17**, 183–202.

Maliga, P. (2002) Engineering the plastid genome of higher plants. *Curr. Opin. Plant. Biol.*, **5**, 164–172.

Malik, S. and Roeder, R.G. (2000) Transcriptional regulation through mediator-like coactivators in yeast and metazoan cells. *Trends Biochem. Sci.*, **25**, 277–283.

Mandel, M. and Higa, A. (1970) Calcium-dependent bacteriophage DNA infection. *J. Mol. Biol.*, **53**, 159–162.

Maniatis, T., Hardison, R.C., Lacy, E., Lauer, J., O'Connell, C., Quon, D., Sim, G.K. and Efstratiadis, A. (1978) The isolation of structural genes from libraries of eucaryotic DNA. *Cell*, **15**, 687–701.

Maniatis, T. and Reed, R. (2002) An extensive network of coupling among gene expression machines. *Nature*, **416**, 499–506.

Mansour, S.L., Thomas, K.R. and Capecchi, M.R. (1988) Disruption of the proto-oncogene *int-2* in mouse embryo-derived stem cells: a general strategy for targeting mutations to non-selectable genes. *Nature*, **336**, 348–352.

Marchuk, D., Drumm, M., Saulino, A. and Collins, F.S. (1991) Construction of T-vectors, a rapid and general system for direct cloning of unmodified PCR products. *Nucleic Acids Res.*, **19**, 1154.

Marmorstein, R., Carey, M., Ptashne, M. and Harrison, S.C. (1992) DNA recognition by GAL4: structure of a protein–DNA complex. *Nature*, **356**, 408–414.

Marmorstein, R. and Harrison, S.C. (1994) Crystal structure of a PPR1–DNA complex: DNA recognition by proteins containing a Zn_2Cys_6 binuclear cluster. *Genes Dev.*, **8**, 2504–2512.

Marsolier, M.C., Prioleau, M.N. and Sentenac, A. (1997) A RNA polymerase III-based two-hybrid system to study RNA polymerase II transcriptional regulators. *J. Mol. Biol.*, **268**, 243–249.

Martegani, E., Vanoni, M., Zippel, R., Coccetti, P., Brambilla, R., Ferrari, C., Sturani, E. and Alberghina, L. (1992) Cloning by functional complementation of a mouse cDNA encoding a homologue of CDC25, a *Saccharomyces cerevisiae* RAS activator. *EMBO J.*, **11**, 2151–2157.

Martin, G.A., Kawaguchi, R., Lam, Y., DeGiovanni, A., Fukushima, M. and Mutter, W. (2001) High-yield, *in vitro* protein expression using a continuous-exchange, coupled transcription/translation system. *Biotechniques*, **31**, 948–953.

Martinez, J., Patkaniowska, A., Urlaub, H., Lührmann, R. and Tuschl, T. (2002) Single-stranded antisense siRNAs guide target RNA cleavage in RNAi. *Cell*, **110**, 563.

Maruyama, M., Kumagai, T., Matoba, Y., Hayashida, M., Fujii, T., Hata, Y. and Sugiyama, M. (2001) Crystal structures of the transposon Tn5-carried bleomycin resistance determinant uncomplexed and complexed with bleomycin. *J. Biol. Chem.*, **276**, 9992–9999.

Mascorrogallardo, J.O., Covarrubias, A.A. and Gaxiola, R. (1996) Construction of a *CUP1* promoter-based vector to modulate gene expression in *Saccharomyces cerevisiae. Gene*, **172**, 169–170.

Mathews, M.B. (1995) Structure, function, and evolution of adenovirus virus-associated RNAs. *Curr. Top. Microbiol. Immunol.*, **199**, 173–187.

Matsumura, M., Signor, G. and Matthews, B.W. (1989) Substantial increase of protein stability by multiple disulphide bonds. *Nature*, **342**, 291–293.

Matsushita, T., Elliger, S., Elliger, C., Podsakoff, G., Villarreal, L., Kurtzman, G.J., Iwaki, Y. and Colosi, P. (1998) Adeno-associated virus vectors can be efficiently produced without helper virus. *Gene Ther.*, **5**, 938–945.

Matzke, A.J. and Chilton, M.D. (1981) Site-specific insertion of genes into T-DNA of the *Agrobacterium* tumor-inducing plasmid: an approach to genetic engineering of higher plant cells. *J. Mol. Appl. Genet.*, **1**, 39–49.

Maundrell, K. (1990) *nmt1* of fission yeast. A highly transcribed gene completely repressed by thiamine. *J. Biol. Chem.*, **265**, 10 857–10 864.

Maundrell, K. (1993) Thiamine-repressible expression vectors pREP and pRIP for fission yeast. *Gene*, **123**, 127–130.

Maxam, A.M. and Gilbert, W. (1977) A new method for sequencing DNA. *Proc. Natl. Acad. Sci. USA*, **74**, 560–564.

Maxam, A.M. and Gilbert, W. (1980) Sequencing end-labeled DNA with base-specific chemical cleavages. *Methods Enzymol.*, **65**, 499–560.

Mayford, M., Bach, M.E., Huang, Y.Y., Wang, L., Hawkins, R.D. and Kandel, E.R. (1996) Control of memory formation through regulated expression of a CaMKII transgene. *Science*, **274**, 1678–1683.

Mayford, M., Wang, J., Kandel, E.R. and O'Dell, T.J. (1995) CaMKII regulates the frequency-response function of hippocampal synapses for the production of both LTD and LTP. *Cell*, **81**, 891–904.

McBride, K.E., Svab, Z., Schaaf, D.J., Hogan, P.S., Stalker, D.M. and Maliga, P. (1995) Amplification of a chimeric *Bacillus* gene in chloroplasts leads to an extraordinary level of an insecticidal protein in tobacco. *Bio/technology*, **13**, 362–365.

McBride, L.J. and Caruthers, M.H. (1983) An investigation of several deoxynucleoside phosphoramidites useful for synthesizing deoxyligonucleotides. *Tetrahedron Lett.*, **24**, 245–248.

McCracken, S., Fong, N., Yankulov, K., Ballantyne, S., Pan, G., Greenblatt, J., Patterson, S.D., Wickens, M. and Bentley, D.L. (1997) The C-terminal domain of RNA polymerase II couples mRNA processing to transcription. *Nature*, **385**, 357–361.

McCreery, T. (1997) Digoxigenin labeling. *Mol. Biotechnol.*, **7**, 121–124.

McPherson, M.J. and Møller, S.G. (eds.). (2000) *PCR Basics: from Background to Bench.* BIOS, Oxford.

Melchers, L.S. and Stuiver, M.H. (2000) Novel genes for disease-resistance breeding. *Curr. Opin. Plant. Biol.*, **3**, 147–152.

Melton, D.A. (1987) Translation of messenger RNA in injected frog oocytes. *Methods Enzymol.*, **152**, 288–296.

Mertz, J.E. and Davis, R.W. (1972) Cleavage of DNA by R1 restriction endonuclease generates cohesive ends. *Proc. Natl. Acad. Sci. USA*, **69**, 3370–3374.

Meselson, M. and Stahl, F.W. (1958) The replication of DNA in *Escherichia coli*. *Proc. Natl. Acad. Sci. USA*, **44**, 671–682.

Meselson, M. and Yuan, R. (1968) DNA restriction enzyme from *E. coli. Nature*, **217**, 1110–1114.

Meselson, M.S. and Radding, C.M. (1975) A general model for genetic recombination. *Proc. Natl. Acad. Sci. USA*, **72**, 358–361.

Messing, J., Gronenborn, B., Müller-Hill, B. and Hans Hopschneider, P. (1977) Filamentous coliphage M13 as a cloning vehicle: insertion of a *Hin*dII fragment of the *lac* regulatory region in M13 replicative form *in vitro. Proc. Natl. Acad. Sci. USA*, **74**, 3642–3646.

Mierendorf, R.C., Percy, C. and Young, R.A. (1987) Gene isolation by screening λgt11 libraries with antibodies. *Methods Enzymol.*, **152**, 458–469.

Mingeot-Leclercq, M.P., Glupczynski, Y. and Tulkens, P.M. (1999) Aminoglycosides: activity and resistance. *Antimicrob. Agents Chemother.*, **43**, 727–737.

Mocharla, H., Mocharla, R. and Hodes, M.E. (1990) Coupled reverse transcription–polymerase chain reaction (RT–PCR) as a sensitive and rapid method for isozyme genotyping. *Gene*, **93**, 271–275.

Moir, D.T. and Dumais, D.R. (1987) Glycosylation and secretion of human α-1-antitrypsin by yeast. *Gene*, **56**, 209–217.

Monteilhet, C., Perrin, A., Thierry, A., Colleaux, L. and Dujon, B. (1990) Purification and characterization of the *in vitro* activity of I-Sce I, a novel and highly specific endonuclease encoded by a group I intron. *Nucleic Acids Res.*, **18**, 1407–1413.

Montgomery, M.K., Xu, S. and Fire, A. (1998) RNA as a target of double-stranded RNA-mediated genetic interference in *Caenorhabditis elegans. Proc. Natl. Acad. Sci. USA*, **95**, 15 502–15 507.

Morgan, R.A. and Anderson, W.F. (1993) Human gene therapy. *Annu. Rev. Biochem.*, **62**, 191–217.

Morgan, T.H. (1910) Sex-limited inheritance in *Drosophila. Science*, **32**, 120–122.

Mouse Genome Sequencing Consortium. (2002) Initial sequencing and comparative analysis of the mouse genome. *Nature*, **420**, 540–562.

Müller-Hill, B., Crapo, L. and Gilbert, W. (1968) Mutants that make more *lac* repressor. *Proc. Natl. Acad. Sci. USA*, **59**, 1259–1264.

Mulligan, R.C. and Berg, P. (1981) Selection for animal cells that express the *Escherichia coli* gene coding for xanthine–guanine phosphoribosyltransferase. *Proc. Natl. Acad. Sci. USA*, **78**, 2072–2076.

Mullis, K.B. (1990) The unusual origin of the polymerase chain reaction. *Sci. Am.*, **262**, 56–61, 64–55.

Mullis, K.B. and Faloona, F.A. (1987) Specific synthesis of DNA *in vitro via* a polymerase-catalyzed chain reaction. *Methods Enzymol.*, **155**, 335–350.

Munro, S. and Pelham, H.R.B. (1987) A C-terminal signal prevents secretion of luminal ER proteins. *Cell*, **48**, 899–907.

Murakawa, G.J., Zaia, J.A., Spallone, P.A., Stephens, D.A., Kaplan, B.E., Wallace, R.B. and Rossi, J.J. (1988) Direct detection of HIV-1 RNA from AIDS and ARC patient samples. *DNA*, **7**, 287–295.

Murby, M., Uhlén, M. and Ståhl, S. (1996) Upstream strategies to minimize proteolytic degradation upon recombinant production in *Escherichia coli*. *Protein Expr. Purif.*, 7, 129–136.

Murray, V. (1989) Improved double-stranded DNA sequencing using the linear polymerase chain reaction. *Nucleic Acids Res.*, 17, 8889.

Myers, E.W., Sutton, G.G., Smith, H.O., Adams, M.D. and Venter, J.C. (2002) On the sequencing and assembly of the human genome. *Proc. Natl. Acad. Sci. USA*, 99, 4145–4146.

Myers, T.W. and Gelfand, D.H. (1991) Reverse transcription and DNA amplification by a *Thermus thermophilus* DNA polymerase. *Biochemistry*, 30, 7661–7666.

Mylin, L.M., Hofmann, K.J., Schultz, L.D. and Hopper, J.E. (1990) Regulated *GAL4* expression cassette providing controllable and high-level output from high-copy galactose promoters in yeast. *Methods Enzymol.*, 185, 297–308.

Nagai, K., Perutz, M.F. and Poyart, C. (1985) Oxygen binding properties of human mutant hemoglobins synthesized in *Escherichia coli*. *Proc. Natl. Acad. Sci. USA*, 82, 7252–7255.

Nagai, N., Hosokawa, M., Itohara, S., Adachi, E., Matsushita, T., Hosokawa, N. and Nagata, K. (2000) Embryonic lethality of molecular chaperone hsp47 knockout mice is associated with defects in collagen biosynthesis. *J. Cell. Biol.*, 150, 1499–1506.

Nagy, A., Rossant, J., Nagy, R., Abramow-Newerly, W. and Roder, J.C. (1993) Derivation of completely cell culture-derived mice from early-passage embryonic stem cells. *Proc. Natl. Acad. Sci. USA*, 90, 8424–8428.

Naim, H.Y. and Roth, M.G. (1994) SV40 virus expression vectors. *Methods Cell Biol.*, 43, 113–136.

Najjar, S.M. and Lewis, R.E. (1999) Persistent expression of foreign genes in cultured hepatocytes: expression vectors. *Gene*, 230, 41–45.

Nakamaye, K.L. and Eckstein, F. (1986) Inhibition of restriction endonuclease *Nci*I cleavage by phosphorothioate groups and its application to oligonucleotide-directed mutagenesis. *Nucleic Acids Res.*, 14, 9679–9698.

Nakano, T., Kodama, H. and Honjo, T. (1994) Generation of lymphohematopoietic cells from embryonic stem cells in culture. *Science*, 265, 1098–1101.

Negrutiu, I., Shillito, R.D., Potrykus, I., Biasini, G. and Sala, F. (1987) Hybrid genes in the analysis of transformation conditions I. Setting up a simple method for direct gene transfer in plant protoplasts. *Plant Mol. Biol.*, 8, 363–373.

Neville, M., Stutz, F., Lee, L., Davis, L.I. and Rosbash, M. (1997) The importin-β family member Crm1p bridges the interaction between Rev and the nuclear pore complex during nuclear export. *Curr. Biol.*, 7, 767–775.

Newton, C.R., Graham, A., Heptinstall, L.E., Powell, S.J., Summers, C., Kalsheker, N., Smith, J.C. and Markham, A.F. (1989) Analysis of any point mutation in DNA. The amplification refractory mutation system (ARMS). *Nucleic Acids Res.*, 17, 2503–2516.

Ngo, H., Tschudi, C., Gull, K. and Ullu, E. (1998) Double-stranded RNA induces mRNA degradation in *Trypanosoma brucei*. *Proc. Natl. Acad. Sci. USA*, 95, 14687–14692.

NIH/CEPH Collaborative Mapping Group. (1992) A comprehensive genetic linkage map of the human genome. *Science*, **258**, 67–86.

Nishikura, K. (2001) A short primer on RNAi: RNA-directed RNA polymerase acts as a key catalyst. *Cell*, **107**, 415–418.

Norrander, J., Kempe, T. and Messing, J. (1983) Construction of improved M13 vectors using oligodeoxynucleotide-directed mutagenesis. *Gene*, **26**, 101–106.

Norton, P.A. and Coffin, J.M. (1985) Bacterial β-galactosidase as a marker of Rous sarcoma virus gene expression and replication. *Mol. Cell. Biol.*, **5**, 281–290.

O'Neil, K.T. and Hoess, R.H. (1995) Phage display: protein engineering by directed evolution. *Curr. Opin. Struct. Biol.*, **5**, 443–449.

O'Reilly, D., Hanscombe, O. and O'Hare, P. (1997) A single serine residue at position 375 of VP16 is critical for complex assembly with Oct-1 and HCF and is a target of phosphorylation by casein kinase II. *EMBO J.*, **16**, 2420–2430.

Ogonuki, N., Inoue, K., Yamamoto, Y., Noguchi, Y., Tanemura, K., Suzuki, O., Nakayama, H., Doi, K., Ohtomo, Y., Satoh, M., Nishida, A. and Ogura, A. (2002) Early death of mice cloned from somatic cells. *Nat. Genet.*, **30**, 253–254.

Okazaki, R., Okazaki, T., Sakabe, K., Sugimoto, K. and Sugino, A. (1968) Mechanism of DNA chain growth. I. Possible discontinuity and unusual secondary structure of newly synthesized chains. *Proc. Natl. Acad. Sci. USA*, **59**, 598–605.

Old, R.W., Woodland, H.R., Ballantine, J.E., Aldridge, T.C., Newton, C.A., Bains, W.A. and Turner, P.C. (1982) Organization and expression of cloned histone gene clusters from *Xenopus laevis* and *X. borealis*. *Nucleic Acids Res.*, **10**, 7561–7580.

Olins, A.L. and Olins, D.E. (1974) Spheroid chromatin units (ν bodies). *Science*, **183**, 330–332.

Olivares, E.C., Hollis, R.P., Chalberg, T.W., Meuse, L., Kay, M.A. and Calos, M.P. (2002) Site-specific genomic integration produces therapeutic Factor IX levels in mice. *Nat. Biotechnol.*, **20**, 1124–1128.

Olsen, M.J., Stephens, D., Griffiths, D., Daugherty, P., Georgiou, G. and Iverson, B.L. (2000) Function-based isolation of novel enzymes from a large library. *Nat. Biotechnol.*, **18**, 1071–1074.

Orel, V. (1995) *Gregor Mendel: the First Geneticist*. Oxford University Press, Oxford.

Orrantia, E. and Chang, P.L. (1990) Intracellular distribution of DNA internalized through calcium phosphate precipitation. *Exp. Cell. Res.*, **190**, 170–174.

Osada, J. and Maeda, N. (1998) Preparation of knockout mice. *Methods Mol. Biol.*, **110**, 79–92.

Pajusola, K., Gruchala, M., Joch, H., Luscher, T.F., Yla-Herttuala, S. and Bueler, H. (2002) Cell-type-specific characteristics modulate the transduction efficiency of adeno-associated virus type 2 and restrain infection of endothelial cells. *J. Virol.*, **76**, 11 530–11 540.

Palmer, B.R. and Marinus, M.G. (1994) The *dam* and *dcm* strains of *Escherichia coli* – a review. *Gene*, **143**, 1–12.

Palmiter, R.D. and Brinster, R.L. (1986) Germ-line transformation of mice. *Annu. Rev. Genet.*, **20**, 465–499.

Palmiter, R.D., Brinster, R.L., Hammer, R.E., Trumbauer, M.E., Rosenfeld, M.G., Birnberg, N.C. and Evans, R.M. (1982) Dramatic growth of mice that develop from eggs microinjected with metallothionein–growth hormone fusion genes. *Nature*, **300**, 611–615.

Palmiter, R.D., Chen, H.Y. and Brinster, R.L. (1982) Differential regulation of metallothionein–thymidine kinase fusion genes in transgenic mice and their offspring. *Cell*, **29**, 701–710.

Panne, D., Muller, S.A., Wirtz, S., Engel, A. and Bickle, T.A. (2001) The *Mcr*BC restriction endonuclease assembles into a ring structure in the presence of G nucleotides. *EMBO J.*, **20**, 3210–3217.

Panne, D., Raleigh, E.A. and Bickle, T.A. (1999) The *Mcr*BC endonuclease translocates DNA in a reaction dependent on GTP hydrolysis. *J. Mol. Biol.*, **290**, 49–60.

Paszkowski, J., Shillito, R.D., Saul, M., Vandak, V., Hohn, T., Hohn, B. and Potrykus, I. (1984) Direct gene transfer to plants. *EMBO J.*, **3**, 2717–2722.

Pauling, L. and Corey, R.B. (1953) A proposed structure for the nucleic acids. *Proc. Natl. Acad. Sci. USA*, **39**, 84–97.

Pauling, L., Corey, R.B. and Branson, H.R. (1951) The structure of proteins: two hydrogen bonded helical configurations of the polypeptide chain. *Proc. Natl. Acad. Sci. USA*, **37**, 205–211.

Pawlowski, W.P. and Somers, D.A. (1998) Transgenic DNA integrated into the oat genome is frequently interspersed by host DNA. *Proc. Natl. Acad. Sci. USA*, **95**, 12 106–12 110.

Pazin, M.J. and Kadonaga, J.T. (1997) SWI2/SNF2 and related proteins: ATP-driven motors that disrupt protein–DNA interactions? *Cell*, **88**, 737–740.

Pennica, D., Holmes, W.E., Kohr, W.J., Harkins, R.N., Vehar, G.A., Ward, C.A., Bennett, W.F., Yelverton, E., Seeburg, P.H., Heyneker, H.L., Goeddel, D.V. and Collen, D. (1983) Cloning and expression of human tissue-type plasminogen activator cDNA in *E. coli*. *Nature*, **301**, 214–221.

Pérez-Martín, J. (1999) Chromatin and transcription in *Saccharomyces cerevisiae*. *FEMS Microbiol. Rev.*, **23**, 503–523.

Perou, C.M., Jeffrey, S.S., van de Rijn, M., Rees, C.A., Eisen, M.B., Ross, D.T., Pergamenschikov, A., Williams, C.F., Zhu, S.X., Lee, J.C., Lashkari, D., Shalon, D., Brown, P.O. and Botstein, D. (1999) Distinctive gene expression patterns in human mammary epithelial cells and breast cancers. *Proc. Natl. Acad. Sci. USA*, **96**, 9212–9217.

Perutz, M.F., Rossmann, M.G., Cullis, A.F., Muirhead, H., Will, G. and North, A.C.T. (1960) Structure of haemoglobin. A three-dimensional Fourier synthesis at 5.5Å resolution, obtained by X-ray analysis. *Nature*, **185**, 416–422.

Phelps, C.J. *et al.* (2002) Production of α1,3-galactosyltransferase-deficient pigs. *Science*, **299**, 411–414.

Pierce, J.C., Sauer, B. and Sternberg, N. (1992) A positive selection vector for cloning high molecular weight DNA by the bacteriophage P1 system: improved cloning efficacy. *Proc. Natl. Acad. Sci. USA*, **89**, 2056–2060.

Pingoud, A. and Jeltsch, A. (2001) Structure and function of type II restriction endonucleases. *Nucleic Acids Res.*, **29**, 3705–3727.

Plasterk, R.H. and Ketting, R.F. (2000) The silence of the genes. *Curr. Opin. Genet. Dev.*, **10**, 562–567.

Polejaeva, I.A., Chen, S.H., Vaught, T.D., Page, R.L., Mullins, J., Ball, S., Dai, Y., Boone, J., Walker, S., Ayares, D.L., Colman, A. and Campbell, K.H. (2000) Cloned pigs produced by nuclear transfer from adult somatic cells. *Nature*, **407**, 86–90.

Portner-Taliana, A., Russell, M., Froning, K.J., Budworth, P.R., Comiskey, J.D. and Hoeffler, J.P. (2000) *In vivo* selection of single-chain antibodies using a yeast two-hybrid system. *J. Immunol. Methods*, **238**, 161–172.

Possee, R.D. (1997) Baculoviruses as expression vectors. *Curr. Opin. Biotechnol.*, **8**, 569–572.

Postic, C., Shiota, M., Niswender, K.D., Jetton, T.L., Chen, Y., Moates, J.M., Shelton, K.D., Lindner, J., Cherrington, A.D. and Magnuson, M.A. (1999) Dual roles for glucokinase in glucose homeostasis as determined by liver and pancreatic beta cell-specific gene knock-outs using Cre recombinase. *J. Biol. Chem.*, **274**, 305–315.

Powell, P.A., Stark, D.M., Sanders, P.R. and Beachy, R.N. (1989) Protection against tobacco mosaic virus in transgenic plants that express tobacco mosaic virus antisense RNA. *Proc. Natl. Acad. Sci. USA*, **86**, 6949–6952.

Power, C.A. and Meyer, A. (2000) Generation of stable cell lines expressing chemokine receptors. *Methods Mol. Biol.*, **138**, 99–104.

Pruchnic, R., Cao, B., Peterson, Z.Q., Xiao, X., Li, J., Samulski, R.J., Epperly, M. and Huard, J. (2000) The use of adeno-associated virus to circumvent the maturation-dependent viral transduction of muscle fibers. *Hum. Gene Ther.*, **11**, 521–536.

Ptashne, M. (1992) *A Genetic Switch: Phage λ and Higher Organisms*. Cell Press–Blackwell, Cambridge, MA.

Pugliese, L., Coda, A., Malcovati, M. and Bolognesi, M. (1993) Three-dimensional structure of the tetragonal crystal form of egg-white avidin in its functional complex with biotin at 2.7 Å resolution. *J. Mol. Biol.*, **231**, 698–710.

Puig, O., Caspary, F., Rigaut, G., Rutz, B., Bouveret, E., Bragado-Nilsson, E., Wilm, M. and Seraphin, B. (2001) The tandem affinity purification (TAP) method: a general procedure of protein complex purification. *Methods*, **24**, 218–229.

Quinlan, R.A., Moir, R.D. and Stewart, M. (1989) Expression in *Escherichia coli* of fragments of glial fibrillary acidic protein: characterization, assembly properties and paracrystal formation. *J. Cell. Sci.*, **93**, 71–83.

Quiocho, F.A., Spurlino, J.C. and Rodseth, L.E. (1997) Extensive features of tight oligosaccharide binding revealed in high-resolution structures of the maltodextrin transport/chemosensory receptor. *Structure*, **5**, 997–1015.

Rackwitz, H.R., Zehetner, G., Murialdo, H., Delius, H., Chai, J.H., Poustka, A., Frischauf, A. and Lehrach, H. (1985) Analysis of cosmids using linearization by phage λ terminase. *Gene*, **40**, 259–266.

Raines, R.T., McCormick, M., Van Oosbree, T.R. and Mierendorf, R.C. (2000) The S. Tag fusion system for protein purification. *Methods Enzymol.*, **326**, 362–376.

Ramalho-Santos, M., Yoon, S., Matsuzaki, Y., Mulligan, R.C. and Melton, D.A. (2002) 'Stemness': transcriptional profiling of embryonic and adult stem cells. *Science*, **298**, 597–600.

Ramirez, A., Bravo, A., Jorcano, J.L. and Vidal, M. (1994) Sequences 5′ of the bovine keratin 5 gene direct tissue- and cell-type-specific expression of a *lacZ* gene in the adult and during development. *Differentiation*, **58**, 53–64.

Rathus, C., Bower, R. and Birch, R.G. (1993) Effects of promoter, intron and enhancer elements on transient gene expression in sugar-cane and carrot protoplasts. *Plant Mol. Biol.*, **23**, 613–618.

Ratner, L. *et al.* (1985) Complete nucleotide sequence of the AIDS virus, HTLV-III. *Nature*, **313**, 277–284.

Ratzkin, B. and Carbon, J. (1977) Functional expression of cloned yeast DNA in *Escherichia coli. Proc. Natl. Acad. Sci. USA*, **74**, 487–491.

Raymond, S. and Weintraub, L. (1959) Acrylamide gel as a supporting medium for zone electrophoresis. *Science*, **130**, 711.

Reece, R.J. and Platt, A. (1997) Signaling activation and repression of RNA polymerase II transcription in yeast. *BioEssays*, **19**, 1001–1010.

Reece, R.J. and Ptashne, M. (1993) Determinants of binding-site specificity among yeast C_6 zinc cluster proteins. *Science*, **261**, 909–911.

Reece, R.J., Rickles, R.J. and Ptashne, M. (1993) Overproduction and single-step purification of GAL4 fusion proteins from *Escherichia coli. Gene*, **126**, 105–107.

Ren, B., Robert, F., Wyrick, J.J., Aparicio, O., Jennings, E.G., Simon, I., Zeitlinger, J., Schreiber, J., Hannett, N., Kanin, E., Volkert, T.L., Wilson, C.J., Bell, S.P. and Young, R.A. (2000) Genome-wide location and function of DNA binding proteins. *Science*, **290**, 2306–2309.

Rigaut, G., Shevchenko, A., Rutz, B., Wilm, M., Mann, M. and Seraphin, B. (1999) A generic protein purification method for protein complex characterization and proteome exploration. *Nat. Biotechnol.*, **17**, 1030–1032.

Roberts, L. (1991) GRAIL seeks out genes buried in DNA sequence. *Science*, **254**, 805.

Roberts, R.J., *et al.* (2003) A nomenclature for restriction enzymes, DNA methyltransferases, homing endonucleases and their genes. *Nucleic Acids Res.*, **31**, 1805–1812.

Rome, J.J., Shayani, V., Newman, K.D., Farrell, S., Lee, S.W., Virmani, R. and Dichek, D.A. (1994) Adenoviral vector-mediated gene transfer into sheep arteries using a double-balloon catheter. *Hum. Gene Ther.*, **5**, 1249–1258.

Rommel, C., Leibiger, I.B., Leibiger, B. and Walther, R. (1994) CT-boxes are involved in control of the rat insulin II gene expression. *FEBS Lett.*, **345**, 17–22.

Rothstein, R.J. (1983) One-step gene disruption in yeast. *Methods Enzymol.*, **101**, 202–211.

Roy, P., Mikhailov, M. and Bishop, D.H. (1997) Baculovirus multigene expression vectors and their use for understanding the assembly process of architecturally complex virus particles. *Gene*, **190**, 119–129.

Roy, P.J., Stuart, J.M., Lund, J. and Kim, S.K. (2002) Chromosomal clustering of muscle-expressed genes in *Caenorhabditis elegans. Nature*, **418**, 975–979.

Rozen, S. and Skaletsky, H.J. (1998) Primer3. Code available at http://www-genome.wi.mit.edu/genome_software/other/primer3.html.

Ruf, S., Hermann, M., Berger, I.J., Carrer, H. and Bock, R. (2001) Stable genetic transformation of tomato plastids and expression of a foreign protein in fruit. *Nat. Biotechnol.*, **19**, 870–875.

Saiki, R.K., Scharf, S., Faloona, F., Mullis, K.B., Horn, G.T., Erlich, H.A. and Arnheim, N. (1985) Enzymatic amplification of β-globin genomic sequences and restriction site analysis for diagnosis of sickle cell anemia. *Science*, **230**, 1350–1354.

Sanger, F., Air, G.M., Barrell, B.G., Brown, N.L., Coulson, A.R., Fiddes, C.A., Hutchison, C.A., Slocombe, P.M. and Smith, M. (1977) Nucleotide sequence of bacteriophage ϕX174 DNA. *Nature*, **265**, 687–695.

Sanger, F. and Coulson, A.R. (1975) A rapid method for determining sequences in DNA by primed synthesis with DNA polymerase. *J. Mol. Biol.*, **94**, 441–448.

Sanger, F., Nicklen, S. and Coulson, A.R. (1977) DNA sequencing with chain-terminating inhibitors. *Proc. Natl. Acad. Sci. USA*, **74**, 5463–5467.

Sanlioglu, S., Monick, M.M., Luleci, G., Hunninghake, G.W. and Engelhardt, J.F. (2001) Rate limiting steps of AAV transduction and implications for human gene therapy. *Curr. Gene Ther.*, **1**, 137–147.

Sato, S., Nakamura, Y., Kaneko, T., Asamizu, E. and Tabata, S. (1999) Complete structure of the chloroplast genome of *Arabidopsis thaliana*. *DNA Res.*, **6**, 283–290.

Schaefer-Ridder, M., Wang, Y. and Hofschneider, P.H. (1982) Liposomes as gene carriers: efficient transformation of mouse L cells by thymidine kinase gene. *Science*, **215**, 166–168.

Schimke, R.T., Kaufman, R.J., Alt, F.W. and Kellems, R.F. (1978) Gene amplification and drug resistance in cultured murine cells. *Science*, **202**, 1051–1055.

Schlötterer, C. (2000) Evolutionary dynamics of microsatellite DNA. *Chromosoma*, **109**, 365–371.

Scholthof, H.B., Scholthof, K.-B.G. and Jackson, A.O. (1996) Plant virus gene vectors for transient expression of foreign proteins in plants. *Annu. Rev. Phytopathol.*, **34**, 299–323.

Schönbrunn, E., Eschenburg, S., Shuttleworth, W.A., Schloss, J.V., Amrhein, N., Evans, J.N. and Kabsch, W. (2001) Interaction of the herbicide glyphosate with its target enzyme 5-enolpyruvylshikimate 3-phosphate synthase in atomic detail. *Proc. Natl. Acad. Sci. USA*, **98**, 1376–1380.

Schubert, R., Panitz, R., Manteuffel, R., Nagy, I., Wobus, U. and Baumlein, H. (1994) Tissue-specific expression of an oat 12S seed globulin gene in developing tobacco seeds: differential mRNA and protein accumulation. *Plant Mol. Biol.*, **26**, 203–210.

Schueler, M.G., Higgins, A.W., Rudd, M.K., Gustashaw, K. and Willard, H.F. (2001) Genomic and genetic definition of a functional human centromere. *Science*, **294**, 109–115.

Schultz, L.D., Hofmann, K.J., Mylin, L.M., Montgomery, D.L., Ellis, R.W. and Hopper, J.E. (1987) Regulated overproduction of the *GAL4* gene product greatly increases expression from galactose-inducible promoters on multi-copy expression vectors in yeast. *Gene*, **61**, 123–133.

Schwartz, B., Ivanov, M.A., Pitard, B., Escriou, V., Rangara, R., Byk, G., Wils, P., Crouzet, J. and Scherman, D. (1999) Synthetic DNA-compacting peptides derived from human sequence enhance cationic lipid-mediated gene transfer *in vitro* and *in vivo*. *Gene Ther.*, **6**, 282–292.

Schwartz, D.C. and Cantor, C.R. (1984) Separation of yeast chromosome-sized DNAs by pulsed field gradient gel electrophoresis. *Cell*, **37**, 67–75.

Schwartz, T., Behlke, J., Lowenhaupt, K., Heinemann, U. and Rich, A. (2001) Structure of the DLM-1-Z-DNA complex reveals a conserved family of Z-DNA-binding proteins. *Nat. Struct. Biol.*, **8**, 761–765.

SenGupta, D.J., Zhang, B., Kraemer, B., Pochart, P., Fields, S. and Wickens, M. (1996) A three-hybrid system to detect RNA–protein interactions *in vivo*. *Proc. Natl. Acad. Sci. USA*, **93**, 8496–8501.

Serebriiskii, I., Khazak, V. and Golemis, E.A. (1999) A two-hybrid dual bait system to discriminate specificity of protein interactions. *J. Biol. Chem.*, **274**, 17080–17087.

Sgaramella, V. and Ehrlich, S.D. (1978) Use of the T4 polynucleotide ligase in the joining of flush-ended DNA segments generated by restriction endonucleases. *Eur. J. Biochem.*, **86**, 531–537.

Shah, D.M., Horsch, R.B., H.J., K., Kishore, G.M., Winter, J.A., Tumer, N.E., Hironaka, C.M., Sanders, P.R., Gasser, C.S., Aykent, S., Siegel, N.R., Rogers, S.G. and Fraley, R.T. (1986) Engineering herbicide tolerance in transgenic plants. *Science*, **233**, 478–481.

Shalon, D., Smith, S.J. and Brown, P.O. (1996) A DNA microarray system for analyzing complex DNA samples using two-color fluorescent probe hybridization. *Genome Res.*, **6**, 639–645.

Sharp, P.A. (1994) Split genes and RNA splicing. *Cell*, **77**, 805–815.

Shi, H., Djikeng, A., Mark, T., Wirtz, E., Tschudi, C. and Ullu, E. (2000) Genetic interference in *Trypanosoma brucei* by heritable and inducible double-stranded RNA. *RNA*, **6**, 1069–1076.

Shi, Q., Wang, Y. and Worton, R. (1997) Modulation of the specificity and activity of a cellular promoter in an adenoviral vector. *Hum. Gene Ther.*, **8**, 403–410.

Shiels, P.G., Kind, A.J., Campbell, K.H., Waddington, D., Wilmut, I., Colman, A. and Schnieke, A.E. (1999) Analysis of telomere lengths in cloned sheep. *Nature*, **399**, 316–317.

Shillito, R.D., Saul, M.W., Paszkowski, J., Müller, M. and Potrykus, I. (1985) High efficiency direct gene transfer to plants. *Bio/technology*, **3**, 1099–1103.

Shin, T., Kraemer, D., Pryor, J., Liu, L., Rugila, J., Howe, L., Buck, S., Murphy, K., Lyons, L. and Westhusin, M. (2002) A cat cloned by nuclear transplantation. *Nature*, **415**, 859.

Shine, J. and Dalgarno, L. (1975) Determinant of cistron specificity in bacterial ribosomes. *Nature*, **254**, 34–38.

Shizuya, H., Birren, B., Kim, U.J., Mancino, V., Slepak, T., Tachiiri, Y. and Simon, M. (1992) Cloning and stable maintenance of 300-kilobase-pair fragments of human DNA in *Escherichia coli* using an F-factor-based vector. *Proc. Natl. Acad. Sci. USA*, **89**, 8794–8797.

Shoemaker, D.D. and Linsley, P.S. (2002) Recent developments in DNA microarrays. *Curr. Opin. Microbiol.*, **5**, 334–337.

Short, J.M., Fernandez, J.M., Sorge, J.A. and Huse, W.D. (1988) λZAP: a bacteriophage lambda expression vector with *in vivo* excision properties. *Nucleic Acids Res.*, **16**, 7583–7600.

Singh, H., LeBowitz, J.H., Baldwin, A.S., Jr. and Sharp, P.A. (1988) Molecular cloning of an enhancer binding protein: isolation by screening of an expression library with a recognition site DNA. *Cell*, **52**, 415–423.

Slater, G.W., Mayer, P. and Drouin, G. (1996) Migration of DNA through gels. *Methods Enzymol.*, **270**, 272–295.

Slatko, B.E. (1996) Thermal cycle dideoxy DNA sequencing. *Mol. Biotechnol.*, **6**, 311–322.

Smith, C.J.S., Watson, C.F., Ray, J., Bird, C.R., Morris, P.C., Schuch, W. and Grierson, D. (1988) Antisense RNA inhibition of polygalacturonasegene expression in transgenic tomatoes. *Nature*, **334**, 724–726.

Smith, D.R., Smyth, A.P. and Moir, D.T. (1990) Amplification of large artificial chromosomes. *Proc. Natl. Acad. Sci. USA*, **87**, 8242–8246.

Smith, G.E., Summers, M.D. and Fraser, M.J. (1983) Production of human β-interferon in insect cells infected with a baculovirus expression vector. *Mol. Cell. Biol.*, **3**, 2156–2165.

Smith, G.P. (1985) Filamentous fusion phage: novel expression vectors that display cloned antigens on the virion surface. *Science*, **228**, 1315–1317.

Smith, L.C. and Wilmut, I. (1989) Influence of nuclear and cytoplasmic activity on the development *in vivo* of sheep embryos after nuclear transplantation. *Biol. Reprod.*, **40**, 1027–1035.

Sommerfelt, M.A. (1999) Retrovirus receptors. *J. Gen. Virol.*, **80**, 3049–3064.

Sorek, R. and Amitai, M. (2001) Piecing together the significance of splicing. *Nat. Biotechnol.*, **19**, 196.

Sorlie, T. *et al.* (2001) Gene expression patterns of breast carcinomas distinguish tumor subclasses with clinical implications. *Proc. Natl. Acad. Sci. USA*, **98**, 10 869–10 874.

Southern, E.M. (1975) Detection of specific sequences among DNA fragments separated by gel electrophoresis. *J. Mol. Biol.*, **98**, 503–517.

St John, T.P. and Davis, R.W. (1981) The organization and transcription of the galactose gene cluster of *Saccharomyces*. *J. Mol. Biol.*, **152**, 285–315.

St-Onge, L., Furth, P.A. and Gruss, P. (1996) Temporal control of the Cre recombinase in transgenic mice by a tetracycline responsive promoter. *Nucleic Acids Res.*, **24**, 3875–3877.

Stalker, D.M., Hiatt, W.R. and Comai, L. (1985) A single amino acid substitution in the enzyme 5-enolpyruvylshikimate-3-phosphate synthase confers resistance to the herbicide glyphosate. *J. Biol. Chem.*, **260**, 4724–4728.

Stanley, J. (1993) Geminiviruses: plant viral vectors. *Curr. Opin. Genet. Dev.*, **3**, 91–96.

Stark, M.J.R. (1987) Multicopy expression vectors carrying the *lac* repressor gene for regulated high-level expression of genes in *Escherichia coli*. *Gene*, **51**, 255–267.

Statham, S. and Morgan, R.A. (1999) Gene therapy clinical trials for HIV. *Curr. Opin. Mol. Ther.*, **1**, 430–436.

Staub, J.M., Garcia, B., Graves, J., Hajdukiewicz, P.T., Hunter, P., Nehra, N., Paradkar, V., Schlittler, M., Carroll, J.A., Spatola, L., Ward, D., Ye, G. and Russell, D.A. (2000) High-yield production of a human therapeutic protein in tobacco chloroplasts. *Nat. Biotechnol.*, **18**, 333–338.

Staub, J.M. and Maliga, P. (1992) Long regions of homologous DNA are incorporated into the tobacco plastid genome by transformation. *Plant Cell*, **4**, 39–45.

Sternberg, N., Hamilton, D., Austin, S., Yarmolinsky, M. and Hoess, R. (1981) Site-specific recombination and its role in the life cycle of bacteriophage P1. *Cold Spring Harbor Symp. Quant. Biol.*, **45**, 297–309.

Stevens, A. (1960) Incorporation of the adenine ribonucleotide into RNA by cell fractions from *E. coli* B. *Biochem. Biophys. Res. Commun.*, **3**, 92–96.

Stewart, F.J., Panne, D., Bickle, T.A. and Raleigh, E.A. (2000) Methyl-specific DNA binding by *Mcr*BC, a modification-dependent restriction enzyme. *J. Mol. Biol.*, **298**, 611–622.

Stirpe, F., Barbieri, L., Battelli, M.G., Soria, M. and Lappi, D.A. (1992) Ribosome-inactivating proteins from plants: present status and future prospects. *Bio/technology*, **10**, 405–412.

Strahl-Bolsinger, S., Hecht, A., Luo, K. and Grunstein, M. (1997) *SIR2* and *SIR4* interactions differ in core and extended telomeric heterochromatin in yeast. *Genes Dev.*, **11**, 83–93.

Strange, R.C., Jones, P.W. and Fryer, A.A. (2000) Glutathione S-transferase: genetics and role in toxicology. *Toxicol. Lett.*, **112/113**, 357–363.

Studier, F.W. and Moffatt, B.A. (1986) Use of bacteriophage T7 RNA polymerase to direct selective high-level expression of cloned genes. *J. Mol. Biol.*, **189**, 113–130.

Studier, F.W., Rosenberg, A.H., Dunn, J.J. and Dubendorff, J.W. (1990) Use of T7 RNA polymerase to direct expression of cloned genes. *Methods Enzymol.*, **185**, 60–89.

Studitsky, V.M., Clark, D.J. and Felsenfeld, G. (1995) Overcoming a nucleosomal barrier to transcription. *Cell*, **83**, 19–27.

Sturtevant, A.H. (1913) The linear arrangement of six sex-linked factors in *Drosophila*, as shown by their mode of association. *J. Exp. Biol.*, **14**, 43–59.

Stutz, F., Neville, M. and Rosbash, M. (1995) Identification of a novel nuclear pore-associated protein as a functional target of the HIV-1 Rev protein in yeast. *Cell*, **82**, 495–506.

Sugamura, K., Asao, H., Kondo, M., Tanaka, N., Ishii, N., Ohbo, K., Nakamura, M. and Takeshita, T. (1996) The interleukin-2 receptor γ chain: its role in the multiple cytokine receptor complexes and T cell development in XSCID. *Annu. Rev. Immunol.*, **14**, 179–205.

Sukharev, S.I., Klenchin, V.A., Serov, S.M., Chernomordik, L.V. and Chizmadzhev, Yu.A. (1992) Electroporation and electrophoretic DNA transfer into cells. The effect of DNA interaction with electropores. *Biophys. J.*, **63**, 1320–1327.

Sulston, J.E. and Horvitz, H.R. (1977) Post-embryonic cell lineages of the nematode, *Caenorhabditis elegans*. *Dev. Biol.*, **56**, 110–156.

Suzuki, K., Hattori, Y., Uraji, M., Ohta, N., Iwata, K., Murata, K., Kato, A. and Yoshida, K. (2000) Complete nucleotide sequence of a plant tumor-inducing Ti plasmid. *Gene*, **242**, 331–336.

Svab, Z., Hajdukiewicz, P. and Maliga, P. (1990) Stable transformation of plastids in higher plants. *Proc. Natl. Acad. Sci. USA*, **87**, 8526–8530.

Svejstrup, J.Q., Li, Y., Fellows, J., Gnatt, A., Bjorklund, S. and Kornberg, R.D. (1997) Evidence for a mediator cycle at the initiation of transcription. *Proc. Natl. Acad. Sci. USA*, **94**, 6075–6078.

Swaminathan, K., Flynn, P.J., Reece, R.J. and Marmorstein, R. (1997) Crystal structure of the PUT3–DNA complex reveals a novel mechanism for DNA recognition by a protein containing a Zn_2Cys_6 binuclear cluster. *Nature Struct. Biol.*, **4**, 751–759.

Szybalska, E.H. and Szybalski, W. (1962) Genetics of human cell lines, IV. DNA-mediated heritable transformation of a biochemical trait. *Proc. Natl. Acad. Sci. USA*, **48**, 2026–2034.

Taanman, J.W., van der Veen, A.Y., Schrage, C., de Vries, H. and Buys, C.H. (1991) Assignment of the gene coding for human cytochrome c oxidase subunit VIb to chromosome 19, band q13.1, by fluorescence *in situ* hybridisation. *Hum. Genet.*, **87**, 325–327.

Takahashi, M., Degenkolb, J. and Hillen, W. (1991) Determination of the equilibrium association constant between Tet repressor and tetracycline at limiting Mg^{2+} concentrations: a generally applicable method for effector-dependent high-affinity complexes. *Anal. Biochem.*, **199**, 197–202.

Takumi, T. (1997) Use of PCR for cDNA library screening. *Methods Mol. Biol.*, **67**, 339–344.

Tamashiro, K.L., Wakayama, T., Akutsu, H., Yamazaki, Y., Lachey, J.L., Wortman, M.D., Seeley, R.J., D'Alessio, D.A., Woods, S.C., Yanagimachi, R. and Sakai, R.R. (2002) Cloned mice have an obese phenotype not transmitted to their offspring. *Nat. Med.*, **8**, 262–267.

Tassabehji, M., Read, A.P., Newton, V.E., Harris, R., Balling, R., Gruss, P. and Strachan, T. (1992) Waardenburg's syndrome patients have mutations in the human homologue of the Pax-3 paired box gene. *Nature*, **355**, 635–636.

Taylor, I.C.A., Workman, J.L., Schuetz, T.J. and Kingston, R.E. (1991) Facilitated binding of GAL4 and heat shock factor to nucleosome templates: Differential function of DNA-binding domains. *Genes Dev.*, **5**, 1285–1298.

Taylor, J.H., Woods, P.S. and Hughes, W.C. (1957) The organization and duplication of chromosomes revealed by autoradiographic studies using tritium-labeled thymidine. *Proc. Natl. Acad. Sci. USA*, **48**, 122–128.

Taylor, J.W., Ott, J. and Eckstein, F. (1985) The rapid generation of oligonucleotide-directed mutations at high frequency using phosphorothioate-modified DNA. *Nucleic Acids Res.*, **13**, 8765–8785.

Teichler Zallen, D. (2000) US gene therapy in crisis. *Trends Genet.*, **16**, 272–275.

Temin, H.M. and Mizutani, S. (1970) RNA-dependent DNA polymerase in virions of Rous sarcoma virus. *Nature*, **226**, 1211–1213.

The *Arabidopsis* Genome Initiative. (2000) Analysis of the genome sequence of the flowering plant *Arabidopsis thaliana*. *Nature*, **408**, 796–815.

The *C. elegans* Sequencing Consortium. (1998) Genome sequence of the nematode *C. elegans*: a platform for investigating biology. *Science*, **282**, 2012–2018.

Thomas, K.R. and Capecchi, M.R. (1987) Site-directed mutagenesis by gene targeting in mouse embryo-derived stem cells. *Cell*, **51**, 503–512.

Thomas, P.S. (1980) Hybridization of denatured RNA and small DNA fragments transferred to nitrocellulose. *Proc. Natl. Acad. Sci. USA*, **77**, 5201–5205.

Thomas, R. (1993) The denaturation of DNA. *Gene*, **135**, 77–79.

Threadgill, D.W., Dlugosz, A.A., Hansen, L.A., Tennenbaum, T., Lichti, U., Yee, D., LaMantia, C., Mourton, T., Herrup, K., Harris, R.C., Barnard, J.A., Yuspa, S.H., Coffey, R.J. and Magnuson, T. (1995) Targeted disruption of mouse EGF receptor: effect of genetic background on mutant phenotype. *Science*, **269**, 230–234.

Timmons, L., Court, D.L. and Fire, A. (2001) Ingestion of bacterially expressed dsR-NAs can produce specific and potent genetic interference in *Caenorhabditis elegans*. *Gene*, **263**, 103–112.

Timson, D.J., Singleton, M.R. and Wigley, D.B. (2000) DNA ligases in the repair and replication of DNA. *Mutat. Res.*, **460**, 301–318.

Tobias, J.W., Shrader, T.E., Rocap, G. and Varshavsky, A. (1991) The N-end rule in bacteria. *Science*, **254**, 1374–1377.

Todd, R.B. and Andrianopoulos, A. (1997) Evolution of a fungal regulatory gene family: the $Zn(II)_2Cys_6$ binuclear cluster DNA binding motif. *Fungal Genet. Biol.*, **21**, 388–405.

Touraev, A., Stöger, E., Voronin, V. and Heberle-Bors, E. (1997) Plant male germ line transformation. *Plant J.*, **12**, 949–956.

Triezenberg, S.J. (1995) Structure and function of transcriptional activation domains. *Curr. Opin. Gen. Dev.*, **5**, 190–196.

Tschopp, J.F., Brust, P.F., Cregg, J.M., Stillman, C.A. and Gingeras, T.R. (1987) Expression of the *lacZ* gene from two methanol-regulated promoters in *Pichia pastoris*. *Nucleic Acids Res.*, **15**, 3859–3876.

Turner, R. and Foster, G.D. (1998) *In vitro* transcription and translation. *Methods Mol. Biol.*, **81**, 293–299.

Uetz, P. *et al.* (2000) A comprehensive analysis of protein–protein interactions in *Saccharomyces cerevisiae*. *Nature*, **403**, 623–627.

Ullmann, A. (1992) Complementation in β-galactosidase: from protein structure to genetic engineering. *Bioessays*, **14**, 201–205.

Van Larebeke, N., Engler, G., Holsters, M., Van den Elsacker, S., Zaenen, I., Schilper-oort, R.A. and Schell, J. (1974) Large plasmid in *Agrobacterium tumefaciens* essential for crown gall-inducing ability. *Nature*, **252**, 169–170.

Vara, J.A., Portela, A., Ortin, J. and Jimenez, A. (1986) Expression in mammalian cells of a gene from *Streptomyces alboniger* conferring puromycin resistance. *Nucleic Acids Res.*, **14**, 4617–4624.

Varshavsky, A. (1992) The N-end rule. *Cell*, **69**, 725–735.

Vashee, S., Simancek, P., Challberg, M.D. and Kelly, T.J. (2001) Assembly of the human origin recognition complex. *J. Biol. Chem.*, **276**, 26 666–26 673.

Vasquez, J.R., Evnin, L.B., Higaki, J.N. and Craik, C.S. (1989) An expression system for trypsin. *J. Cell. Biochem.*, **39**, 265–276.

Vassalli, J.D. and Saurat, J.H. (1996) Cuts and scrapes? Plasmin heals! *Nat. Med.*, **2**, 284–285.

Venter, J.C. *et al.* (2001) The sequence of the human genome. *Science*, **291**, 1304–1351.

Vieira, J. and Messing, J. (1982) The pUC plasmids, an M13mp7-derived system for insertion mutagenesis and sequencing with synthetic universal primers. *Gene*, **19**, 259–268.

Vinograd, J., Lebowitz, J., Radloff, R., Watson, R. and Laipis, P. (1965) The twisted circular form of polyoma viral DNA. *Proc. Natl. Acad. Sci. USA*, **53**, 1104–1111.

Vinson, C.R., LaMarco, K.L., Johnson, P.F., Landschulz, W.H. and McKnight, S.L. (1988) *In situ* detection of sequence-specific DNA binding activity specified by a recombinant bacteriophage. *Genes Dev.*, **2**, 801–806.

Voinnet, O. (2001) RNA silencing as a plant immune system against viruses. *Trends Genet.*, **17**, 449–459.

von Freiesleben, U., Krekling, M.A., Hansen, F.G. and Lobner-Olesen, A. (2000) The eclipse period of *Escherichia coli. EMBO J.*, **19**, 6240–6248.

Wach, A., Brachat, A., Pöhlmann, R. and Philippsen, P. (1994) New heterologous modules for classical or PCR-based gene disruptions in Saccharomyces cerevisiae. *Yeast*, **10**, 1793–1808.

Wadhwa, P.D., Zielske, S.P., Roth, J.C., Ballas, C.B., Bowman, J.E. and Gerson, S.L. (2002) Cancer gene therapy: scientific basis. *Annu. Rev. Med.*, **53**, 437–452.

Wakayama, T., Perry, A.C., Zuccotti, M., Johnson, K.R. and Yanagimachi, R. (1998) Full-term development of mice from enucleated oocytes injected with cumulus cell nuclei. *Nature*, **394**, 369–374.

Walker, J., Crowley, P., Moreman, A.D. and Barrett, J. (1993) Biochemical properties of cloned glutathione S-transferases from *Schistosoma mansoni* and *Schistosoma japonicum. Mol. Biochem. Parasitol.*, **61**, 255–264.

Walker, P.A., Leong, L.E., Ng, P.W., Tan, S.H., Waller, S., Murphy, D. and Porter, A.G. (1994) Efficient and rapid affinity purification of proteins using recombinant fusion proteases. *Bio/technology*, **12**, 601–605.

Wallace, R.B., Schold, M., Johnson, M.J., Dembek, P. and Itakura, K. (1981) Oligonucleotide directed mutagenesis of the human β-globin gene: a general method for producing specific point mutations in cloned DNA. *Nucleic Acids Res.*, **9**, 3647–3656.

Wallace, R.B., Shaffer, J., Murphy, R.F., Bonner, J., Hirose, T. and Itakura, K. (1979) Hybridization of synthetic oligodeoxyribonucleotides to ΦX174 DNA: The effect of single base pair mismatch. *Nucleic Acids Res.*, **6**, 3543–3557.

Wang, A., Deems, R.A. and Dennis, E.A. (1997) Cloning, expression, and catalytic mechanism of murine lysophospholipase I. *J. Biol. Chem.*, **272**, 12 723–12 729.

Wang, M.M. and Reed, R.R. (1993) Molecular cloning of the olfactory neuronal transcription factor Olf-1 by genetic selection in yeast. *Nature*, **364**, 121–126.

Wang, R.F., Cao, W.W. and Johnson, M.G. (1992) A simplified, single tube, single buffer system for RNA-PCR. *Biotechniques*, **12**, 702–704.

Wang, W. and Malcolm, B.A. (1999) Two-stage PCR protocol allowing introduction of multiple mutations, deletions and insertions using QuikChange site-directed mutagenesis. *Biotechniques*, **26**, 680–682.

Watanabe, T., Suzuki, K., Oyanagi, W., Ohnishi, K. and Tanaka, H. (1990) Gene cloning of chitinase A1 from *Bacillus circulans* WL-12 revealed its evolutionary relationship to *Serratia* chitinase and to the type III homology units of fibronectin. *J. Biol. Chem.*, **265**, 15 659–15 665.

Waterston, R.H., Lander, E.S. and Sulston, J.E. (2002) On the sequencing of the human genome. *Proc. Natl. Acad. Sci. USA*, **99**, 3712–3716.

Watson, J.D. and Crick, F.C.H. (1953a) Molecular structure of nucleic acids. A structure for deoxyribose nucleic acid. *Nature*, **171**, 737–738.

Watson, J.D. and Crick, F.H.C. (1953b) Genetical implications of the structure of deoxyribonucleic acid. *Nature*, **171**, 964–967.

Watson, R. and Yamazaki, H. (1973) Alkali hydrolysis of RNA of *Escherichia coli* deposited on filter paper disks. *Anal. Biochem.*, **51**, 312–314.

Weinmann, A.S., Yan, P.S., Oberley, M.J., Huang, T.H. and Farnham, P.J. (2002) Isolating human transcription factor targets by coupling chromatin immunoprecipitation and CpG island microarray analysis. *Genes Dev.*, **16**, 235–244.

Weiss, B. and Richardson, C.C. (1967) Enzymatic breakage and joining of deoxyribonucleic acid, I. Repair of single-strand breaks in DNA by an enzyme system from *Escherichia coli* infected with T4 bacteriophage. *Proc. Natl. Acad. Sci. USA*, **57**, 1021–1028.

Weiss, S.B. (1960) Enzymatic incorporation of ribonucleotide triphosphates into the interpolynucleotide linkages of ribonucleic acid. *Proc. Natl. Acad. Sci. USA*, **46**, 1020–1030.

Wells, J.A., Vasser, M. and Powers, D.B. (1985) Cassette mutagenesis: an efficient method for generation of multiple mutations at defined sites. *Gene*, **34**, 315–323.

White, J., Maltais, L. and Nebert, D. (1998) Networking nomenclature. *Nat. Genet.*, **18**, 209.

Wigler, M., Perucho, M., Kurtz, D., Dana, S., Pellicer, A., Axel, R. and Silverstein, S. (1980) Transformation of mammalian cells with an amplifiable dominant-acting gene. *Proc. Natl. Acad. Sci. USA*, **77**, 3567–3570.

Wigler, M., Silverstein, S., Lee, L.S., Pellicer, A., Cheng, Y. and Axel, R. (1977) Transfer of purified herpes virus thymidine kinase gene to cultured mouse cells. *Cell*, **11**, 223–232.

Wild, J., Hradecna, Z. and Szybalski, W. (2001) Single-copy/high-copy (SC/HC) pBAC/oriV novel vectors for genomics and gene expression. *Plasmid*, **45**, 142–143.

Wilks, H.M., Hart, K.W., Feeney, R., Dunn, C.R., Muirhead, H., Chia, W.N., Barstow, D.A., Atkinson, T., Clarke, A.R. and Holbrook, J.J. (1988) A specific, highly active malate dehydrogenase by redesign of a lactate dehydrogenase framework. *Science*, **242**, 1541–1544.

Willard, H.F. (2000) Artificial chromosomes coming to life. *Science*, **290**, 1308–1309.

Williams, G.J., Domann, S., Nelson, A. and Berry, A. (2003) Modifying the stereochemistry of an enzyme-catalyzed reaction by directed evolution. *Proc. Natl. Acad. Sci. USA*, **100**, 3143–3148.

Williams, N. (2002) Dolly clouds cloning hopes. *Curr. Biol.*, **12**, R79–R80.

Wilmut, I., Beaujean, N., de Sousa, P.A., Dinnyes, A., King, T.J., Paterson, L.A., Wells, D.N. and Young, L.E. (2002) Somatic cell nuclear transfer. *Nature*, **419**, 583–586.

Wilmut, I., Schnieke, A.E., McWhir, J., Kind, A.J. and Campbell, K.H. (1997) Viable offspring derived from fetal and adult mammalian cells. *Nature*, **385**, 810–813.

Wilson, I.A., Niman, H.L., Houghten, R.A., Cherenson, A.R., Connolly, M.L. and Lerner, R.A. (1984) The structure of an antigenic determinant in a protein. *Cell*, **37**, 767–778.

Winge, D.R. (1998) Copper-regulatory domain involved in gene expression. *Prog. Nucleic Acid Res. Mol. Biol.*, **58**, 165–195.

Winge, D.R., Jensen, L.T. and Srinivasan, C. (1998) Metal-ion regulation of gene expression in yeast. *Curr. Opin. Chem. Biol.*, **2**, 216–221.

Wingert, L. and Von Hippel, P.H. (1968) The conformation dependent hydrolysis of DNA by micrococcal nuclease. *Biochim. Biophys. Acta*, **157**, 114–126.

Winzeler, E.A. *et al.* (1999) Functional characterization of the *S. cerevisiae* genome by gene deletion and parallel analysis. *Science*, **285**, 901–906.

Wittwer, C.T., Herrmann, M.G., Moss, A.A. and Rasmussen, R.P. (1997) Continuous fluorescence monitoring of rapid cycle DNA amplification. *Biotechniques*, **22**, 130–138.

Wolff, J.A., Malone, R.W., Williams, P., Chong, W., Acsadi, G., Jani, A. and Felgner, P.L. (1990) Direct gene transfer into mouse muscle *in vivo*. *Science*, **247**, 1465–1468.

Wong, T.K. and Neumann, E. (1982) Electric field mediated gene transfer. *Biochem. Biophys. Res. Commun.*, **107**, 584–587.

Wood, V. *et al.* (2002) The genome sequence of *Schizosaccharomyces pombe*. *Nature*, **415**, 871–880.

Worrall, A.F. (1994) Site-directed mutagenesis by the cassette method. *Methods Mol. Biol.*, **30**, 199–210.

Wrestler, J.C., Lipes, B.D., Birren, B.W. and Lai, E. (1996) Pulsed-field gel electrophoresis. *Methods Enzymol.*, **270**, 255–272.

Wright, G., Carver, A., Cottom, D., Reeves, D., Scott, A., Simons, P., Wilmut, I., Garner, I. and Colman, A. (1991) High level expression of active human α_1-antitrypsin in the milk of transgenic sheep. *Bio/technology*, **9**, 830–834.

Wu, D.Y., Ugozzoli, L., Pal, B.K. and Wallace, R.B. (1989) Allele-specific enzymatic amplification of β-globin DNA for diagnosis of sickle cell anemia. *Proc. Natl. Acad. Sci. USA*, **86**, 2757–2760.

Wu, R. and Taylor, E. (1971) Nucleotide sequence analysis of DNA. II. Complete nucleotide sequence of the cohesive ends of bacteriophage λ DNA. *J. Mol. Biol.*, **57**, 491–511.

Wurm, F.M., Gwinn, K.A. and Kingston, R.E. (1986) Inducible overproduction of the mouse c-myc protein in mammalian cells. *Proc. Natl. Acad. Sci. USA*, **83**, 5414–5418.

Wyckoff, E. and Hsieh, T.S. (1988) Functional expression of a *Drosophila* gene in yeast: genetic complementation of DNA topoisomerase II. *Proc. Natl. Acad. Sci. USA*, **85**, 6272–6276.

Wysocki, L.J. and Gefter, M.L. (1989) Gene conversion and the generation of antibody diversity. *Annu. Rev. Biochem.*, **58**, 509–531.

Xu, C.W., Mendelsohn, A.R. and Brent, R. (1997) Cells that register logical relationships among proteins. *Proc. Natl. Acad. Sci. USA*, **94**, 12 473–12 478.

Yadav, N.S., Vanderleyden, J., Bennett, D.R., Barnes, W.M. and Chilton, M.-D. (1982) Short direct repeats flank the T-DNA on a nopaline Ti plasmid. *Proc. Natl. Acad. Sci. USA*, **79**, 6322–6326.

Yamamoto, T. and Yokota, T. (1980) Construction of a physical map of a kanamycin (Km) transposon, Tn5, and a comparison to another Km transposon, Tn903. *Mol. Gen. Genet.*, **178**, 77–83.

Yanisch-Perron, C., Vieira, J. and Messing, J. (1985) Improved M13 phage cloning vectors and host strains: nucleotide sequences of the M13mp18 and pUC19 vectors. *Gene*, **33**, 103–119.

Yansura, D.G. and Henner, D.J. (1990) Use of *Escherichia coli trp* promoter for direct expression of proteins. *Methods Enzymol.*, **185**, 54–60.

Yates, J.L., Warren, N. and Sugden, B. (1985) Stable replication of plasmids derived from Epstein–Barr virus in various mammalian cells. *Nature*, **313**, 812–815.

Yee, S.P. and Rigby, P.W. (1993) The regulation of myogenin gene expression during the embryonic development of the mouse. *Genes Dev.*, **7**, 1277–1289.

Yie, Y., Wei, Z. and Tien, P. (1993) A simplified and reliable protocol for plasmid DNA sequencing: fast miniprep and denaturation. *Nucleic Acids Res.*, **21**, 361.

Yip, T.T. and Hutchens, T.W. (1996) Immobilized metal ion affinity chromatography. *Methods Mol. Biol.*, **59**, 197–210.

Yocum, R.R., Hanley, S., West, R. and Ptashne, M. (1984) Use of *lacZ* fusions to delimit regulatory elements of the inducible divergent *GAL1–GAL10* promoter in *Saccharomyces cerevisiae*. *Mol. Cell. Biol.*, **4**, 1985–1998.

Yoshida, M. and Seiki, M. (1987) Recent advances in the molecular biology of HTLV-1: trans-activation of viral and cellular genes. *Annu. Rev. Immunol.*, **5**, 541–559.

Yu, J. *et al.* (2002) A draft sequence of the rice genome (*Oryza sativa* L. ssp. *indica*). *Science*, **296**, 79–92.

Zaenen, I., Van Larebeke, N., Van Montagu, M. and Schell, J. (1974) Supercoiled circular DNA in crown-gall inducing *Agrobacterium* strains. *J. Mol. Biol.*, **86**, 109–127.

Zaitlin, M. and Palukaitis, P. (2000) Advances in understanding plant viruses and virus diseases. *Annu. Rev. Phytopathol.*, **38**, 117–143.

Zambryski, P., Joos, H., Genetello, C., Leemans, J., Van Montagu, M. and Schell, J. (1983) Ti plasmid vector for the introduction of DNA into plant cells without altering their normal regeneration capacity. *EMBO J.*, **2**, 2143–2150.

Zamenhof, S., Brawerman, G. and Chargaff, E. (1952) On the desoxypentose nucleic acids from several microorganisms. *Biochim. Biophys. Acta*, **9**, 402–405.

Zamore, P.D., Tuschl, T., Sharp, P.A. and Bartel, D.P. (2000) RNAi: double-stranded RNA directs the ATP-dependent cleavage of mRNA at 21 to 23 nucleotide intervals. *Cell*, **101**, 25–33.

Zenke, F.T., Engles, R., Vollenbroich, V., Meyer, J., Hollenberg, C.P. and Breunig, K.D. (1996) Activation of Gal4p by galactose-dependent interaction of galactokinase and Gal80p. *Science*, **272**, 1662–1665.

Zhang, Y.L. and Cramer, W.A. (1993) Intramembrane helix–helix interactions as the basis of inhibition of the Colicin E1 ion channel by its immunity protein. *J. Biol. Chem.*, **268**, 10 176–10 184.

Zhou, M.Y., Clark, S.E. and Gomez-Sanchez, C.E. (1995) Universal cloning method by TA strategy. *Biotechniques*, **19**, 34–35.

Zimmerman, L., Parr, B., Lendahl, U., Cunningham, M., McKay, R., Gavin, B., Mann, J., Vassileva, G. and McMahon, A. (1994) Independent regulatory elements in the nestin gene direct transgene expression to neural stem cells or muscle precursors. *Neuron*, **12**, 11–24.

Zimmerman, S.B., Little, J.W., Oshinsky, C.K. and Gellert, M. (1967) Enzymatic joining of DNA strands: a novel reaction of diphosphopyridine nucleotide. *Proc. Natl. Acad. Sci. USA*, **57**, 1841–1848.

Zinder, N.D. and Boeke, J.D. (1982) The filamentous phage (Ff) as vectors for recombinant DNA – a review. *Gene*, **19**, 1–10.

Zoller, M.J. and Smith, M. (1983) Oligonucleotide-directed mutagenesis of DNA fragments cloned into M13 vectors. *Methods Enzymol.*, **100**, 468–500.

Zufferey, R., Dull, T., Mandel, R.J., Bukovsky, A., Quiroz, D., Naldini, L. and Trono, D. (1998) Self-inactivating lentivirus vector for safe and efficient *in vivo* gene delivery. *J. Virol.*, **72**, 9873–9880.

Index

Analysis of Genes and Genomes   Richard J. Reece
© 2004 John Wiley & Sons, Ltd   ISBNs: 0-470-84379-9 (HB); 0-470-84380-2 (PB)